722

Lecture Notes in Computer Science

Lecture Notes in Computer Science

722

Edited by G. Goos and J. Hartmanis

Advisory Board: W. Brauer   D. Gries   J. Stoer

Alfonso Miola (Ed.)

Design and Implementation of Symbolic Computation Systems

International Symposium, DISCO '93
Gmunden, Austria, September 15-17, 1993
Proceedings

Springer-Verlag
Berlin Heidelberg New York
London Paris Tokyo
Hong Kong Barcelona
Budapest

Alfonso Miola (Ed.)

Design and Implementation of Symbolic Computation Systems

International Symposium, DISCO '93
Gmunden, Austria, September 15-17, 1993
Proceedings

Springer-Verlag

Berlin Heidelberg New York
London Paris Tokyo
Hong Kong Barcelona
Budapest

Series Editors

Gerhard Goos
Universität Karlsruhe
Postfach 69 80
Vincenz-Priessnitz-Straße 1
D-76131 Karlsruhe, Germany

Juris Hartmanis
Cornell University
Department of Computer Science
4130 Upson Hall
Ithaca, NY 14853, USA

Volume Editor

Alfonso Miola
Dipartimento di Informatica e Sistemistica, Università di Roma "La Sapienza"
Via Salaria, 113, I-00198 Roma, Italia

CR Subject Classification (1991): D.1, D.2,1, D.2.10, D.3, I.1, I.2.2-3, I.2.5,
I.3.5-6

ISBN 3-540-57235-X Springer-Verlag Berlin Heidelberg New York
ISBN 0-387-57235-X Springer-Verlag New York Berlin Heidelberg

© Springer-Verlag Berlin Heidelberg 1993
Printed in Germany

Typesetting: Camera-ready by author
Printing and binding: Druckhaus Beltz, Hemsbach/Bergstr.
45/3140-543210 - Printed on acid-free paper

Foreword

This volume contains the proceedings of the Third International Symposium on Design and Implementation of Symbolic Computation Systems, DISCO '93.

The growing importance of systems for symbolic computation has essentially influenced the decision to organize the DISCO conference series. DISCO '93 takes place in Gmunden, Austria, September 15 - 17, 1993, as an international event in the field, organized and sponsored by the Research Institute for Symbolic Computation (University J. Kepler, Linz, Austria) and by the Dipartimento di Informatica e Sistemistica (University "La Sapienza", Roma, Italy).

DISCO '93 focuses mainly on the most innovative methodological and technological aspects of hardware and software system design and implementation for symbolic and algebraic computation, automated reasoning, geometric modeling and computation, and automatic programming.

The international research communities have recognized the relevance of the proposed objectives and topics which are generally not well covered in other conferences in the areas of symbolic and algebraic computation.

DISCO '93 includes papers on theory, languages, software environments, architectures and in particular, papers on the design and the development of significant running systems.

The general objective of DISCO '93 is to present an up-to-date view of the field, while encouraging the scientific exchange among academic, industrial and user communities on the development of systems for symbolic computation. Therefore it is devoted to researchers, developers and users from academia, scientific institutions, and industry who are interested in the most recent advances and trends in the field of symbolic computation.

The Program Chairman received 56 submissions for DISCO '93 and organized the reviewing process in cooperation with the Program Committee. Each paper was sent to two Program Committee members and then carefully reviewed by at least three independent referees, including Program Committee members. The Program Committee met on April 13 to 14, 1993 at the Dipartimento Informatica e Sistemistica. Università di Roma "La Sapienza" (Italy), to reach the final decision on acceptance of the submitted papers. The resulting DISCO '93 Scientific Program corresponds well to the initial objectives.

Among the submissions, 22 papers were selected as full contributions for presentation at the conference, as well as in this volume, under classified sections. Six further papers were selected as contributions for a presentation at the conference, concerning work in progress or running systems relevant to the themes of the symposium. These papers are included in a separate section of the present volume.

All my personal appreciation goes, in particular to Franz Lichtenberger, the Symposium Chairman, and to both the Program Committee and the Organizing Committee members for their indefatigable and valuable cooperation.

On behalf of the Program Committee, I would like to thank the authors of the submitted papers for their significant response to our Call for Papers, the invited speakers for having agreed to make their outstanding contributions to DISCO '93, and the referees for their cooperation in timely and precisely reviewing the papers.

Roma, July 1993   Alfonso Miola

Symposium Officers

General Chairman

F. Lichtenberger (Austria)

Program Committee

J. Fitch (UK), C. M. Hoffmann (USA), H. Hong (Austria), C. Kirchner (France), A. Kreczmar (Poland), A. Miola (Chairman, Italy), M. Monagan (Switzerland), E. G. Omodeo (Italy), F. Pfenning (USA), M. Wirsing (Germany)

Organizing Committee

- Research Insitute for Symbolic Computation, Johannes Kepler University, Austria
- Dipartimento di Informatica e Sistemistica, Università di Roma "La Sapienza", Roma, Italy

List of Referees

L. C. Aiello

M. P. Bonacina

A. Bossi

A. Bouhoula

M. Bronstein

R. Bündgen

H. J. Bürckert

D. Cantone

O. Caprotti

M. Casini Schaerf

P. Ciancarini

B. Ciciani

G. Cioni

F. D'Amore

R. De Nicola

J. Derzinger

J. Despeyroux

P. Di Blasio

F. Donini

D. Dranidis

S. Gastinger

W. Gehrke

E. Giovannetti

M. Grabowsky

D. Gruntz

R. Hennicker

C. Hintermeyer

T. Jebelean

B. Kacewicz

F. Kluzniak

F. Kroeger

M. Lenzerini

C. Limongelli

V. Manca

L. Mandel

A. Marchetti Spaccamela

V. Marian

M. Mehlich

T. Mora

A. Muech

A. Neubacher

F. Nickl

C. Palamidessi

F. Parisi Presicce

A. Pettorossi

A. Pierantonio

F. Pirri

A. Policriti

M. Proietti

B. Reus

G. Rossi

M. Rusinovitch

P. Santas

M. Schaerf

W. Schreiner

K. Siegl

A. Skowron

R. Stabl

T. Streicher

K. Sutner

M. Temperini

M. Turski

L. Unycryn

S. Valentini

I. Walukiewicz

W. Windsteiger

J. Winkowski

P. Zimmermann

Contents

Automated Reasoning

Software Systems

System Description

Mathematica: A System for Doing Mathematics by Computer?

Bruno Buchberger

Research Institute for Symbolic Computation (RISC),
Johannes Kepler University
A-4040 Linz, Austria
Tel. +43 (7236) 3231-41
FAX +43 (7236) 3231-30
buchberger@risc.uni-linz.ac.at

Abstract

The title of my talk coincides with the title of Stephen Wolfram's book on his Mathematica system except that in the title of my talk there is a question mark. The content of my talk is my personal answer to this question.

We start from analyzing what it means to do mathematics and introduce what we call the "creativity spiral in mathematics": "Doing mathematics", in my view, is iterating the cycle "observation – proof – programming – application".

Our main point is that Mathematica supports well three passes of this spiral, namely "observation – programming – application", and it helps a little in some simple forms of proof. However, Mathematica does not yet seem to be the right setting for proving in the broad sense of the word as understood by the mathematics community. We give some arguments for this and develop some ideas how a future system for doing mathematics might look like.

Mathematica: A System for Doing Mathematics by Computer?

Bruno Buchberger

Research Institute for Symbolic Computation (RISC)
Johannes Kepler University
A-4040 Linz, Austria
Tel. ++43 (7236) 3231-41
FAX ++43 (7236) 3231-30
buchberg@risc.uni-linz.ac.at

Abstract

The title of my talk coincides with the title of Stephen Wolfram's book on his Mathematica system except that in the title of my talk there is a question mark. The content of my talk is my personal answer to this question.

We start from analyzing what it means to do mathematics and introduce what we call the "creativity spirale in mathematics": "Doing mathematics", in my view, is iterating the cycle "observation – proof – programming – application" .

Our main point is that Mathematica supports well three passes of this spirale, namely "observation – programming – application" and it helps a little in some simple forms of proof. However, Mathematica does not yet seem to be the right setting for proving in the broad sense of the word as understood by the mathematics community. We give some arguments for this and develop some ideas how a future system for doing mathematics might look like.

Proving the Correctness of Algebraic Implementations by the ISAR System

Bernhard Bauer *, Rolf Hennicker**

* Institut für Informatik, Technische Universität München,
Arcisstr. 16, D-8000 München 2,
E-mail: bauer@informatik.tu-muenchen.de
** Institut für Informatik, Ludwig-Maximilians-Universität München,
Leopoldstr. 11b, D-8000 München 40,
E-mail: hennicke@informatik.uni-muenchen.de

Abstract. We present an interactive system, called ISAR, which provides an environment for correctness proofs of algebraic implementation steps. The correctness notion of implementation is based on behavioural semantics and the underlying proof procedure of the system is based on the principle of context induction (which is a particular instance of structural induction). The input of the ISAR system is an implementation step consisting of an abstract specification to be implemented, a concrete specification used as a basis for the implementation and an implementation construction. If all steps of the (interactive) proof procedure are performed the system has proved the correctness of the implementation step.

1 Introduction

Much work has been done in the field of algebraic specifications to provide formal concepts for the development of correct programs from given specifications. However, in order to be useful in practice, a formal theory of correct program development is not sufficient: Formal implementation notions should be supplied by appropriate proof methods and, even more important, by tools providing mechanical support for proving the correctness of implementation steps.

In this paper an interactive system for algebraic implementation proofs, called ISAR, is presented which sets out from the observational view of software development: The basic assumption is that a software product is a correct implementation if it satisfies the desired input/output behaviour, independently from the internal properties of a program which may not satisfy a given specification. This covers well known practical examples like the implementation of sets by lists (since lists do not satisfy the characteristic set equations but lists have the same behaviour as sets if only membership tests $x \in S$ are observable) or the familiar implementation of stacks by arrays with pointer (since arrays with pointer do not satisfy the usual stack equation $pop(push(x, s)) = s$ but they have the same behaviour as stacks if only the top elements of stacks are observed).

A formalization of this intuitive idea is presented in [Hennicker 90, 92] where an implementation relation for specifications is defined based on behavioural semantics in the sense of [Nivela, Orejas 88], [Reichel 85]. In particular, in [Hennicker 90, 92] a proof theoretic characterization and a proof method, called *context induction*, is presented for proving behavioural implementation relations. The characterization of implementations says that a specification *SP1* is a *behavioural implementation* of a specification *SP* if and only if for all *observable contexts* c (over the signature of *SP*) and for all axioms t = r of *SP* the "observable" equations c[t] = c[r] are deducible from the axioms of the implementation *SP1* (for all ground instantiations over the signature of *SP*).

It is the basic idea of the ISAR system to prove this condition by context induction, i.e. by structural induction on the set of observable contexts, in order to show that SP1 is an implementation of SP. The underlying algorithm of the ISAR system providing a procedure for context induction proofs was developed in [Hennicker 92].

Usually implementations of an abstract specification are constructed on top of existing (concrete) specifications of standard data structures like lists, arrays, trees etc. In order to document the construction of the implementation, the input of the ISAR system is an *implementation step* which consists of three parts: an abstract specification SP-A to be implemented, a concrete specification SP-C used as a basis for the implementation and a construction of the implementation. Such constructions are represented by appropriate enrichments and/or renamings performed on top of SP-C. An implementation step is called *correct* if the application of the implementation construction to SP-C yields a behavioural implementation of SP-A.

In order to prove the correctness of an implementation step the ISAR system first normalizes all specification expressions. Then the *context induction prover*, the kernel of the system, is called for proving the implementation relation for the normalized specifications. Thereby all contexts and all proof obligations to be considered for the implementation proof are automatically generated. For the proof of equations the system is connected to the TIP system which is a narrowing-based inductive theorem prover (cf. [Fraus, Hußmann 91]). All steps of an implementation proof can be guided by appropriate interaction with the user. In particular, as usual when dealing with induction proofs, it is often necessary to find an appropriate generalization of the actual induction assertion if a nesting of context induction (implemented by a recursive call of the context induction prover) is performed. For that purpose the ISAR system generates automatically a set of particular contexts each context representing a generalization of the actual induction assertion. Then the user may select an appropriate context representing an assertion which is general enough for achieving successful termination of the proof algorithm.

As we will show by an example for the construction of generalized induction assertions it may be necessary to define additional function symbols which generalize (some) functions of the abstract specification. (For instance for the proof of the array pointer implementation of stacks a generalization of the *pop* operation by an operation *iterated_pop: nat, stack → stack* is used where *iterated_pop(n, s)* performs n pop operations on a stack *s*.) Such function generalizations can be added as "hints" to an implementation step. Hints cannot be generated automatically. In this case the intuition of the system user is needed.

2 Basic Concepts

In this section we summarize the theoretical foundations of the ISAR system. In particular the notions of behavioural specifications and behavioural implementations are defined. Most definitions and results of this section can be found in [Hennicker 90] or (slightly revised) in [Hennicker 92].

2.1 Algebraic Preliminaries

First, we briefly review the basic notions of algebraic specifications which will be used in the following (for more details see e.g. [Ehrig, Mahr 85]). A (many sorted) *signature* Σ is a pair (S, F) where S is a set of *sorts* and F is a set of *function symbols* (also called *functions* for short). To every function symbol $f \in F$ a functionality $s_1, \ldots, s_n \to s$ with $s_1, \ldots, s_n \in S$ is associated. If $n = 0$ then f is called *constant* of sort s. A *signature morphism* $\rho: \Sigma \to \Sigma'$ between two signatures $\Sigma = (S, F)$ and $\Sigma' = (S', F')$ is a pair $(\rho_{sorts}, \rho_{functs})$ of mappings $\rho_{sorts}: S \to S'$, $\rho_{functs}: F \to F'$ such that for all $f \in F$ with functionality $s_1, \ldots, s_n \to s$, $\rho_{functs}(f)$ has functionality $\rho_{sorts}(s_1), \ldots, \rho_{sorts}(s_n) \to \rho_{sorts}(s)$. A signature $\Sigma' = (S', F')$ is called *subsignature* of Σ (written $\Sigma' \subseteq \Sigma$) if $S' \subseteq S$ and $F' \subseteq F$.

The *term algebra* $W_\Sigma(X)$ of all Σ-*terms* with variables of X (where $X = (X_s)_{s \in S}$ is an S-sorted family of sets of variables) is defined as usual. In particular, for any sort $s \in S$, $W_\Sigma(X)_s$ denotes the set of *terms* of sort s. If $X = \emptyset$ then $W_\Sigma(\emptyset)$ is abbreviated by W_Σ and W_Σ is called *ground term algebra*. We assume that any signature $\Sigma = (S, F)$ is *inhabited*, i.e. for each sort $s \in S$ there exists a ground term $t \in (W_\Sigma)_s$. A *substitution* $\sigma: X \to W_\Sigma(X)$ is a family of mappings $(\sigma_s: X_s \to W_\Sigma(X)_s)_{s \in S}$. For any term $t \in W_\Sigma(X)$, the *instantiation* $\sigma(t) =_{def} t[\sigma(x_1)/x_1, \ldots, \sigma(x_n)/x_n]$ is defined by replacing all variables $x_1, \ldots, x_n$ occurring in t by the terms $\sigma(x_1), \ldots, \sigma(x_n)$. A substitution $\sigma: X \to W_\Sigma$ is called *ground substitution*.

2.2 Behavioural Specifications

The syntax of behavioural specifications is defined similarly to [Nivela, Orejas 88] and [Reichel 85] where a distinguished set of sorts of a specification is declared as observable:

A *behavioural specification* $SP = (\Sigma, Obs, E)$ consists of a signature $\Sigma = (S, F)$, a subset $Obs \subseteq S$ of *observable sorts* and a set E of *axioms*. Any behavioural specification is assumed to contain the observable sort *bool*, two constants *true, false*: $\to bool$ (denoting the truth values) and the axiom *true* $\neq$ *false*. The axioms of E\ {*true* $\neq$ *false*} are equations $t = r$ with terms $t, r \in W_\Sigma(X)$.

Specifications can be reused for the definition of new behavioural specifications by the operators *enrich* for enriching a given specification by some sorts, functions and axioms, + for the combination of two specifications and *rename* for renaming the sorts and functions of a specification. More precisely we define for any behavioural specification SP = (Σ, Obs, E) with signature Σ = (S, F):

enrich SP **by sorts** S1 **observable sorts** Obs1 **functions** F1 **axioms** E1 =$_{def}$
  ((S ∪ S1, F ∪ F1), Obs ∪ Obs1, E ∪ E1),

SP + SP1 =$_{def}$ (Σ ∪ Σ1, Obs ∪ Obs1, E ∪ E1) where SP1 = (Σ1, Obs1, E1),

rename SP **by** ρ =$_{def}$ (Σ1, ρ$_{sorts}$(Obs), ρ$_{ax}$(E))
  where ρ: Σ → Σ1 is a bijective signature morphism and ρ$_{ax}$ is the
  extension of ρ to Σ-formulas.

Note that the enrich operator is only defined if (S ∪ S1, F ∪ F1) forms a signature, Obs1 is a subset of S ∪ S1 and E1 are axioms over the signature (S ∪ S1, F ∪ F1). Moreover, note that in contrast to specification building operators in the sense of ASL (cf. [Wirsing 86]) the above operators are only defined syntactically in order to express textual abbreviations.

As an example, the following behavioural specification STACK describes the usual data structure of stacks with a constant *empty*, denoting the empty stack, an operation *push* for adding an element to a stack, an operation *top* for selecting the top element of a stack and an operation *pop* for removing the top element. STACK is an enrichment of BOOL and of an arbitrary specification ELEM of the elements of a stack. The sort *elem* for the elements is declared as observable while the sort *stack* is not observable. Hence, the behaviour of stacks can only be observed via their top elements.

  spec STACK = **enrich** BOOL + ELEM **by**
    sorts stack
    observable sorts elem
    functions empty: → stack,
         push: elem, stack → stack,
         top: stack → elem,
         pop: stack → stack
    axioms top(push(e, s)) = e,
        pop(push(e, s)) = s **endspec**

2.3 Behavioural Implementations

The definition of the behavioural implementation concept is based on the assumption that from the software user's point of view a software product is a correct implementation if it satisfies the desired input/output behaviour. Hence a behavioural specification SP1 = (Σ1, Obs1, E1) is called behavioural implementation of SP = (Σ, Obs, E) if SP1 respects the observable properties of SP. A precise formal definition of this intuitive notion using behavioural semantics (cf. e.g. [Nivela, Orejas 88], [Reichel 85]) is given in [Hennicker 90, 92]. Since we are interested in automated implementation proofs we will present here only the following proof theoretic characterization of behavioural implementations (cf. [Hennicker 90, 92]) which is the theoretical background of the ISAR system. The characterization uses the notion of a Σ-*context* which is any term $c[z_S]$ over the signature Σ of SP which contains a distinguished variable z_S of some sort s ∈ S (where z_S occurs exactly once in $c[z_S]$). If the (result) sort, say s_0, of $c[z_S]$ belongs to Obs then $c[z_S]$ is called *observable Σ-context*. The application of a context $c[z_S]$ to a term t of sort s is defined by the substitution of z_S by t. Instead of $c[t/z_S]$ we also write briefly c[t]. In particular, for any sort s, the variable z_S is itself a Σ-context (called *trivial context*) of sort s and $z_S[t] = t$.

2.1 Proposition   Let SP1 = (Σ1, Obs1, E1) and SP = (Σ, Obs, E) be behavioural specifications such that Σ ⊆ Σ1 and Obs ⊆ Obs1.
SP1 is a behavioural implementation of SP, if and only if for all observable Σ-contexts $c[z_S]$ and for all axioms (t = r) ∈ E (such that t and r are of sort s),
      SP1 ⊢ σ(c[t]) = σ(c[r]) holds for all ground substitutions σ: X → $W_Σ$.

In the above proposition SP1 ⊢ σ(c[t]) = σ(c[r]) means that the equation σ(c[t]) = σ(c[r]) is deducible from the axioms of the implementation SP1 by the usual axioms and rules of the equational calculus, cf. e.g. [Ehrig, Mahr 85]. (The additional derivation rule (R) of [Hennicker 92] is only relevant for the necessity of the implementation condition if the implementation is inconsistent.) Since behavioural semantics in [Hennicker 90, 92] is restricted to term generated algebras it is enough to consider all ground instantiations σ(c[t]) = σ(c[r]) of the equations c[t] = c[r]. Note that the "non observable" axioms t = r of an abstract specification SP need not to be satisfied by an implementation. Only the observable consequences of those axioms (formally expressed by applications of observable contexts) have to be satisfied by an implementation (for all ground substitutions). For instance, an implementation of the specification STACK not necessarily has to satisfy the non observable stack equation *pop(push(e, s)) = s* but it has to satisfy all (ground instantiations of) applications of observable contexts to this equation as e.g. σ(*top(pop(push(e, s)))*) = σ(*top(s)*) with σ: X → $W_Σ$.

3 The ISAR System

3.1 Implementation Proofs by Context Induction

Proposition 2.1 provides the starting point for an automatization of implementation proofs since it gives a proof theoretic characterization of behavioural implementations where certain equations have to be derived from the axioms of the implementation. In particular, the proposition says that it is sufficient to show that all ground instantiations $\sigma(c[t]) = \sigma(c[r])$ of the equations $c[t] = c[r]$ are valid in the implementation SP1. Hence it is enough to prove that the equations $c[t] = c[r]$ are *inductive theorems* of SP1 where an equation e is called inductive theorem of SP1 if all ground instantiations of e are theorems of SP1, i.e. SP1 $\vdash \sigma(e)$ for all ground substitutions $\sigma: X \rightarrow W_{\Sigma 1}$ (cf. [Padawitz 88]). (Note that the inductive theorem property is slightly stronger than the condition of Proposition 2.1 since there only ground substitutions σ w.r.t. the subsignature $\Sigma \subseteq \Sigma 1$ are considered.) Then, for the proof of inductive theorems one may use theorem provers like the prover of Boyer and Moore (cf. [Boyer, Moore 88]) or the Larch Prover (cf. [Garland, Guttag 89]).

However, things are not that easy because, in general, infinitely many observable contexts exist and therefore one has to prove usually infinitely many equations $c[t] = c[r]$. Hence, for proving that SP1 is a behavioural implementation of SP, it is enough to show that the following property $P(c[z_s])$ is valid for all observable Σ-contexts $c[z_s]$:

$$P(c[z_s]) = true \iff_{def} \text{ for all axioms } (t = r) \in E \text{ (such that t, r are of sort s),}$$
$$\text{the equation } c[t] = c[r] \text{ is an inductive theorem of SP1.}$$

Since observable Σ-contexts are particular terms (over the signature of the abstract specification) the syntactic subterm ordering defines a Noetherian relation on the set of observable Σ-contexts and therefore we can apply *context induction* (which is a particular instance of structural induction, cf. [Burstall 69]) for showing the validity of $P(c[z_s])$ for all observable Σ-contexts $c[z_s]$.

It turns out that in many cases implementation proofs by context induction work quite schematically although usually a lot of different cases of contexts have to be distinguished. Hence it is the aim of the ISAR system to provide a tool which automates (to a certain extent) implementation proofs by context induction. The principle idea for executing implementation proofs by the ISAR system is the following one: (For a detailed description of the underlying algorithm of the system we refer to [Hennicker 92].)

Let SP and SP1 be as above. In the first step (which is the base of the context induction) it has to be shown that $P(z_s)$ is valid for all trivial observable Σ-contexts z_s (which just are variables of observable sort s). According to the definition of the property P this means that one has to prove that all "observable" axioms $t = r$ of SP (with terms t and r of observable sort) are inductive theorems of SP1. For the proof of the equations the ISAR system is connected to the TIP system (cf. [Fraus, Hußmann 91]) which is a narrowing-based inductive theorem prover.

In the second step, the induction step is performed for all contexts of the form $f(...,c[z_s],...)$ where f is a function symbol of SP with observable result sort and $c[z_s]$

ranges over all Σ-contexts of sort, say s_i. Then, for the proof of the actual induction assertion it has to be shown that for all Σ-contexts $c[z_s]$ of sort s_i, $P(f(...,c[z_s],...))$ holds, i.e. for all axioms $t = r$ of SP (such that t, r are of sort s) the equation $f(...,c[t],...) = f(...,c[r],...)$ is an inductive theorem of SP1. For that purpose two cases are distinguished:

Case 1: s_i is an observable sort of SP. Then, by hypothesis of the context induction, $P(c[z_s])$ holds, i.e. $c[t] = c[r]$ is an inductive theorem of SP1 for all axioms $t = r$ of SP. Hence $f(...,c[t],...) = f(...,c[r],...)$ is also an inductive theorem of SP1 for all axioms $t = r$ of SP, i.e. $P(f(...,c[z_s],...))$ holds.

Case 2: s_i is not an observable sort of SP. Then, the hypothesis of the context induction cannot be applied for $c[z_s]$ and therefore a nested context induction (over all Σ-contexts $c[z_s]$ of sort s_i) is performed for proving the property $P(f(...,c[z_s],...))$ for all Σ-contexts $c[z_s]$ of sort s_i.

In the ISAR system each nesting of context induction is implemented by a recursive call of the *context induction prover* where the actual parameter is a (fixed) context $c0[z_{s_i}]$ which represents the actual induction assertion "$P(c0[c[z_s]])$ is valid for all Σ-contexts $c[z_s]$ of sort s_i". (For instance, the induction assertion "$P(f(...,c[z_s],...))$ is valid for all Σ-contexts $c[z_s]$ of sort s_i" is represented by the context $f(...,z_{s_i},...)$.) Initially the context induction prover is called for all trivial contexts z_{s_0} of observable sort $s_0 \in S$. Then, if all steps of the proof procedure are performed the principle of context induction implies that the property $P(c[z_s])$ is proved for all observable Σ-contexts $c[z_s]$ and hence it is shown that SP1 is a behavioural implementation of SP. Obviously, the proof algorithm of the ISAR system cannot be complete (in the sense that all valid implementation relations can be proved by ISAR) because context induction and inductive theorem proving are not complete.

In order to achieve successful termination of the implementation proof it is often necessary to find an appropriate generalization of the actual induction assertion. Therefore automated reasoning has to be supplemented by interaction with the user who may select before each nesting of context induction a context which represents a generalization of the actual induction assertion. For instance, any subcontext $c0'[z_{s_i}]$ of a context $c0[z_{s_i}]$ represents a generalization of the assertion represented by $c0[z_{s_i}]$ because it is easy to show (using the congruence rule of the equational calculus) that in this case $P(c0'[c[z_s]])$ implies $P(c0[c[z_s]])$ for all Σ-contexts $c[z_s]$ of sort s_i. Hence, instead of $c0[z_{s_i}]$ any subcontext $c0'[z_{s_i}]$ can be correctly used as an actual parameter of a recursive call of the proof algorithm.

The ISAR system generates automatically before each nesting of context induction all subcontexts of the context $c0[z_{s_i}]$ which represents the actual induction assertion and the user may choose an appropriate one. Besides the generalization of contexts by subcontexts there are two further constructions of context generalizations performed by the ISAR system: The first one allows to abstract from a context $c0[z_s]$ by replacing subterm(s) of c0[z_S] (which do not contain z_S) by variables. The second one allows to construct new contexts by applying rewrite steps to the original context c0[z_S]. An example for the generation of context generalizations can be seen in the implementation proof of Section 3.3.

3.2 Implementation Steps

Usually the implementation of an abstract specification is constructed on top of already existing specifications of concrete data structures like lists, arrays, trees etc. (see e.g. [Ehrig et al. 82], [Sannella, Tarlecki 88] for formalizations of implementation constructions). In order to document the construction of the implementation the input of the ISAR system is the description of an *implementation step* consisting of three parts: an abstract specification SP = (Σ, Obs, E) to be implemented, a concrete specification SP-C = (Σ-C, Obs-C, E-C) used as a basis for the implementation and a construction of the implementation on top of SP-C. The implementation construction can be defined by some enrichment and/or renaming of SP-C. For instance, the following implementation step performs first a renaming of SP-C w.r.t. a signature morphism ρ and then an enrichment Δ = **sorts** S1 **observable sorts** Obs1 **functions** F1 **axioms** E1 of the renamed version of SP-C:

```
implementation step SP_by_SP-C =
    SP is implemented by SP-C
    via renaming ρ, enrichment Δ
endimplstep
```

Such an implementation step is called *correct* if
      enrich (rename SP-C **by** ρ**) by** Δ is a behavioural implementation of SP.
The correctness of implementation steps when performing first an enrichment and then a renaming is defined analogously. Before starting an implementation proof the *normalizer* of the ISAR system computes normal forms of all specifications according to the definition of the operators *enrich, rename* and + (cf. Section 2.2).
In some implementation proofs it may be necessary to use particular lemmas (i.e. theorems which are valid in the implementing specification) and even auxiliary function definitions which are used for the construction of contexts which represent sufficiently general induction assertions. Such lemmas and function definitions can be added as "hints" to an implementation step. Hints cannot be generated automatically. In this case the intuition of the system user is necessary. However, it seems that in most examples the auxiliary functions can be created just by generalizations of those abstract functions which would be iteratively used in recursive calls of the proof procedure (cf. the generalization of the *pop* operation by the operation *iterated_pop* below.)
As an example we consider an implementation step which implements stacks on top of

the following specification of (dynamic) arrays. For simplicity only those array operations are defined which are necessary for the example: *vac* denotes the empty array, *put* inserts an element into an array at a particular index and *get* delivers the actual value for a given index. The indices are natural numbers which are specified in the underlying specification NAT. It is assumed that the specification ELEM of the array elements contains a constant *constelem* and a conditional function *ifelem . then . else . fi: bool, elem, elem → elem*.

```
spec ARRAY=
    enrich BOOL + NAT + ELEM by
      sorts array
      observable sorts elem
      functions vac : -> array,
                put : array, nat, elem -> array,
                get : nat, array -> elem
        axioms
          get(k, put(a, l, e)) =
                ifelem eq_nat(k, l) then e else get(k, a) fi,
          get(k, vac) = constelem    endspec
```

Then the implementation of stacks on top of arrays is defined by the following implementation step:

```
implementation step STACK_by_ARRAY =
    STACK is implemented by ARRAY
    via enrichment
            sorts stack
            functions   pair : array, nat -> stack,
                        empty : -> stack,
                        push : elem, stack -> stack,
                        top : stack -> elem,
                        pop : stack -> stack
            axioms
                empty = pair(vac, 0),
                push(e, pair(a, p)) =
                        pair(put(a, p+1, e), p +1),
                top(pair(a, p)) = get(p, a),
                pop(pair(a, p)) = pair(a, p-1)
    hints
        auxiliary functions
            iterated_pop : nat, stack -> stack
        axioms
            iterated_pop(zero, s) = s,
            iterated_pop(n+1, s) = iterated_pop(n, pop(s))
        lemmas
            iterated_pop(i, pair(a,n)) = pair(a, n-i)
endimplstep
```

In the above implementation construction stacks are implemented by their familiar array pointer representation, i.e. by pairs consisting of an array and a pointer (a natural number) which points to the top element of a stack. The stack operations *empty, push, top* and *pop* are implemented as usual. For instance, the *pop* operation simply decrements the pointer without deleting the entry at the last top position. Hence the abstract stack equation *pop(push(e, s)) = e* is not valid in the implementation. But we will see that nevertheless the implementation step is correct since the implementation satisfies all observable consequences of the abstract stack equations. The usefulness and necessity of the hints will be seen in the following when the implementation proof is performed by the system.

3.3 An Example Session with the ISAR System

After the ISAR system is called we first give a command for reading the file where the implementation step STACK_by_ARRAY together with the specifications STACK and ARRAY is stored. After syntactical analysis all specifications and the implementation step are normalized and it is checked whether the signature and the observable sorts of the abstract specification are included in the (normalized) implementation because this is the precondition of our implementation definition. It is possible to display all specifications and the implementation step in their normal form and in their structured form using the commands list norm or list struct.

Now the implementation proof can be started by calling the context induction prover. In the following we will show how the system performs the implementation proof and we will give detailed comments - written in *italics* - which do not belong to the output of the system.

****** CONTEXT-INDUCTION ******

Observable Sorts of STACK:
bool, elem

***** PROOF-OBLIGATION: ******
FOR ALL CONTEXTS c OF SORT bool, elem AND FOR ALL AXIOMS t=r OF STACK IT IS TO SHOW:
IMPLEMENTATION |- c[t] = c[r]

(This is exactly the property P(c) defined in Section 3.1 (since induction is allowed for the derivation). In the next step the axioms of STACK are considered explicitely. Thereby the axioms of the underlying specifications BOOL and ELEM are omitted because these axioms belong already to the implementation. The variables X29, X30, ... are system variables.)

THE FOLLOWING EQUATIONS HAVE TO BE PROVED
FOR ALL CONTEXTS c OF SORT bool, elem:
  c[top(push(X29, X30))] = c[X29]   *(if the context variable z_s has sort s = elem)*
  c[pop(push(X31, X32))] = c[X32]   *(if the context variable z_s has sort s = stack)*

* BASE OF THE CONTEXT INDUCTION FOR CONTEXTS OF SORT: bool, elem
* DEPTH: 0
  ONLY THE OBSERVABLE AXIOMS HAVE TO BE PROVED

(For the base of the context induction the property $P(z_{elem})$ has to be proved where z_{elem} is the trivial context of sort elem. Note that $P(z_{bool})$ is trivially satisfied since no axiom of sort bool is considered.)

* PROOF OBLIGATIONS:
  (1): **top(push(X29, X30)) = X29**

ISAR/PROOF>all
(The command "all" says that all equations of the proof obligation - here only one exists - are selected to be proved. For the proof of the equation the ISAR system calls the TIP system.)

  top(push(X29, X30)) = X29
 # EQUATION PROVED BY TIP-SYSTEM.

(The proof of this equation is performed by induction on the structure of X30 where the function "pair" is used as a constructor for stacks. In general, constructors can be chosen interactively during the TIP proof or they can be declared previously in the implementing specification. Constructor completeness can be checked by the TIP system and a listing of the TIP proof can be displayed if required.)

* BASE OF THE CONTEXT INDUCTION OF DEPTH 0 FINISHED

*CONTEXT INDUCTION STEP FOR CONTEXTS OF SORT: bool, elem
* DEPTH: 0

(It is enough to perform the induction step for contexts with outermost function symbol top. For all other function symbols of STACK with result bool or elem the argument sorts are observable as well and hence the induction step is trivial.)

* THE FOLLOWING ABSTRACT FUNCTION WITH OBSERVABLE RESULT-SORT
* HAS TO BE CONSIDERED
  (1): top : stack -> elem

ISAR/PROOF>all
  Selected function: top : stack -> elem

*(At this point the actual induction assertion to be proved by a nested context induction is: "$P(top(c))$ is valid for all contexts c of sort stack". However in the induction step of the nested context induction contexts of the form $top(pop(c))$ have to be considered but it is not possible to prove $P(top(pop(c)))$ using the hypothesis $P(top(c))$. Hence one could try to start a second nesting of context induction for proving $P(top(pop(c)))$. But then the same situation occurs leading to a further iteration of context induction for proving $P(top(pop(pop(c))))$ etc. Consequently, the implementation proof would not terminate. A possible solution is to code the iteration of the pop operation by an auxiliary function iterated_pop: nat, stack $\rightarrow$ stack such that iterated_pop(n, s) pops a stack s n-times. The operation iterated_pop is specified in the hints of the implementation step.
Now having the hints available the ISAR system generates the following set of contexts. Context (1) represents the original induction assertion which was not successful. Context (2) represent a more general induction assertion (namely: "$P(c)$ is valid for all contexts c of sort stack") but since the non observable STACK axiom pop(push(e, s)) = s is not valid in the implementation this assertion is too general. Context (3) is*

equivalent to context (1) by the first axiom of the hints. Context (4) is a generalization of context (3) where the term zero is replaced by a variable. We will see that the selection of (4) will be successful.)

```
* GENERATED CONTEXTS:
  (1): top(z_stack)
  (2): z_stack
  (3): top(iterated_pop(zero, z_stack))
  (4): top(iterated_pop(X145, z_stack))
```

ISAR/PROOF>sel 4

(Now the assertion represented by context (4) is displayed.)

***** PROOF-OBLIGATION: ******
FOR ALL CONTEXTS c OF SORT stack AND FOR ALL AXIOMS t=r OF STACK
IT IS TO SHOW:
IMPLEMENTATION |- top(iterated_pop(X145,c[t])) =
                  top(iterated_pop(X145,c[r]))

```
THE FOLLOWING EQUATIONS HAVE TO BE PROVED
FOR ALL CONTEXTS c OF SORT stack:
  top(iterated_pop(X145, c[pop(push(X31, X32))]])) =
  top(iterated_pop(X145, c[X32]))
```

* BASE OF THE CONTEXT INDUCTION FOR CONTEXTS OF SORT: stack
* DEPTH: 1

(Depth 1 indicates the depth of the nesting of context induction which is actually performed.)

* PROOF OBLIGATIONS:
```
  (1): top(iterated_pop(X145, pop(push(X31, X32)))) =
       top(iterated_pop(X145, X32))
```

ISAR/PROOF>all

```
  top(iterated_pop(X145, pop(push(X31, X32)))) =
  top(iterated_pop(X145, X32))
# EQUATION PROVED BY TIP-SYSTEM.
```

(For the proof of this equation the lemma stated as a hint in the implementation step is used. The proof is again performed by induction on the structure of X32 using the constructor "pair".)

* BASE OF THE CONTEXT INDUCTION OF DEPTH 1 FINISHED

* CONTEXT INDUCTION STEP FOR CONTEXTS OF SORT: stack
* DEPTH 1

* THE FOLLOWING ABSTRACT FUNCTIONS WITH RESULT-SORT stack
* HAVE TO BE CONSIDERED
```
  (1): push : elem, stack -> stack
  (2): pop : stack -> stack
```

ISAR/PROOF>all
  Selected function: push : elem, stack -> stack
* WHICH ARGUMENT?
ISAR/PROOF>all
  push : elem, **stack** -> stack

(Here the system does not select the first argument sort of push since elem is an observable sort and hence the induction step is trivial. For the second argument sort the actual induction assertion to be proved is: "P(top(iterated_pop(X145, push(X729, c)))) is valid for all contexts c of sort stack." Thereby the induction hypothesis of the nested context induction can be

applied, i.e. one has to show that for all contexts c of sort stack, P(top(iterated_pop(X145,c))) implies P(top(iterated_pop (X145, push(X729, c)))). For this it is enough to show for newly introduced constants constant1_stack and constant2_stack of sort stack that the proof obligation below can be derived from the implementation if the following additional hypothesis is added to the axioms of the implementation:)

```
* ADDITIONAL HYPOTHESIS OF THE CONTEXT INDUCTION:
    top(iterated_pop(X145, constant1_stack)) =
    top(iterated_pop(X145, constant2_stack))
* PROOF OBLIGATION:
    top(iterated_pop(X145, push(X729, constant1_stack))) =
    top(iterated_pop(X145, push(X729, constant2_stack)))
# EQUATION PROVED BY TIP-SYSTEM.
```

(The proof is performed by induction over X145 using the constructors "zero" and "succ" of the natural numbers. The base of the induction uses the equation top(push(X29, X30)) = X29 which has been proved previously and therefore is automatically added to the lemmas of the implementation. The induction step uses the equation top(iterated_pop(X145, pop(push(X31, X32)))) = top(iterated_pop(X145, X32)) (which has also been proved before) and the hypothesis of the context induction.)

```
* THE FOLLOWING ABSTRACT FUNCTIONS WITH RESULT-SORT stack
* HAVE TO BE CONSIDERED
    (1): push : elem, stack -> stack * proved
    (2): pop : stack -> stack

    Selected function: pop : stack -> stack

* ADDITIONAL HYPOTHESIS OF THE CONTEXTT INDUCTION:
    top(iterated_pop(X145, constant1_stack)) =
    top(iterated_pop(X145, constant2_stack))
* PROOF OBLIGATION:
    top(iterated_pop(X145, pop(constant1_stack))) =
    top(iterated_pop(X145, pop(constant2_stack)))
# EQUATION PROVED BY TIP-SYSTEM.
```

(The proof uses the second axiom of iterated_pop and the hypothesis.)

```
* CONTEXT INDUCTION OF DEPTH 1 FINISHED

* CONTEXT INDUCTION OF DEPTH 0 FINISHED

* END OF THE IMPLEMENTATION PROOF
***** ALL PROOF OBLIGATIONS PROVED! *****
```

3.4 The Structure of the ISAR System

The ISAR system is written in the programming language PASCAL in order to be compatible with the TIP system which verifies all proof obligations generated by ISAR. The main modules of ISAR are

- a *scanner* and a *parser* with mixfix-parser for the syntactical analysis of specifications and implementation steps,
- a *normalizer* for flattening structured specifications and implementation steps,
- a *context generator* to produce automatically contexts which represent generalizations of the actual induction assertion,

- a *TIP-interface* for the exchange of informations between the ISAR and the TIP system and
- the proof-modules of the *TIP-system* for the verification of the proof obligations.

For lack of space we cannot give here a more elaborated description of the internal structure and technical details of the ISAR system. Interested readers may consult [Bauer 93].

4 Concluding Remarks

The development of the ISAR system is the consequent third step after having introduced the context induction principle for proving behavioural implementations in [Hennicker 90] and after the investigation of a proof procedure for context induction proofs in [Hennicker 92]. The proof techniques of the ISAR system and the underlying implementation concept are based on behavioural semantics which is a major difference for instance to the ISDV system (cf. [Beierle, Voß 85]) where implementations and abstract specifications are related by representation homomorphisms. Recently, in [Bidoit, Hennicker 92] a method was developed for proving observational theorems over a behavioural specification with the Larch Prover (cf. [Garland, Guttag 88]) where (under particular assumptions) an explicit use of context induction could be avoided by using the partioned by deduction rule of LP. It is an interesting objective of future research how an environment for proving behavioural implementations could be built on top of LP.

The actual version of the ISAR system is restricted to equational specifications but it is intended to provide an extension to conditional equational specifications (with observable premises of the axioms) and to implementations of parameterized specifications. In particular, for dealing with parameterized implementations the solution is very simple: One just has to guarantee that proofs of equations by the TIP system do not use induction on parameter sorts.

An important direction of future development concerns the use of ISAR for the representation of reusable software components and for retrieving reusable components from a component library. In fact an implementation step as it is processed by ISAR represents a two level reusable component in the sense of [Wirsing 88] where the specification to be implemented represents the abstract description of the component's behaviour and the implementation represents the realisation of the component. (Actual research deals with an extension of the implementation notion such that object-oriented classes with imperative method definitions can be used as implementations.) Concerning the retrieval of components the ISAR system provides already a tool which allows to check whether the reuse of components which are retrieved from a component library by syntactic signature matching (cf. [Chen et al. 93]) is semantically correct with respect to a given goal specification. As a consequence, we suggest to combine both techniques, syntactic signature matching and correctness proofs by ISAR, in order to obtain a complete retrieval system for reusable software components.

Acknowledgement We would like to thank Martin Wirsing for several valuable suggestions for the design of the ISAR system. This work is partially sponsored by the German BMFT project KORSO.

References

[Bauer 93] An interactive system for algebraic implementation proofs: The ISAR system from the user's point of view. Universität München, Technical Report (to appear), 1993.

[Beierle, Voß 85] C. Beierle, A. Voß: Algebraic specification and implementation in an integrated software development and verification system. MEMO SEKI-12, FB Informatik, Universität Kaiserslautern, 1985.

[Bidoit, Hennicker 92] How to prove observational theorems with LP. Proc. of the First International Workshop on Larch, July 1992, Boston, USA, Springer Verlag Workshop in Computing Series, 1993. Also in: Laboratoire d'Informatique de l' Ecole Normale Supérieure, Paris, LIENS-92-23, 1992.

[Boyer, Moore 88] R. S. Boyer, J. S. Moore: A computational logic handbook. Academic Press, New York, 1988.

[Burstall 69] R. M. Burstall: Proving properties of programs by structural induction. *Comp. Journal* 12, 41-48, 1969.

[Chen et al. 93] P. S. Chen, R. Hennicker, M. Jarke: On the retrieval of reusable software components. In: R. Prieto-Diaz, W. B. Frakes (eds.): Advances in Software Reuse. *Selected Papers from the Second International Workshop on Software Reusability*. Lucca, Italy, 1993. IEEE Computer Society Press, Los Alamitos, California, Order Number 3130, 99-108, 1993.

[Ehrig et al. 82] H. Ehrig, H.-J. Kreowski, B. Mahr, P. Padawitz: Algebraic Imple-mentation of Abstract Data Types. *Theoretical Computer Science* 20, 209-263, 1982.

[Ehrig, Mahr 85] H. Ehrig, B. Mahr: Fundamentals of algebraic specification 1, EATCS Monographs on Theoretical Computer Science 6, Springer, Berlin, 1985.

[Fraus, Hußmann 91] U. Fraus, H. Hußmann: A narrowing-based theorem prover. Extended Abstract. In: Proc. RTA '91, Rewriting Techniques and its Applications, Lecture Notes in Computer Science 488, 435-436, 1991.

[Garland, Guttag 88] S. J. Garland, J. V. Guttag: An overview of LP, the Larch Prover. In: Proc. RTA '89, Rewriting Techniques and its Applications, Lecture Notes in Computer Science 355, 137-151, 1989.

[Hennicker 90] R. Hennicker: Context Induction: a proof principle for behavioural abstractions. In: A. Miola (ed.): Proc. DISCO '90, International Symposium on Design and Implementation of Symbolic Computation Systems, Capri, April 1990. Lecture Notes in Computer Science 429, 101-110, 1990.

[Hennicker 92] A semi-algorithm for algebraic implementation proofs. *Theoretical Computer Science* 104, Special Issue, 53-87, 1992.

[Nivela, Orejas 88] Mª P. Nivela, F. Orejas: Initial behaviour semantics for algebraic specifi- cations. In: D. T. Sannella, A. Tarlecki (eds.): Proc. 5th Workshop on Algebraic Specifi- cations of Abstract Data Types, Lecture Notes in Computer Science 332, 184-207, 1988.

[Padawitz 88] P. Padawitz: Computing in Horn clause theories. EATCS Monographs on Theoretical Computer Science 16, Springer, Berlin, 1988.

[Reichel 85] H. Reichel: Initial restrictions of behaviour. IFIP *Working Conference*, The Role of Abstract Models in Information Processing, 1985.

[Sannella, Tarlecki 88] D. T. Sannella, A. Tarlecki: Toward formal development of programs from algebraic specifications: implementation revisited. *Acta Informatica* 25, 233-281, 1988.

[Wirsing 86] M. Wirsing: Structured algebraic specifications: a kernel language. *Theoretical Computer Science* 42, 123-249, 1986.

[Wirsing 88] Algebraic description of reusable software components. In: Proc. COMPEURO '88, Comp. Society Order Number 834, 300-312, 1988.

Sketching Concepts and Computational Model of TROLL *light*

Martin Gogolla*, Stefan Conrad*, and Rudolf Herzig*

Technische Universität Braunschweig, Informatik, Abt. Datenbanken
Postfach 3329, D-38023 Braunschweig, GERMANY
e-mail: gogolla@idb.cs.tu-bs.de

Abstract

The specification language TROLL *light* is intended to be used for conceptual modeling of information systems. It is designed to describe the Universe of Discourse (UoD) as a system of concurrently existing and interacting objects, i.e., an object community.

The first part of the present paper introduces the various language concepts offered by TROLL *light*. TROLL *light* objects have observable properties modeled by attributes, and the behavior of objects is described by events. Possible object observations may be restricted by constraints, whereas event occurrences may be restricted to specified life-cycles. TROLL *light* objects are organized in an object hierarchy established by sub-object relationships. Communication among objects is supported by event calling.

The second part of our paper outlines a simplified computational model for TROLL *light*. After introducing signatures for collections of object descriptions (or templates as they are called in TROLL *light*) we explain how single states of an object community are constructed. By parallel occurrence of a finite set of events the states of object communities change. The object community itself is regarded as a graph where the nodes are the object community states reachable from an initial state and the edges represent transitions between states.

1 Introduction

Formal program specification techniques have been receiving more and more attention in recent years. Several approaches to specification are studied nowadays, for example: Specification of functions (VDM, Z [Jon86, BHL90]), abstract data types [EM85, EGL89, EM90, Wir90, BKL+91], predicate logic and extensions like temporal and modal logic, semantic data models [HK87], and process specification (CCS [Mil80], CSP [Hoa85], petri nets [Rei85]). But for different reasons, none of the above approaches seems to meet in isolation all the requirements needed for the conceptual modeling of information systems: The specification of functions and abstract data

---

*Work reported here has been partially supported by the CEC under Grant No. 6112 (COMPASS) and BMFT under Grant No. 01 IS 203 D (KORSO).

types does not adequately support persistent objects or the interactive nature of information systems, predicate logic seems to be too general, semantic data models do not take into account the evolution of information systems, and process specifications do not reflect structural aspects.

The paradigm of object-orientation seems to offer more natural ways to model systems and to develop modular software. The two areas formal specification and object-orientation are brought together by object-oriented specification. In particular, object-oriented specification tries to take advantage of several specification techniques by combining their capabilities. For this extremely current research field semantic foundations of object-oriented specifications are given for instance in [EGS90, SE91, Mes92b]. But in addition to semantic foundations, concrete specification languages for objects are needed as well. As far as we know OBLOG [SSE87, CSS89, SSG+91, SGG+91, SRGS91] was the first proposal. Based on experience with this language and on results achieved in the ESPRIT BRWG IS-CORE the language TROLL [JSHS91, JSS91] was developed. Other specification languages for objects are ABEL [DO91], CMSL [Wie91], MONDEL [BBE+90], OS [Bre91], and Π [Gab93].

We did not want to re-invent the wheel, and therefore took the specification language TROLL as a basis for our language. The richness of concepts in TROLL allows us to model the Universe of Discourse (UoD) as adequately as possible. With this aim in mind even some partially redundant language concepts are offered by TROLL. Thus it was necessary to evaluate which concepts are redundant and could be disregarded for our language. Other concepts of TROLL, like temporal logic, have been excluded for pragmatical reasons (so as to simplify animation and to make verification manageable). Finally, we obtained a language with a small number of concepts and hence called it TROLL *light*.

However, TROLL *light* is not just a subset of TROLL. Some details have been added or modified in order to round off TROLL *light*. This was necessary because we needed a clear and balanced semantic basis for our specification language. In particular we want to stress the fact that in TROLL *light* classes are understood as composite objects having the class extension as sub-objects. Therefore in contrast to TROLL an extra notion of class is not needed in TROLL *light*. This leads to a more orthogonal use of object descriptions. Over and above that concepts like class attributes, meta-classes, or heterogeneous classes are inherent in TROLL *light* while they had to be introduced in TROLL by additional language features. Second TROLL *light* incorporates a query calculus providing a general declarative query facility for object-oriented databases. For instance, terms of this calculus may be used in object specifications to describe derivation rules for attributes, or to query object communities in an ad hoc manner.

The paper is organized as follows. Section 2 shortly introduces the various language concepts offered by TROLL *light*. A more detailed presentation of TROLL *light* can be found in [CGH92]. In Section 3 a simplified computational model for TROLL *light* is outlined. The last section discusses future work and gives some concluding remarks.

2 Concepts of TROLL *light*

TROLL *light* is a language for describing static and dynamic properties of objects. As in TROLL object descriptions are called templates in TROLL *light*. Because of their pure declarative nature templates may be compared with the notion of class found in object-oriented programming languages. In the context of databases however, classes are also associated with class extensions so that we settled on a fresh designation. Templates show the following structure.

```
TEMPLATE name of the template
    DATA TYPES    data types used in current template
    TEMPLATES     other templates used in current template
    SUBOBJECTS    slots for sub-objects
    ATTRIBUTES    slots for attributes
    EVENTS        event generators
    CONSTRAINTS   restricting conditions on object states
    VALUATION     effect of event occurrences on attributes
    DERIVATION    rules for derived attributes
    INTERACTION   synchronization of events in different objects
    BEHAVIOR      description of object behavior by a CSP-like process
END TEMPLATE;
```

To give an example for templates let us assume that we have to describe authors. For every author the name, the date of birth, and the number of books sold by year have to be stored. An author may change her name only once in her life. An appropriate TROLL *light* specification would hence be:

```
TEMPLATE Author
    DATA TYPES    String, Date, Nat;
    ATTRIBUTES    Name:string; DateOfBirth:date;
                  SoldBooks(Year:nat):nat;
    EVENTS        BIRTH create(Name:string, DateOfBirth:date);
                  changeName(NewName:string);
                  storeSoldBooks(Year:nat, Number:nat);
                  DEATH destroy;
    VALUATION     [create(N,D)] Name=N, DateOfBirth=D;
                  [changeName(N)] Name=N;
                  [storeSoldBooks(Y,NR)] SoldBooks(Y)=NR;
    BEHAVIOR      PROCESS authorLife1 =
                    ( storeSoldBooks -> authorLife1 |
                      changeName -> authorLife2 |
                      destroy -> POSTMORTEM );
                  PROCESS authorLife2 =
                    ( storeSoldBooks -> authorLife2 |
                      destroy -> POSTMORTEM );
                  ( create -> authorLife1 );
END TEMPLATE;
```

Data types are assumed to be specified with a data specification language. In the KORSO project we use SPECTRUM [BFG+92] as a reference language, but other proposals like ACT ONE [EFH83], PLUSS [Gau84], Extended ML [ST86], or OBJ3 [GW88] will do their job just as good. With the DATA TYPES section the signature of data types is made known to the current template. For example, referring to Nat means that the data sort nat, operations like + : nat x nat -> nat, and predicates like <= : nat x nat are visible in the current template definition. Note that we employ a certain convention concerning the naming of data types, templates and associated sorts. We use names starting with an upper case letter to denote data types and templates whereas the corresponding sort names start with a lower case letter.

Attributes denote observable *properties* of objects. They are specified in the ATTRIBUTE section of a template by $a[(\lceil p_1 : \rceil s_1, \ldots, \lceil p_n : \rceil s_n)] : d$, where a is an attribute name generator, d is a sort expression determining the range of an attribute, and $s_1, \ldots, s_n$ ($n \geq 0$) denote optional parameter sorts (data or object sorts). To stress the meaning of parameter sorts, optional parameter names p_i might be added. The sort expression d may be built over both data sorts and object sorts by applying predefined type constructors. We have decided to include the type constructors tuple, set, bag, list, and union. Of course other choices can be made. Thereby, *complex* attributes can be specified, e.g., data-valued, object-valued, multi-valued, composite, alternative attributes, and so on. The interpretation of all sort expressions contains the undefined element $\perp$, and therefore all attributes are *optional* by default. Attribute names may be provided with parameters. For example, by the declaration SoldBooks(Year:nat):nat a possibly infinite number of attribute names like SoldBooks(1993) is introduced. We demand that in a given state of an object only a finite number of attributes takes values different from $\perp$ such that only these attributes have to be stored. A parametrized attribute $a(s_1, \ldots, s_n) : s$ can also be viewed as an attribute $a : \text{set}(\text{tuple}(s_1, \ldots, s_n, s))$, but clearly the formerly inherent functional dependency would have to be described by an explicit constraint. The same works for parametrized sub-object constructors to be discussed later.

Incidents possibly appearing in an object's life are modeled by *events*. Events are specified in the EVENT section of a template by $e[(\lceil p_1 : \rceil s_1, \ldots, \lceil p_n : \rceil s_n)]$, where e denotes an event generator and $s_1, \ldots, s_n$ ($n \geq 0$) represent optional parameter sort expressions. To underline the meaning of parameters, optional parameter names p_i may be added. Event parameters are used to define the effects of an event occurrence on the current state of an object (see the explanation of the VALUATION section below), or they are used to describe the transmission of data during communication (see the explanation of the INTERACTION section below). Special events in an object's life cycle are the BIRTH and DEATH events with which an object's life is started or ended. Several birth or death events may be specified for the same object. A template may have no death event, but we require a it to have at least one birth event.

The effect of events on attribute values is specified in the VALUATION section of a template definition by *valuation rules*, having the following form.

$$[\{precondition\}] \; [event\_descr] \; attr\_term = term$$

Such a rule says that immediately after an event occurrence belonging to the event descriptor *event_descr*, the attribute given by the attribute term *attr_term* has the value determined by destination *term*. The applicability of such a rule may be restricted by a *precondition*, i.e., the valuation rule is applied only if the precondition is true. It is important to note that the precondition as well as both terms are evaluated in the state *before* the event occurrence. Thereby all these items may have free variables which however must appear in the event descriptor. An *event descriptor* consists of an event generator and a (possibly empty) list of variables representing the parameters of an event. By a concrete occurrence of such an event, these variables are instantiated with the actual event parameters in such a way that the precondition and the other terms can be evaluated.

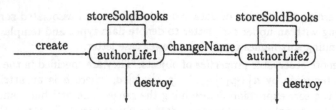

Figure 1: Behavior of authors

To summarize we can say that a valuation is a proposition stating that after the occurrence of a certain event some attributes shall have some special values. The following frame rule is assumed: Attributes which are not caught by a suitable valuation rule of a given event remain unchanged. Before the birth event, all attributes of an object are assumed to be undefined. Thus if the event in question is a birth event, some attributes may remain undefined. It is important to note that an attribute can only be affected by local events, i.e., events which are specified in the same template in which the attribute is specified.

In the BEHAVIOR section the possible life-cycles of objects are restricted to admissible ones by the means of *behavior patterns*. Using an abstract characterization, behavior patterns are given by an event-driven machine, which will be called *o-machine* (o for object) in the sequel. This machine has a finite number of *behavioral* states (cf. [SM92]). However, regarding an object together with its attributes we shall generally get an infinite number of object states. To achieve a compact notation within the BEHAVIOR section such an o-machine is represented by process descriptions. By process descriptions *event sequences*, which an object must go through, can be specified as well as *event dependent branchings*. Event sequences always lead back into processes, and event alternatives, which produce the same o-machine state transitions, can be listed separated by commas. The possibility of providing every state transition with a precondition allows for *guarded events* (cf. CSP [Hoa85]). We have used the keyword POSTMORTEM within the behavior specification to denote that the object vanishes. In Figure 1 the behavior of authors is visualized by the corresponding o-machine representation. Within an object description the behavior section may be missing. In that case life-cycles are unrestricted, i.e., it is only required for life-cycles to start with a birth event and possibly end with a death event.

After dealing with the TROLL *light* features for simple objects we now turn to *composite objects*. In order to combine several authors in a higher-level object, *classes* are usually introduced as containers in object-oriented databases. Here, TROLL *light* does not support an explicit class concept. Classes are viewed as composite objects instead, and therefore classes have to be described by templates as already mentioned in the last section. However, the means of describing the relationship between a container object and the contained objects must be added. This is done by introducing sub-object relationships denoting (exclusive) *part-of* relationships. The following example gives the format of container objects for authors.

```
TEMPLATE AuthorClass
    DATA TYPES   String, Date, Nat;
    TEMPLATES    Author;
    SUBOBJECTS   Authors(No:nat):author;
```

```
ATTRIBUTES    DERIVED NumberOfAuthors:nat;
EVENTS        BIRTH createClass;
                    addObject(No:nat, Name:string, DateOfBirth:date);
                    removeObject(No:nat);
              DEATH destroyClass;
CONSTRAINTS   NumberOfAuthors<10000;
DERIVATION    NumberOfAuthors=CNT(Authors);
INTERACTION   addObject(N,S,D) >> Authors(N).create(S,D);
              removeObject(N) >> Authors(N).destroy;
END TEMPLATE;
```

Within the TEMPLATES section, other (existing) *templates* can be made known to the current template. We assume templates to induce corresponding object sorts. Hence referring to Author means that the object sort author, and the attributes and the event generators of Author are visible in AuthorClass.

An object of sort authorClass will hold finitely many author objects as private components or *sub-objects*. In order to be able to distinguish several authors from a *logical* point of view, an explicit *identification mechanism* is needed. One way to install such a mechanism would be to assign a unique name for each sub-object, e.g., MyAuthor, YourAuthor, ... Indeed, such a name allocation could be expressed in TROLL *light* as follows

    SUBOBJECTS MyAuthor:author; YourAuthor:author; ...;

But clearly, in the case of a large number of authors such a procedure is not practicable. A solution is given by parametrized sub-object constructors as shown in the example. As with parametrized attributes, a possibly infinite number of logical sub-object names like Authors(42) can be defined by the sub-object name declaration for authors, but not the author objects themselves. The declaration only states that in context of an object of sort authorClass, author objects are identified by a natural number. In semantic data models the parameter No would be called a *key*. But the parameters need not be related to any attributes of the objects they identify. Each logical sub-object name corresponds to an object of the appropriate object sort. Analogously to attributes, we demand that in each state only a finite number of defined sub-objects exists.

Interaction rules are specified in the INTERACTION part of the template definition. During the lifetime of an author class there will be events concerning the insertion or deletion of authors. In AuthorClass, the insertion of an author should always be combined with the creation of a new author object. In the template definition this is expressed by an event calling rule addObject(N,S,D) >> Authors(N).create(S,D), where N denotes a natural number, S a string, and D a date. The general event calling scheme is expressed as

    [{*precondition*}] [*src_obj_term*.]*src_event_descr* >> [*dest_obj_term*.]*dest_event_term* .

Such a rule states that whenever an event belonging to the source event descriptor *src_event_descr* occurs in the source object denoted by *src_obj_term*, and the *precondition* holds, then the event denoted by the destination event term *dest_event_term* must occur simultaneously in the destination object denoted by *dest_obj_term*. If one of the object terms is missing, then the corresponding event descriptor (respectively event term) refers to the current object. The source object term is not allowed to have free

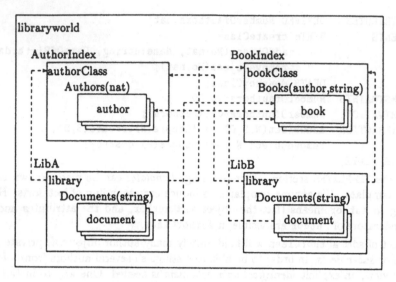

Figure 2: Instance of sort `libraryworld`

variables, but the destination object term, the destination event term, and the precondition are allowed to have free variables, which however have to occur in the source event descriptor. As already mentioned in the explanation of valuation rules, these free variables are instantiated by an event occurrence corresponding to the event descriptor in such a way that the precondition and the other terms can be evaluated. In almost all cases the objects to be called for must be alive. The one and only exception is when a parent object calls for the birth of a sub-object. In this case it is even a requisite that the destination object does not exist already.

In addition to the mentioned language features TROLL *light* offers some further, more sophisticated concepts which will be mentioned only briefly: Possible object states can be restricted by explicit *constraints*, which are specified in the `CONSTRAINTS` section of a template. Constraints are given by closed formulas. For example, an author container of sort `authorClass` may only be populated by less than 10000 authors. Derivable attributes can be specified by stating *derivation rules* which, in fact, are closed terms. In the example we declared the attribute `NumberOfAuthors` of `authorClass` as derived. We employed the function `CNT` which counts the elements of a multi-valued term. Please remember that in the derivation rule `Authors` is a term of sort `set(tuple(nat,author))`. So `NumberOfAuthors` does not need to be set by explicit valuation rules.

Up to now we have only dealt with the description of objects (i.e., templates). In order to attain object instances we must choose one template as the *schema template*, and one fixed object of the corresponding object sort as the *schema object*. Because the schema object is not a sub-object of any other specified object it has no logical object name. Therefore, we give the special name INIT to this object. Then we can make use of templates to derive all other object names by means of a term algebra construction. This will be shown in the next section.

In Figure 2 we give an example of a possible embedding of an object of sort `authorClass`. The considered world consists of an object representing a *library world* which has two libraries, an author index and a book index as sub-objects. The author index and the book index are modeled as class objects having authors respectively books as sub-objects. The libraries have (disjoint sets of) documents as sub-objects. In the figure, objects are represented by rectangles with the corresponding object sorts standing in the upper left corner and the logical sub-object name standing outside. Thereby, sub-object relationship is represented by putting the rectangles of the sub-objects into the rectangle of their common superobject. A similiar approach for representing object structures graphically can be found [KS91].

Within the figure dashed lines are used to state another kind of relationship. In the specification these relationships are given by *object-valued* attributes. For instance, each document refers to a corresponding book, and each book refers to (a list of) authors. Thereby, *object sharing* is possible because several books can have common authors. Furthermore, object-valued attributes can change in the course of time whereas sub-object relationships are fixed. Therefore, object-valued attributes provide a flexible mechanism to model arbitrary relationships between objects.

3 Simplified Computational Model for TROLL *light*

In this section we will outline a simplified computational model for TROLL *light*. Indeed this model is just one of many possible models. Unluckily it is beyond the scope of this paper to discuss a general notion of model for specification of templates. Our computational model serves as a basis for an animation system of the language. For this reason, we use concrete set- and graph-theoretical concepts to describe it. We will not go into detailed technicalities, but we will give an overview and a general taste how the computational model works by means of examples. We assume that a collection of TROLL *light* templates together with one distinguished (schema) template is given. We first define the data part of our model.

Definition 3.1: Data signature and data algebra

A *data signature* $\Sigma_D = (S_D, \Omega_D, \Pi_D)$ is given by

- a set S_D of data sorts,
- a family of sets of operation symbols $\Omega_D = < \Omega_{D,ws} >_{ws \in S_D^* \times S_D}$, and
- a family of sets of predicate symbols $\Pi_D = < \Pi_{D,w} >_{w \in S_D^*}$.

We assume a data signature is interpreted by one fixed term-generated data algebra $DA \in \text{ALG}_{\Sigma_D}$. The set $E_{DA} = \{t_L = t_R \mid t_L, t_R \in T_{\Sigma_D}, t_L^{DA} = t_R^{DA}\}$ denotes the equations induced on Σ_D-terms by DA. □

The algebra DA is assumed to be specified with a data specification language. Next we define signatures for template collections where the names of object sorts, and the names and parameters of sub-object constructors, attributes, events, and o-machine states are given.

Definition 3.2: Template collection signature

A *template collection signature* $\Sigma_O = (S_O, SUB_O, ATT_O, EVT_O, OMS_O)$ is given by

- a set S_O of object sorts — from now on we assume $S := S_D \cup S_O$ —,
- a family of sets of sub-object symbols $SUB_O = < SUB_{O,ows} >_{ows \in S_O \times S^* \times S_O}$,
- a family of sets of attribute symbols $ATT_O = < ATT_{O,ows} >_{ows \in S_O \times S^* \times S}$,
- a family of sets of event symbols $EVT_O = < EVT_{O,ow} >_{ow \in S_O \times S^*}$, and
- a family of finite sets of o-machine states $OMS_O = < OMS_o >_{o \in S_O}$.

The notation $u : o \times s_1 \times \ldots \times s_n \to s$ stands for $u \in SUB_{O,os_1 \ldots s_n s}$. We assume an analogous notation for attribute symbols, event symbols, and o-machine states. □

Example 3.3:

For our running example we have the following template collection signature.

$$
\begin{array}{lll}
S_O & = \{ & \text{author, authorClass} \ \} \\
SUB_O & = \{ & \text{Authors : authorClass} \times \text{nat} \to \text{author} \ \} \\
ATT_O & = \{ & \text{Name : author} \to \text{string,} \\
& & \text{DateOfBirth : author} \to \text{date,} \\
& & \text{SoldBooks : author} \times \text{nat} \to \text{nat,} \\
& & \text{NumberOfAuthors : authorClass} \to \text{nat} \ \} \\
EVT_O & = \{ & \text{create : author} \times \text{string} \times \text{date,} \\
& & \text{changeName : author} \times \text{string,} \\
& & \text{storeSoldBooks : author} \times \text{nat} \times \text{nat,} \\
& & \text{destroy : author,} \\
& & \text{createClass : authorClass,} \\
& & \text{addObject : authorClass} \times \text{nat} \times \text{string} \times \text{date,} \\
& & \text{removeObject : authorClass} \times \text{nat,} \\
& & \text{destroyClass : authorClass} \ \} \\
OMS_O & = \{ & \text{authorClasslife : authorClass,} \\
& & \text{authorLife1, authorLife2 : author} \ \}
\end{array}
$$

□

After having fixed the syntax of our template collections we can define a name space for possible objects to be considered.

Definition 3.4: Universe of possible object names

We construct the *universe of possible object names* by considering the term algebra generated by data operations in Σ_D, sub-object operations in SUB_O, and a special constant INIT for the object sort of the distinguished (schema) template. Additionally we factorize this term algebra by the equations E_{DA}:

$$UNIV := < T_{\Sigma_D + SUB_O + \text{INIT}, E_{DA}, o} >_{o \in S_O}.$$ □

Example 3.5:

For our running example template AuthorClass is the schema template and therefore INIT is a constant for object sort authorClass.

$UNIV_{authorClass} = \{ \text{INIT} \}$

$UNIV_{author} = \{ \ldots, \text{Authors(INIT,23), Authors(INIT,42), } \ldots \}$ □

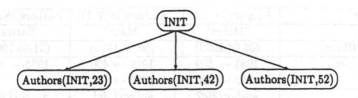

Figure 3: Objects in example object community state

Up to now we have only considered the template signatures. Now we will assume that families of constraints, valuation formulas, derivation rules, interaction rules, and behavior descriptions constitute the template specifications. We first define how object community states, i.e., snapshots describing the shape of the object community after a number of events occurred, look like.

Definition 3.6: Object community state

An *object community state* is a finite tree (N, E) with root INIT (provided $N \neq \emptyset$) such that the conditions given next are satisfied.

- The nodes N are given as an S_O-indexed family of finite sets with $N_s \subseteq UNIV_s$.

- The edges are determined by $E = \{(t, u(t, t_1, ..., t_n)) \mid t, u(t, t_1, ..., t_n) \in N\}$.

- With the above it is possible to define $I_s := \begin{cases} DA_s & \text{if } s \in S_D \\ N_s & \text{if } s \in S_O \end{cases}$.

- A sub-object symbol $u : o \times s_1 \times ... \times s_n \rightarrow s$ is interpreted as a mapping
  $u_I : N_o \times I_{s_1} \times ... \times I_{s_n} \rightarrow N_s \cup \{\bot\}$.
  $u_I : (t_o, t_1, ..., t_n) \mapsto \begin{cases} u(t_o, t_1, ..., t_n) & \text{if } u(t_o, t_1, ..., t_n) \in N_s \\ \bot & \text{otherwise} \end{cases}$
  Each u_I is already fixed by the choice of the nodes N.

- An attribute symbol $a : o \times s_1 \times ... \times s_n \rightarrow s$ is interpreted as a mapping $a_I : N_o \times I_{s_1} \times ... \times I_{s_n} \rightarrow I_s \cup \{\bot\}$. In contrast to u_I the interpretation of a non-derived attribute is free.

- Each object has a fixed o-machine state determined by a family of functions $< \delta_o >_{o \in S_O}$ with $\delta_o : N_o \rightarrow OMS_o$ for each $o \in S_O$.

- The above interpretation must obey the definitions given for derived attributes and must fulfill the specified static constraints.

The initial object community state S_0 is defined by $S_0 = (\emptyset, \emptyset)$. The set of all object community states is denoted by STATES. □

Example 3.7:

Consider the object community state for the running example depicted in Figure 3. Here only the sub-object relationships and their tree structure but not the attributes are shown. More general graph structures with shared objects can be realized with object-valued attributes.

author	Authors(INIT,52)	Authors(INIT,23)	Authors(INIT,42)
Name$_I$	'Mailer'	'Mann'	'Sartre'
DateOfBirth$_I$	(01.04.1929)	(06.06.1875)	(21.06.1905)
SoldBooks$_I$	1981 $\mapsto$ 300	1979 $\mapsto$ 400	1985 $\mapsto$ 600
	1983 $\mapsto$ 600	1980 $\mapsto$ 700	
δ_{author}	authorLife2	authorLife1	authorLife1
authorClass	INIT		
NumberOfAuthors$_I$	3		
$\delta_{authorClass}$	authorClasslife		

Figure 4: Attributes and o-machine states in example object community state

The attribute and o-machine state values for the example state are characterized by the table in Figure 4. In the state there is an author which has already changed his name from 'Kundera' to 'Mailer' and consequently his o-machine state is 'authorLife2' disallowing a second name change.

If we had a constraint restricting authors to be born in the 20th century (for example by requiring CONSTRAINTS DateOfBirth>=(01.01.1900)), then the above would not be a valid object community state. On the other hand we even get a valid object community state if δ_{author} maps Authors(INIT,52) to 'authorLife1'. $\square$

So far we have not considered events. The occurrence of events changes the state of the object community. Due to our mechanism for event calling the occurrence of one event may force other events to occur as well. Therefore state transitions will in general be induced by sets of events.

Definition 3.8: Event and closed event set

A possible *event* in an object community state (N, E) consists of an event symbol $e : o \times s_1 \times ... \times s_n$ together with appropriate actual parameters $(t_o, t_1, ..., t_n) \in (N_o \times I_{s_1} \times ... \times I_{s_n})$ where the first parameter — the object for which the event takes place — is written in dot notation before the event symbol: $t_o.e(t_1, ..., t_n)$. A finite set of events $\{\underline{e}_1, ..., \underline{e}_n\}$ is called *closed* with respect to event calling, if there does not exist another event $\underline{e}$ and an interaction rule such that an event $\underline{e}_i$ calls for $\underline{e}$.

The set of all possible closed events sets is denoted by EVENTS. $\square$

Definition 3.9: Object community state transitions

The specification of the templates determine a relation

    TRANSITION $\subseteq$ STATES $\times$ EVENTS $\times$ STATES

given as follows. Let an object community state S and a closed event set $\{\underline{e}_1, ..., \underline{e}_n\}$ be given. $(S, \{\underline{e}_1, ..., \underline{e}_n\}, S') \in$ TRANSITION, if the following conditions are satisfied.

- The object community state S' differs from S only by modifications forced by evaluating valuation rules of events from the event set (modification of attribute functions) and by birth and death events from the event set (modification of the sub-object structure).

- There is at most one event per object in the event set.
- Each event fits into its object's life-cycle determined by the behavior patterns.
- In the state S' the o-machine states have changed in accordance with the behavior patterns.

Recall, if S' is a valid object community state, then the constraints are satisfied by S'.
□

Example 3.10:

We give examples of transitions for the object community state in Example 3.7 and also examples for event sets which do not induce such transitions. One possible transition is induced by the closed event set:

{ INIT.addObject(37,'Mann',(27.03.1871)),

   Authors(INIT,37).create('Mann',(27.03.1871)) }.

The state belonging to the transition is characterized as follows:

- $N'_{authorClass} = N_{authorClass}$
- $N'_{author} = N_{author} \cup \{$ Authors(INIT,37) $\}$
- $Name'_I = Name_I \cup \{$ Authors(INIT,37) $\mapsto$ 'Mann' $\}$
- $DateOfBirth'_I = DateOfBirth_I \cup \{$ Authors(INIT,37) $\mapsto$ (27.03.1871) $\}$
- $SoldBooks'_I = SoldBooks_I$
- $NumberOfAuthors'_I : INIT \mapsto 4$
- $\delta'_{author} = \delta_{author} \cup \{$ Authors(INIT,37) $\mapsto$ authorLife1 $\}$

After this, another transition may be performed by executing the closed event set:

{ INIT.removeObject(52), Authors(INIT,52).destroy }

As examples for events which cannot be elements of event sets inducing a state transition for the state in Example 3.7 we mention the following events.

- Authors(INIT,52).changeName('Kundera')
- Authors(INIT,52).create('Kundera',(01.04.1929))
- INIT.addObject(23,'Mann',(06.06.1875))
- INIT.removeObject(13)
- INIT.createClass

A sequence of event sets — indeed, this is one sequence of many possible ones — leading to the state in Example 3.7 would start with { INIT.create } and then one could sequentially create the authors 'Kundera', 'Mann', and 'Sartre'. Afterwards one could have the five events updating the SoldBooks attribute. The last step could be the event changing the Name attribute of author 'Kundera' to 'Mailer'. This sequence of event sets has a rather non-parallel nature. But parallel occurrence of events is supported as well. Consider a situation where authors 'Kundera' and 'Mann' already exist and have not changed their name. In this situation the following closed event set could induce a state transition.

{ INIT.addObject(42,'Sartre',(21.06.1905)),

   Authors(INIT,42).create('Sartre',(21.06.1905)),

   Authors(INIT,52).changeName('Mailer'),

   Authors(INIT,23).storeSoldBooks(1980,700) }

These four events occur concurrently in different objects.   □

We now come to our last and most important notion, namely the object community. An object community is the collection of all object community states reachable from the initial state together with the transitions between these states. Thus an object community reflects structure and behavior of the described system.

Definition 3.11: Object community

The semantics of a template collection specification, i.e., the specified *object community*, is a directed, connected graph where the nodes are the object community states and edges represent closed event sets. All nodes must be reachable from the initial state S_0.   □

In general, the object community will be an arbitrary directed graph and not a tree or dag structure. From the initial state subsequent states are reached by performing a birth event for the (schema) object INIT. Not all possible object community states contribute to the object community but only those which are reachable from the initial state. Our computational model implies for instance severe restrictions on the interpretation of attributes: $a_I(t_o, t_1, ..., t_n)$ yields values different from $\perp$ only for a finite number of arguments, because only a finite number of events lies between the initial state and a reachable object community state.

4   Conclusion and Future Work

Although we have now fixed the language, a lot of work remains to be done. Special topics for our future investigations will be formal semantics of TROLL *light*, animation, certification, and an integrated development environment.

Here, we have presented a computational model for TROLL *light* serving as a basis for implementation. But what we desire is an abstract *formal semantics* for TROLL *light* which is compositional. The approach presented in [Mes92a, Mes92b] seems to us very promising since our computational model already reflects the idea to have systems with structured states and transitions between these states. On the other hand the entity algebra approach of [Reg90] or the Berlin projection specifications [EPPB+90] may be suitable tools as well. A detailed overview on algebraic semantics of concurrency can be found in [AR93].

Animation is another aspect of our future studies. Formal specifications cannot be proved to be correct with regard to some informal description of the system to be specified. Only validation can help to ensure that a formally specified system behaves as required. Animation of specifications seems to be an important way to support validation in an efficient manner. Therefore, we develop and implement an animation system for TROLL *light* specifications supporting designers in validating object communities against their specification.

Another aspect we work on is *certification*. *Consistency* is an essential prerequisite for specifications to be used. For instance, animation, as well as proving properties of specifications, requires consistent specifications. Therefore, requirements for specifications to be consistent have to be worked out. *Verification* is needed to prove the properties of specifications. Writing all intended properties of some objects into their template specification seems to be unrealistic, because in this way specifications would become larger and larger and finally nobody would be able to read and understand such specifications.

We also design an *integrated development environment* [VHG+93] for our specification language. In principle, this environment will support all phases of software development: Information analysis, formal specification, animation, certification, and transformation into executable code. Of course we do not expect that at the end of our project there will be a complete development environment. Consequently we are going to implement the important parts of such a system.

References

[AR93]   E. Astesiano and G. Reggio. Algebraic Specification of Concurrency. In M. Bidoit and C. Choppy, editors, *Recent Trends in Data Type Specification — Proc. 8th Workshop on Specification of Abstract Data Types*, pages 1–39. Springer, Berlin, LNCS 655, 1993.

[BBE+90]   G. v. Bochmann, M. Barbeau, M. Erradi, L. Lecomte, P. Mondain-Monval, and N. Williams. Mondel: An Object-Oriented Specification Language. Département d'Informatique et de Recherche Opérationnelle, Publication 748, Université de Montréal, 1990.

[BFG+92]   M. Broy, C. Facchi, R. Grosu, R. Hettler, H. Hussmann, D. Nazareth, F. Regensburger, and K. Stølen. The Requirement and Design Specification Language SPECTRUM — An Informal Introduction (Version 0.3). Technical Report TUM–I9140, Technische Universität München, 1992.

[BHL90]   D. Bjorner, C.A.R. Hoare, and H. Langmaack, editors. *VDM'90: VDM and Z — Formal Methods in Software Development*. Springer, LNCS 428, 1990.

[BKL+91]   M. Bidoit, H.-J. Kreowski, P. Lescanne, F. Orejas, and D. Sannella, editors. The Compass Working Group: *Algebraic System Specification and Development*. Springer, Berlin, LNCS 501, 1991.

[Bre91]   R. Breu. *Algebraic Specification Techniques in Object Oriented Programming Environments*. Springer, LNCS 562, 1991.

[CGH92]   S. Conrad, M. Gogolla, and R. Herzig. TROLL light: A Core Language for Specifying Objects. Informatik-Bericht 92–02, Technische Universität Braunschweig, 1992.

[CSS89]   J.-F. Costa, A. Sernadas, and C. Sernadas. OBL-89 Users Manual (Version 2.3). Internal report, INESC, Lisbon, 1989.

[DO91]   O.-J. Dahl and O. Owe. Formal Development with ABEL. Technical Report 159, University of Oslo, 1991.

[EFH83]   H. Ehrig, W. Fey, and H. Hansen. ACT ONE: An Algebraic Specification Language with Two Levels of Semantics. Technical Report 83-03, Technische Universität Berlin, 1983.

[EGL89]   H.-D. Ehrich, M. Gogolla, and U.W. Lipeck. *Algebraische Spezifikation abstrakter Datentypen – Eine Einführung in die Theorie.* Teubner, Stuttgart, 1989.

[EGS90]   H.-D. Ehrich, J. A. Goguen, and A. Sernadas. A Categorial Theory of Objects as Observed Processes. In J.W. de Bakker, W.P. de Roever, and G. Rozenberg, editors, *Foundations of Object-Oriented Languages (Proc. REX/FOOL Workshop, Noordwijkerhood (NL))*, pages 203–228. Springer, LNCS 489, 1990.

[EM85]   H. Ehrig and B. Mahr. *Fundamentals of Algebraic Specification 1: Equations and Initial Semantics.* Springer, Berlin, 1985.

[EM90]   H. Ehrig and B. Mahr. *Fundamentals of Algebraic Specification 2: Modules and Constraints.* Springer, Berlin, 1990.

[EPPB⁺90] H. Ehrig, F. Parisi-Presicce, P. Boehm, C. Rieckhoff, C. Dimitrovici, and M. Große-Rhode. Combining Data Type and Recursive Process Specifications Using Projection Algebras. *Theoretical Computer Science*, 71:347–380, 1990.

[Gab93]   P. Gabriel. The Object-Based Specification Language Π: Concepts, Syntax, and Semantics. In M. Bidoit and C. Choppy, editors, *Recent Trends in Data Type Specification — Proc. 8th Workshop on Specification of Abstract Data Types*, pages 254–270, Berlin, 1993. Springer, LNCS 655.

[Gau84]   M.-C. Gaudel. A First Introduction to PLUSS. Technical Report, Université de Paris-Sud, Orsay, 1984.

[GW88]   J.A. Goguen and T. Winkler. Introducing OBJ3. Research Report SRI-CSL-88-9, SRI International, 1988.

[HK87]   R. Hull and R. King. Semantic Database Modelling: Survey, Applications, and Research Issues. *ACM Computing Surveys*, 19(3):201–260, 1987.

[Hoa85]   C.A.R. Hoare. *Communicating Sequential Processes.* Prentice-Hall, Englewood Cliffs (NJ), 1985.

[Jon86]   C.B. Jones. *Systematic Software Developing Using VDM.* Prentice-Hall, Englewood Cliffs (NJ), 1986.

[JSHS91]   R. Jungclaus, G. Saake, T. Hartmann, and C. Sernadas. Object-Oriented Specification of Information Systems: The TROLL Language. Informatik-Bericht 91–04, Technische Universität Braunschweig, 1991.

[JSS91]   R. Jungclaus, G. Saake, and C. Sernadas. Formal Specification of Object Systems. In S. Abramsky and T. Maibaum, editors, *Proc. TAPSOFT'91, Brighton*, pages 60–82. Springer, Berlin, LNCS 494, 1991.

[KS91]   G. Kappel and M. Schrefl. Object/Behavior Diagrams. In *Proc. 7th Int. Conf. on Data Engineering, Kobe (Japan)*, pages 530–539, 1991.

[Mes92a]   J. Meseguer. A Logical Theory of Concurrent Objects and its Realization in the Maude Language. In G. Agha, P. Wegener, and A. Yonezawa, editors, *Research Directions in Object-Based Concurrency.* MIT Press, 1992. To appear.

[Mes92b]   J. Meseguer. Conditional Rewriting as a Unified Model of Concurrency. *Theoretical Computer Science*, 96(1):73–156, 1992.

[Mil80]   R. Milner. *A Calculus of Communicating Systems.* Springer, Berlin, 1980.

[Reg90]   G. Reggio. Entities: An Institution for Dynamic Systems. In H. Ehrig, K.P. Jantke, F. Orejas, and H. Reichel, editors, *Recent Trends in Data Type Specification*, pages 246–265. Springer, LNCS 534, 1990.

[Rei85]   W. Reisig. *Petri Nets: An Introduction*. Springer, Berlin, 1985.

[SE91]   A. Sernadas and H.-D. Ehrich. What is an Object, after all? In R.A. Meersman, W. Kent, and S. Khosla, editors, *Object-Oriented Databases: Analysis, Design & Construction (DS-4), Proc. IFIP WG 2.6 Working Conference, Windermere (UK) 1990*, pages 39–70. North-Holland, 1991.

[SGG⁺91]   C. Sernadas, P. Gouveia, J. Gouveia, A. Sernadas, and P. Resende. The Reification Dimension in Object-Oriented Database Design. In D. Harper and M. C. Norrie, editors, *Proc. Int. Workshop on Specification of Database Systems*, pages 275–299. Springer, 1991.

[SM92]   S. Shlaer and S.J. Mellor. *Object Life Cycles: Modeling the World in States*. Yourdon Press computing series, Prentice-Hall, Englewood Cliffs (NJ), 1992.

[SRGS91]   C. Sernadas, P. Resende, P. Gouveia, and A. Sernadas. In-the-large Object-Oriented Design of Information Systems. In F. Van Assche, B. Moulin, and C. Rolland, editors, *Proc. Object-Oriented Approach in Information Systems*, pages 209–232. North Holland, 1991.

[SSE87]   A. Sernadas, C. Sernadas, and H.-D. Ehrich. Object-Oriented Specification of Databases: An Algebraic Approach. In P.M. Stocker and W. Kent, editors, *Proc. 13th Int. Conf. on Very Large Data Bases (VLDB)*, pages 107–116. Morgan-Kaufmann, Palo Alto, 1987.

[SSG⁺91]   A. Sernadas, C. Sernadas, P. Gouveia, P. Resende, and J. Gouveia. OBLOG — Object-Oriented Logic: An Informal Introduction. Technical report, INESC, Lisbon, 1991.

[ST86]   D.T. Sannella and A. Tarlecki. Extended ML: An Institution-Independent Framework for Formal Program Development. In *Proc. Workshop on Category Theory and Computer Programming*, pages 364–389. Springer, LNCS 240, 1986.

[VHG⁺93]   N. Vlachantonis, R. Herzig, M. Gogolla, G. Denker, S. Conrad, and H.-D. Ehrich. Towards Reliable Information Systems: The KORSO Approach. In C. Rolland, editor, *Proc. 5th Int. Conf. Advanced Information Systems Engineering*. Springer, LNCS Series, 1993. *To appear*.

[Wie91]   R. Wieringa. Equational Specification of Dynamic Objects. In R.A. Meersman, W. Kent, and S. Khosla, editors, *Object-Oriented Databases: Analysis, Design & Construction (DS-4), Proc. IFIP WG 2.6 Working Conference, Windermere (UK) 1990*, pages 415–438. North-Holland, 1991.

[Wir90]   M. Wirsing. Algebraic Specification. In J. Van Leeuwen, editor, *Handbook of Theoretical Computer Science, Vol. B*, pages 677–788. Elsevier, North-Holland, 1990.

Analogical Type Theory

Bo Yi,   Jiafu Xu

Institute of Computer Software, Nanjing University
22th, Hankou Road, Nanjing 210008, P. R. China
Email: bmanju@ica.beijing.canet.cn

Abstract: This paper proposes an analogical type system based on ITT. Based on the intuitive meaning of analogy, a set of rules are introduced and justified to deal with analogy. Analogies on types as well as terms are introduced in our system. By an analogy between types, we mean a pair of mappings satisfying coherent conditions on those types. Terms with analogous types are analogical if their focus points meet with each other under the analogy of the types. Analogical theorem proving and term derivation methods are also illustrated as examples of applications of our system.

1 Introduction

As one of the main kinds of reasoning methodologies, analogy is a creative activity in human's brain to accumulate knowledge through similar things or situations. Analogy is not only used in common sense reasoning, but also widely applied to formal inferences. In computer science, analogical methodology is closely related to induction, explanation-based learning, object-oriented programming and software reuse. As the base of abstraction, analogy can serve as a base of methodologies mentioned above. Though analogical methodology is still immature, its applications have been explored in theorem proving[14, 8, 7], machine learning, and program derivation [3, 4].

In [14], we have proposed an analogical calculus LK_A based on Gentzen's natural deduction system LK[11]. Two formulas are analogical in LK_A if their equivalence can be proved in LK_A from a set of correspondings A_c. Lacking intuitionistic meanings, LK could not be directly used as a model of computation in computer science. This motivates the construction of an analogical type theory from Martin-lof's intuitionistic type theory (ITT) [9, 10]. The principle of propositions as types proposed by W. A. Howard [5] and then introduced in programming by Martin-lof [9] can be described as follows: A task (or its specification) can be represented by a proposition. If there is a solution to the task, then, there is a proof which is an evidence of the truth of the proposition. Now, all proofs of the proposition form a set which serves as a type of that proposition. By the way, a term of a type is a

This project is partially supported by National Science Fundations of China.

method to solve the task represented by the type. If a type is constructive, a term in the type is computable, and it can serve as a program to solve that task. This indicates that programming for a task is to prove that the type of the task is inhabited, and find out a term, which is regarded as a program, to support the acknowledgement about its inhabitation. But there is no automated method to derive such a term. A formal proof guided by human is difficult and verbose. It seems helpful for ITT to embrace analogy method to guide a formal proof automatically with existing ones.

In this paper, we proposed an analogical ITT (named AITT) by adding some judgements and rules to an ITT to deal with analogy. In section 2, we descuss the intuitive meaning of analogy on types. Based on the meaning, three kinds of judgements and their introduction rules are introduced in section 3. Some derived rules for special type constructors are discussed in section 4. They make an analogical proof constructive as well. To illustrate applications of AITT, some examples are given in section 5.

2 Analogy on Types

Any two things may seem to be analogical, depending on whatever aspects in your mind. Analogical aspects are called view-ports of analogy[12]. When two things are analogical in your mind, the aim of analogical analysis is not to find whether they are analogical, but to give out a concrete analogical correspondence between these two things. The given analogical correspondence should be precisely, explicitly defined, and, of course, coherent, as mentioned by George Polya[6]. As refined by one of the authors in [13], an analogical correspondence between two systems is a part-to-part or one-to-one association between them, and the association ought to be coherent with the structures of the systems. We can say that an analogical correspondence is an extensional representation of a view-port of an analogy.

In this article, a system is a type (a set or a proposition) under intuitionistic meaning as in [9]. Thus, an analogy correspondence, or A_c, is a pair of mappings from one type to another. The coherence of mappings is that there is an isomorphism between the quotient sets derived from the pair of natural mappings on the types. It is illustrated in Figure 1.

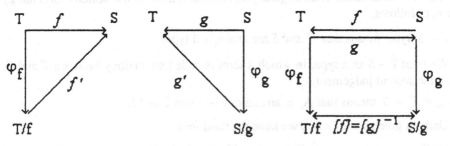

Figure 1. Analogy on types

where φ_f and φ_g are natural mappings under f and g, f' and g' are the unique injections satisfying $f = f' \varphi_f$ and $g = g' \varphi_g$. The mappings $[f]: T/f \to S/g$ and $[g]: S/g \to T/f$ are defined by $\varphi_g f'$ and $\varphi_f g'$ respectively. With coherence, we hope that $[f]$ and $[g]$ are bijections and $[f]^{-1} = [g]$. We give that

Definition 1: $< f, g >$ is an analogy between T and S, if $[f] = [g]^{-1}$.

In this way, a term t in T is analogous to an s in S under $< f, g >$ if $[f]$ maps the equivalence class of t to that of s, and vice versa.

By the definition above, we can prove that

Theorem 1: $< f, g >$ is an analogy between T and S if and only if $f g f = f : T \to S$ and $g f g = g : S \to T$. (Proof is omitted for saving space.)

This theorem gives us an applicable conditions on the mappings f and g themselves. It will be utilized in the introduction rule of analogy in section 3.

It is easy to show that the last diagram in Figure 1 is commutative, i.e.

Theorem 2: $\varphi_g f = [f] \varphi_f$ and $[g] \varphi_g = \varphi_f g$ if $< f, g >$ is an analogy between T and S.

The convers proposition is not true, since commutativity can only guarantee that $[f]$ and $[g]$ be bijective but not inverse mappings of each other. So the definition of analogy on types is stronger than that of commutativity. But, commutativity shows an intuitive meaning of coherence of a correspondence and makes it possible to construct analogical types from analogous base types. The construction will be discussed in section 4.

In the following sections, we assume that the equality $=$ is extensionally predefined as an equivalence relation on each type, and will be used without further explanation. In the next section, we will define three kinds of analogical judgements and basic rules which are justified from the intuitive meaning mentioned above.

3 Analogical Judgements and Basic Rules

We have three kinds of analogical judgements, based on the notions of analogy above, as follows.

1). $T \sim S$ *type* means that T and S are analogical types.

We treat $T \sim S$ as a type, in which a term is called an analogy between T and S. Thus, the second judgement

2). $A_c \in T \sim S$ means that A_c is an analogy between T and S.

Under a given analogy A_c, we have the third form

3). $t \in T \sim_{Ac} s \in S$, which indicates that t in T is analogous to s in S under A_c.

In ITT, logical inclusion $P \supset Q$ is represented by a type $T{\to}S$, where T and S are type versions of propsitions P and Q. Thus, when $T \to S$ and $S \to T$, there exist some kinds of similarities which may serve as an aspect of analogy between T and S. In this way, the first form of judgements ($T \sim S$ *type*) is derivable if and only if $T \to S$ and $S \to T$ are both types. Furthermore, $T \to S$ is a type if T and S are both types, so, we have the rule

$\sim$ Formation:

$$\frac{T \quad type \qquad S \quad type}{T \sim S \quad type}$$

and the substitution rule

$\sim$ Type Substitution:

$$\frac{T = T' \quad type \qquad S = S' \quad type \qquad T' \sim S' \quad type}{T \sim S \quad type}$$

says that T and S are analogical types if their equivalant types T' and S' are.

A precisely defined analogy between T and S is a pair of functions $< f, g >$ satisfying the coherence conditions mentioned in section 2. It can be introduced by the following rule.

$\sim$ Introduction:

$$\frac{f \in T{\to}S \quad g \in S{\to}T \quad fg\,fx = fx \in S\,[x \in T] \quad gfgx = gx \in T\,[x \in S]}{< f, g > \in T \sim S}$$

The rule says that, for any t in T (s in S), its f-image (g-image) is invariant under folding of f and g (g and f). The f-image of S , and g-image of T are called the *kernel* of S under f , and the kernel of T under g , respectively. In other words, T, S can be divided into equivalence classes with elements in the kernel of T, S as representative elements, respectively. It is illustrated in Figure 2

a. Analogical mappings   b. Non-analogical mappings

Figure 2.

Thus, an analogy can also be regarded as a pair of bijective mappings on the kernels, and one is the inverse of the other. A term $t_0 \in T$, for which there exists an $s_0 \in S$ satisfying $f t_0 = s_0$, $g s_0 = t_0$, is called a *focus* term of T. Indeed, s_0 is also a focus in S, which is corresponding to t_0 in T. If f maps t to s_0, and t_0 is a corresponding focus to s_0, we say that t_0 is the focus of t. The focus of s in S is defined similarly. It is easy to show that any element in a kernel is a focus.

There are two elimination rules for $\sim$ type:

$\sim$ Elimination

$$\frac{<f,g> \in T \sim S \qquad t \in T}{f t \in S} \qquad\qquad \frac{<f,g> \in T \sim S \qquad s \in S}{g s \in T}$$

These rules are the basis of analogy derivation. The term $f t$ or $g s$ are analogically derived from the analogy $< f, g >$ and the known term t or s, respectivly. If A_c is an analogy correspondence between T and S, the judgement below

$$t \in T \sim_{Ac} s \in S$$

means t in T is analogous to s in S according to the analogy A_c between T and S. From figure 2.a, the analogical terms can be defined by the rule

$\sim$ Term Introduction:

$$\frac{<f,g> \in T \sim S \qquad t \in T \quad s \in S \qquad g f t = g s \in T}{t \in T \sim_{<f,g>} s \in S}$$

The rule can be justified as follows: t and s are analogical if their focus elements are corresponding under f and g. Since $f t$ and $g s$ are foci of S and T respectively, $g f t = g s$ and $f g s = f t$ mean that $f t$ and $g s$ are corresponding foci, so t and s are analogical under $< f, g >$. $f g s = f t$ can be omitted from the rule's condition part because it can be derived from others.

The following rule, which is a special case of the rule above, defines the one-to-one correspondence on kernels. It is simpler and easier to apply in some special cases.

$\sim$ Term Introduction 1:

$$\frac{<f,g> \in T \sim S \qquad t = g s \in T \quad s = f t \in S}{t \in T \sim_{<f,g>} s \in S}$$

The subscription of $\sim$ will be omitted where no ambiguity occurs in the context. The rule below

$\sim$ Term Introduction 2:

$$\frac{t \in T \qquad s \in S}{t \in T \sim s \in S}$$

is another special case where there exists an $A_c = <f, g>$, such that $f t = s$ and $g s = t$. For instance, let $f x = s$ and $g y = t$ for any $x \in T$, $y \in S$. It is easy to prove that the A_c is exactly an analogy between T and S. In this rule, we omit the subscription A_c since it is partially defined by this introduction and will be expanded in other parts of a proof.

There is also a substitution rule for analogical terms

~ Term Substitution:

$$\frac{t = t' \in T \qquad s = s' \in S \qquad t' \in T \sim s' \in S}{t \in T \sim s \in S}$$

which means that terms are analogical if their equal terms are.

4 Construction of Analogy

In ITT, there are four forms of rules for each type constructor. They are *Formation*, *Introduction*, *Elimination* and *Computation* [1]. For each type constructor, there are analogical rules in AITT corresponding to the original forms, by which we can construct an analogy from analogical base types. We discuss in detail those rules for function types, which are most important and interesting in AITT. The formation rule for analogical function types is simple.

$\rightarrow \sim \rightarrow$ Formation:

$$\frac{A \sim B \qquad C \sim D}{A \rightarrow C \sim B \rightarrow D}$$

The Introduction rule for analogical function types is listed as follows.

$\rightarrow \sim \rightarrow$ Introduction:

$$\frac{<\phi_1, \psi_1> \in A \sim B \qquad <\phi_2, \psi_2> \in C \sim D}{<\lambda f. \phi_2 f \psi_1, \lambda g. \psi_2 g \phi_1> \in A \rightarrow C \sim B \rightarrow D}$$

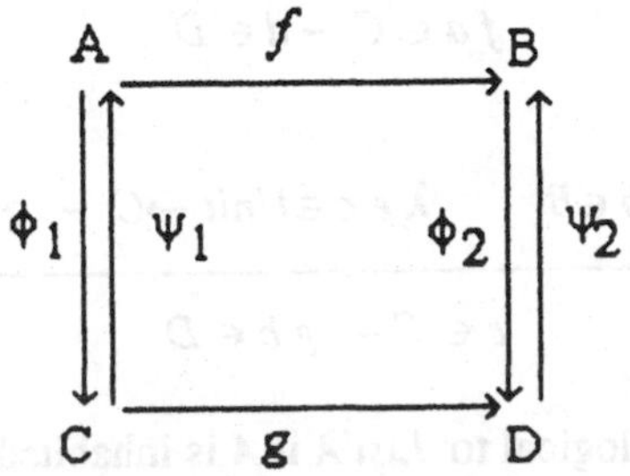

Figure 3   Analogy between function types

To justify this rule, (refer to Figure 3. above), let $\Phi = \lambda f .\phi_2 f \psi_1$, $\Psi = \lambda g.\psi_2 g$ ϕ_1. Then $\Phi f = \phi_2 f \psi_1$, $\Psi g = \psi_2 g \phi_1$. To ensure that $< \Phi, \Psi >$ is really an analogy between $A \to C$ and $B \to D$, we must check that $\Phi \Psi \Phi = \Phi$ and $\Psi \Phi$ $\Psi = \Psi$, according to theorem 1. For any $f : A \to C$, $\Phi \Psi \Phi f = \phi_2 \Psi \Phi f \psi_1$ by unfolding the first Φ, and then it is unfolded to $\phi_2 \psi_2 \Phi f \phi_1 \psi_1$ and $\phi_2 \psi_2 \phi_2 f$ $\psi_1 \phi_1 \psi_1$. But $\phi_2 \psi_2 \phi_2 = \phi_2$ and $\psi_1 \phi_1 \psi_1 = \psi_1$ since $< \phi_1, \psi_1>$ is an analogy between A and B, $< \phi_2 , \psi_2 >$ is that of C and D. So, $\Phi \Psi \Phi f = \Phi f$. The proof of $\Psi \Phi \Psi = \Psi$ is parallel. We restate this as a theorem.

Theorem 3: $\Phi \Psi \Phi = \Phi$ and $\Psi \Phi \Psi = \Psi$, (Φ and Ψ are defined above).

It is evident that analogical functions in function types make the diagram in figure 3 commutative, which coincides with the intuititive meaning of analogy between types. The elimination and computation rules for analogical function types are simple, we only listed them out.

$\to \sim \to$ Elimination (Application):

$$\frac{a \in A \sim b \in B \qquad f \in A \to C \sim g \in B \to D}{fa \in C \sim gb \in D}$$

$\to \sim \to$ Computation:

$$\frac{a \in A \sim b \in B \qquad c \in C \sim d \in D [x \in A \sim y \in B]}{\lambda x.c \ a = c(a/x) \in C \sim \lambda y.d \ b = d(b/y) \in D}$$

In ITT, *Unit* is a type in which there exactly inhabited one element u. A type T can be regarded as a functor type from *Unit* to T, since type T and *Unit* $\to T$ are isomorphic. A *Unit* can be freely appended to a type T to match with anothor function type, as in the following rules, for instance.

Unit Extension Left:

$$\frac{a \in A \sim u \in Unit \qquad f \in A \to C \sim \lambda e.d \in Unit \to D}{fa \in C \sim d \in D}$$

Unit Extension Right:

$$\frac{u \in Unit \sim b \in B \qquad \lambda e.c \in Unit \to C \sim g \in B \to D}{c \in C \sim gb \in D}$$

In this way, *Nat* is analogical to *List A* if A is inhabited since for any given a in A, $< l_a , len >$ is an analogy between *Nat* and *List A*, where $l_a 0 = [ ]$, l_a succ

$n = cons\ a\ l_a\ n$, for any n in *Nat*, and $len\ [\ ] = 0$, $len\ cons\ a\ l = succ\ len\ l$, for any a in A and l in *List A*.

The Π type is a dependent version of function type. Rules for Π which are similar to $\rightarrow$'s are listed as follows.

$\Pi \sim \Pi$ Formation:

$$\frac{A \sim B \qquad C \sim D\,[\,x \in A \sim y \in B\,]}{\Pi\, x{:}A.C \sim \Pi\, y{:}B.D}$$

$\Pi \sim \Pi$ Introduction:

$$\frac{<\phi_1, \psi_1> \,\in A \sim B \qquad <\phi_2, \psi_2> \,\in C \sim D\,[\,x \in A \sim_{<\phi_1,\psi_1>} y \in B\,]}{<\lambda f.\,\phi_2 f \psi_1\, y,\ \lambda g.\,\psi_2 g\, \phi_1\, x> \,\in \Pi\, x{:}A.C \sim \Pi\, y{:}B.D}$$

$\Pi \sim \Pi$ Application:

$$\frac{a \in A \sim b \in B \qquad f \in \Pi\, x{:}A.C \sim g \in \Pi\, y{:}B.D}{f a \in C\,(a/x) \sim g b \in D\,(b/y)}$$

$\Pi \sim \Pi$ Computation:

$$\frac{a \in A \sim b \in B \qquad c \in \Pi\, x{:}A.C \sim d \in \Pi\, y{:}B.D\,[\,x \in A \sim y \in B\,]}{\lambda\, x.c\ a = c\,(a/x) \in C\,(a/x) \sim \lambda\, y.d\ b = d(b/y) \in D(b/y)}$$

The rules for $\times$ and $+$ are simple. We just list introductions for them, other rules are mirror to their original ones.

$\times \sim \times$ Introduction:

$$\frac{<\phi, \psi_1> \,\in A \sim B \qquad <\phi_2, \psi_2> \,\in C \sim D}{<\phi_1 \times \phi_2,\ \psi_1 \times \psi_2> \,\in A \times C \sim B \times D}$$

where $(f \times g)(a, c) = (f a, g c)$

$+ \sim +$ Introduction:

$$\frac{<\phi_1, \psi_1> \in A \sim B \quad <\phi_2, \psi_2> \in C \sim D \quad \phi_1\,(a/x) = \phi_2\,(a/y)[(a,a) \in A \times C] \quad \psi_1(b/x) = \psi_2\,(b/y)[(b,b) \in B \times D]}{<\phi_1 + \psi_2,\ \psi_1 + \psi_2> \in A + C \sim B + D}$$

where $(f + g)inl(a) = i\,nl(fa)$ and $(f + g)inr(c) = inr(gc)$. The last two premises are the second coherent condition, in which ϕ_1 and ϕ_2 should map a common element of A and C to a common element of B and D. For ψ_1 and ψ_2, there is a parallel condition. Rules for Σ are similar to those for $\times$. The formation and introduction are listed here.

$\Sigma \sim \Sigma$ Formation:

$$\frac{A \sim B \qquad C \sim D \;[\,x \in A \sim y \in B\,]}{\Sigma x{:}A.C \;\sim\; \Sigma y{:}B.D}$$

$\Sigma \sim \Sigma$ Introduction:

$$\frac{<\phi_1,\psi_1> \in A \sim B \quad a \in A \sim_{<\phi,\psi>} b \in B \quad <\phi_2,\psi_2> \in C \sim D\,[x \in A \sim_{<\phi1,\psi1>} y \in B]}{<\phi_1 \times \phi_2,\; \psi_1 \times \psi_2> \in \Sigma x{:}A.C \sim \Sigma y{:}B.D}$$

The justification of these rules is not difficult and omitted for saving space.

5   Analogical Derivation

Derivational analogy proposed by J. G. Carbonell[2] is more reasonable than direct analogy. In this section, we illustrate analogical derivation methods based on AITT by examples. When we have types T and S with similar structures, a proof of T (or $t \in T$) can be obtained by proving $T \sim S$ (or $t \in T \sim s \in S$) guided by a proof of S (or $s \in S$) automatically on commom structures of T and S. If there is no common structure at all, the proof is reduced to T ($t \in T$) itself, which can be proved manually as in ITT. By equality of types, we can make some structure transformations to match the structure of a source, then apply analogical rules automatically with the guidence of the source. This is different from direct structure mapping[7] or higher order unification[8] in which no changes can be made on the structure of a proposition. The only thing we need to do is to prove those judgements above an application of a rule $\sim$ Formation or $\sim$ Term Introduction. In fact, these rules give out corresponding types and terms directly when their upper judgements are proved. They can be thought of as sub-tasks to be solved, and can introduce new analogical source if necessary. To deal with this, we give out source introduction rules:

Source Type Introduction:

$$\frac{S \; type \qquad S \sim T \; type}{T \; type}$$

Source Term Introduction:

$$\frac{s \in S \qquad s \in S \sim t \in T}{t \in T}$$

By the way, AITT can be thought of as an analogical analysis tool to find the concrete analogical correspondence between two types, or analogical terms of the types. It can be shown by the following example.

Example 1: The judgement $3 \in Nat \sim [a, b, c] \in List\ A$ is provable in AITT, where Nat and $List\ A$ are types defined in [1] and $a, b,$ and c are in type A. We can derive a concrete analogical correspondence from a proof of the judgement.

To prove this judgement in AITT, we apply term substitution to it, which is deduced to three judgements

1. $succ\ 2 = 3 \in Nat$,
2. $cons\ a\ [b, c] = [a, b, c] \in List\ A$ and
3. $succ\ 2 \in Nat \sim cons\ a\ [b, c] \in List\ A$.

The first two are provable in ITT and thus also in AITT. Applying $\rightarrow \sim \rightarrow$ elimination to the third, we have

3.1 $2 \in Nat \sim [b, c] \in List\ A$, and

3.2 $\lambda n.\ succ\ n \in Nat{\rightarrow}Nat \sim \lambda l.\ cons\ a\ l \in List\ A \rightarrow List\ A$.

The first can be deduced to $0 \in Nat \sim [\ ] \in List\ A$ by using the same rules above. Then it is split by $\sim$ term introduction to $0 \in Nat$ and $[\ ] \in List\ A$, which are axioms of Nat and $List$ types. Thus, we get a corresponding between 0 in Nat and $[\ ]$ in $List\ A$. To prove 3.2 above, we apply a Unit Extension to it and then obtain

3.2.1 $\lambda u.\lambda n\ succ\ n \in Unit \rightarrow Nat \rightarrow Nat \sim \lambda x.\lambda l\ .cons\ x\ l \in A{\rightarrow}List\ A{\rightarrow}List\ A$, and

3.2.2 $u \in Unit \sim a \in A$.

They are also split into judgements in ITT, since there is no further deducible common structure.

As a result of the proof, correspondings in this example are

$u \in Unit$ to $a \in List\ A$, and

$\lambda u.\lambda n.succ\ n \in Unit \rightarrow Nat \rightarrow Nat \sim \lambda x.\lambda l.cons\ x\ l \in A \rightarrow List\ A \rightarrow List\ A$

which are derived from the proof of 3.2., and

$u \in Unit \sim b \in A$

$u \in Unit \sim c \in A$ and

$0 \in Nat \sim [\ ] \in List\ A$ which are derived from a proof of judgement 3.1. above.

Now, we can derive an analogy from the correspondence by non-canonical terms which are applications of elimination schemes on Nat and $List\ A$:

$< \lambda n.Natelim(n, [\ ], \lambda x.\lambda l.cons\ x\ l), \lambda l.Listelim(l, 0, \lambda u.\lambda n.succ\ n) > \in Nat \sim List\ A$

When a proof $s \in S$ is given, and $< \phi, \psi >$ is an analogy between S and T, a corresponding term in T can be obtained by applying ϕ to s. If analogy between S and T is not explicitly given, we can derive such a term by proving $s \in S \sim y$ $\in T$, where y is a new variable which does not occur freely in T and S. y will be instantiated to a term analogous to s in the proof guided by the given proof of s

∈ *S*, when the structures of *T* and *S* are analogical. Such a term can serve as a program for the task (specified by) *T*, It can be illustrated in the following example.

Example 2: Suppose we have known that the function + is defined as $\lambda n.\lambda m.Natelim(n, m, \lambda k.succ\ k) \in Nat \rightarrow Nat \rightarrow Nat$, and given out an example of function $F \in List\ A \rightarrow List\ A \rightarrow List\ A$ which we intend to be analogous to an instance of +.

$3 + 2 = 5 \in Nat \sim F\ [a, b, c]\ [d, e] = [a, b, c, d, e] \in List\ A$.

We want to synthesize the function *F*.

To do this, we replace + by its definition and instantiate *F* to an elimination form on *List A*, according to the analogy between *Nat* and *List A*.

$Natelim\ (3, 2, \lambda k.succ\ k) = 5 \in Nat \sim Listelim\ ([a, b, c], [d, e], \lambda x.\lambda l.F_1 x\ l) = [a, b, c, d, e] \in List\ A$, where F_1 is a variable to be instantiated further.

With the guidence of *Nat computation*, we obtain the judgements below

1. $3 \in Nat \sim [a, b, c] \in List\ A$.

2. $2 \in Nat \sim [d, e] \in List\ A$.

3. $\lambda k.succ\ k \in Nat \rightarrow Nat \sim \lambda x.\lambda l.F_1 x\ l \in A \rightarrow List\ A \rightarrow List\ A$ and

4. $succ\ Natelim\ (2, 2, \lambda k.succ\ k) = 5 \in Nat \sim F_1\ a\ Listelim\ (\ [b, c], [d, e], \lambda x.\lambda l.F_1 x\ l) = [a, b, c, d, e] \in List\ A$.

From proofs of 1. and 2. , which should be easy as described in example 1, we derive a corresponding

$\lambda u.\lambda n\ succ\ n \in Unit \rightarrow Nat \rightarrow Nat \sim \lambda x.\lambda l\ .cons\ x\ l \in A \rightarrow List\ A \rightarrow List\ A$,

which is used to instantiate F_1 in judgement 3. Now , judegment 4 has been instantiated too, and can be easily proved with the guidence of a proof of *succ Natelim* $(2, 2, \lambda k.succ\ k) = 5 \in Nat$ which is a part of the original proof of $3 + 2 = 5 \in Nat$.

Thus, function *F* is synthesized out as

$\lambda l_1.\lambda l_2 Listelim\ ([b, c], [d, e], \lambda x.\lambda l.cons\ x\ l) \in List\ A \rightarrow List\ A \rightarrow List\ A$

This can be regarded as a program derived from an example analogous to an instance of a known function.

6 Conclusions

The correctness is important in program derivation. AITT, inherited from ITT and LK_A, is consistent. A term of a target type can be derived correctly by proving analogy between its example(s) and instance(s) of the source in which the source type is extensionally analogous to the target type. The derived term can be thought of as a program if the target type is a specification of a task. What we should do next is to implement a proof checker based on AITT, and find a method to organize judgements and their proofs so that the proof checker can choose a good source from the proof base for a given target.

7 Acknowledgement

Authors are grateful to Prof. Daoxu Chen, Dr. Jianguo Lu and Dr. Chengxiang Zhai for fruitul discussions.

8 References

1. Roland C. Backhouse, Paul Chisholm. Grant Malcolm and Erik Saaman, Do-it-your-self type theory, Formal aspects of Computing, No 1, pp. 19-84, 1989.
2. J. Carbonell, Derivational Analogy: A Theory of Reconstructive Problem Solving and Expertise Acquisition, in R.S.Michalski et al(ed.), Machine Intelligence: An Artificial Intelligence Approach, Morgan Kaufmann, 1986.
3. Dershowitz, N., Programming by Analogy, in: R. S. Michalski, J. G. Carbonell and T. M. Mitchell (Eds), Machine Learning II: An Artificial Intelligence Approach, Morgan Kaufmann, Los Altos, CA, 1986, pp. 395-423.
4. Dietzen, S. R. and W. L. Scherlis, Analogy in Program Development, in: The Role of Language in Problem Solving 2 , J. C. Boudreaux, B. W. Hamill, and R. Jernigan(Eds), North-Holland, 1987, pp. 95-115.
5. W. A. Howard, The formulae-as-types notion of construction, in 'To H. B. Curry: essay on conbinatory logic, λ-calculus and formalism', ed. Hindely J. R. and Selidin J. P., Academic Press, New York, 1980.
6. George Polya, Induction and Analogy, Princeton Hall, 1957
7. Masateru, Harao, Analogical Reasoning for Natural Deduction Theorem Proving, in (Edited by Zhongzhi Shi) Automated Reasoning, IWAR 92 (Beijing' China) pp. 220-228
8. Jianguo, Lu, Analogical Program Derivation Based on Type Theory, To appear in Theoretical Computer Science Vol.113.
9. Per Martin-Lof, Constructive Mathematics and Computer Programming, in Logic, Methodology and Philosophy of Science, pp. 153-175, North Holland, 1982.
10. Bengt Nordstrom, Kent Petersson and Jan M. Smith, Programming in Mation-lof's Type Theory, Clarendon Press (Oxford) 1990.
11. Takeuti, G. Proof Theory, North-Holland, (2nd. edition), 1987.
12. Bo, Yi and Jiafu, Xu, A Survey on Analogy Reasoning,Computer Science (in Chinese), 1989, No. 4, pp. 1-8
13. Bo, Yi, Analogy Model and Analogy Correspondence:A Formal Theory on Analogy, Doctorial Thesis of Nanjing University, December 1989.
14. Bo, Yi and Jiafu, Xu, Analogy Calculus, To appear in Theoretical Computer Science vol. 113, May 1993.

Improving the Multiprecision Euclidean Algorithm

Tudor Jebelean
RISC-LINZ
Johannes Kepler University, A-4040 Linz, Austria
tjebelea@risc.uni-linz.ac.at

Abstract. We improve the implementation of Lehmer-Euclid algorithm for multiprecision integer GCD computation by partial cosequence computation on pairs of double digits, enhanced condition for exiting the partial cosequence computation, and approximative GCD computation. The combined effect of these improvements is an experimentally measured speed-up by a factor of 2 over the currently used implementation.

Introduction

The speed of multiprecision arithmetic has a decisive influence on the speed of computer algebra systems. By performing a series of measurements upon typical algebraic algorithms [2], we found out that most of the computation time is spent in integer/rational operations, and this proportion increases when the length of the input coefficients increases. Also, the absolute total time increases dramatically, exclusively on behalf of integer/rational operations. For example, in Gröbner Bases computation [1], when increasing the coefficient length from 1 to 10 decimal digits, the proportion of rational operations grows from 62% to 98%, and the total computation time grows by a factor of 25. Also, we noticed that GCD computation is the most time consuming operation on long integers (25% to 70% of the total computing time). [11] also notes that 80% of Gröbner Bases computing time is spent in long integer arithmetic.

For long integers which occur in typical applications (up to 100 computer words), it is generally accepted that the best GCD scheme is Lehmer-Euclid multiprecision Euclidean algorithm (see [9, 8, 3]), although recent research on generalizing the binary GCD algorithm of [14] (see [13, 16, 7, 6]) found better schemes. FFT based algorithms (see [12, 10]), although having a better theoretical complexity, are not efficient for integers in this range.

We present here three ways of improving Lehmer-Euclid algorithm:

- partial cosequence computation using pairs of double digits,
- a better method for detecting the end of the partial cosequence computation,
- approximative GCD computation.

---

* Acknowledgements: Austrian Ministry for Science and Research, project 613.523/ 3-27a/ 89 (Gröbner Bases) and doctoral scholarship; POSSO project(Polynomial Systems Solving – ESPRIT III BRA 6846).

The combined effect of these improvements is a speed-up by a factor of 2 over the (currently used) straight forward implementation of Lehmer's GCD algorithm.

1 The multiprecision Euclidean algorithm

We present here the Euclidean algorithm and the multiprecision version of it [9] and we notice two useful properties of the cosequences. More details can be found in [8, 3]. We adopt here (with small modifications) the notations used in [3].

Let $A > B > 0$ be integers.

The Euclidean algorithm consists in computing the *remainder sequence* of (A, B): $(A_0, A_1, \ldots, A_n, A_{n+1})$ defined by the relations

$$A_0 = A, \quad A_1 = B, \quad A_{i+2} = A_i \bmod A_{i+1}, \quad A_{n+1} = 0. \tag{1}$$

It is well known that

$$A_n = GCD(A, B).$$

The extended Euclidean algorithm consists in computing additionally the *quotient sequence* $(Q_1, \ldots, Q_n)$ defined by

$$Q_{i+1} = \lfloor A_i / A_{i+1} \rfloor,$$

and the *first* and *second cosequences* $(U_0, U_1, \ldots, U_{n+1})$ and $(V_0, V_1, \ldots, V_{n+1})$:

$$(U_0, V_0) = (1, 0), \quad (U_1, V_1) = (0, 1),$$
$$(U_{i+2}, V_{i+2}) = (U_i, V_i) - Q_{i+1}(U_{i+1}, V_{i+1}).$$

Then the following hold:

$$A_{i+2} = A_i - Q_{i+1} A_{i+1}, \tag{2}$$

$$A_i = A U_i + B V_i. \tag{3}$$

It is useful to note that the signs of the elements of each cosequence alternate:

$$\text{if } i \text{ even, then } U_{i+1}, V_i \leq 0, \quad U_i, V_{i+1} \geq 0, \tag{4}$$

which leads to

$$(|U_{i+2}|, |V_{i+2}|) = (|U_i|, |V_i|) + Q_{i+1}(|U_{i+1}|, |V_{i+1}|). \tag{5}$$

These relations are useful for implementing the algorithm when the upper value of U_i, V_i is the maximum value which can be contained in a computer word (e.g. $2^{32} - 1$), because then the signs are difficult to handle, since the sign-bit cannot be used.

Another useful relation is

$$|V_i| \geq Q_1 |U_i|, \quad \text{for all } i \geq 1. \tag{6}$$

Indeed, the relation can be verified directly for $i = 1, 2$ and the induction step is:

$$|V_{i+2}| = Y_i + Q_{i+1}Y_{i+1} \geq Q_1 X_i + Q_{i+1}Q_1 X_{i+1} = Q_1|U_{i+2}|.$$

When A and B are multiprecision integers (several computer words are needed for storing the values), the divisions in (1) are quite time consuming. Lehmer [9] noticed that several steps of the Euclidean algorithm can be simulated by using only simple precision divisions. The idea is to apply the extended algorithm to

$$a = \lfloor A/2^h \rfloor, \quad b = \lfloor B/2^h \rfloor,$$

where $h \geq 0$ is chosen such that $a > b > 0$ are simple precision.

Then one gets the remainder sequence $(a_0, a_1, ..., a_k, a_{k+1})$, the quotient sequence $(q_1, ..., q_k)$ and the cosequences $(u_0, ..., u_{k+1})$ and $(v_0, ..., v_{k+1})$, for some $k \geq 0$ such that

$$q_i = Q_i, \quad \text{for all } i \leq k. \tag{7}$$

This process is called *digit partial cosequence calculation* (DPCC). When (7) is true, we say *the k-length quotient sequences of (a,b) and (A,B) match*. Then also

$$(u_i, v_i) = (U_i, V_i), \quad \text{for all } i \leq k + 1,$$

hence by (3):

$$A_k = u_k A + v_k B, \quad A_{k+1} = u_{k+1}A + v_{k+1}B, \tag{8}$$

and the process can be reiterated with the most significant digits of A_k, A_{k+1}.

However, the condition (7) cannot be tested directly, since Q_i are not known, hence one must have a condition upon a_i, q_i, u_i, v_i which ensures (7).

Lehmer's original algorithm computed the quotient sequences of $(a, b + 1)$ and $(a + 1, b)$ and compared the quotients at each step.

Collins [3] developed a better condition which needs only the computation of the sequences (q_i), (a_i) and (v_i). That is

$$a_{i+1} \geq |v_{i+1}| \quad \text{and} \quad a_{i+1} - a_i \geq |v_{i+1} - v_i|, \quad \text{for all } i \leq k. \tag{9}$$

This condition has the advantage that only one quotient has to be computed, and only one of the cosequences. As noted by G. H. Bradley (acc. to [8]), u_k, u_{k+1} can be found at the end of partial cosequence computation, by using

$$u_k a + v_k b = a_k, \quad u_{k+1}a + v_{k+1}b = a_{k+1}. \tag{10}$$

As a comparison base for our improvements, we implemented this version of the multiprecision Euclidean algorithm using the GNU multiprecision library [5] and GNU optimizing compiler on a Digital DECstation 5000/200. The benchmarks for integers with lengths varying from 5 to 100 digits (32 bit words) are shown in Fig.1(columns S). 1000 random integers were tested for each length. These experimental settings will remain unchanged for all the benchmarks presented in this paper.

2 The double digit algorithm

Recovering A_k, A_{k+1} by (8) involves 4 multiplications of a single digit number by a multidigit number, and this is the most time consuming part of the whole computation. Experimentally, for 32 bit digits, one notices that the final cofactors are usually shorter than 16 bits. Hence, if digit partial cosequences computation (DPCC) would be performed on double digits, then the recovering step would require the same computational effort, but will occur (roughly) two times less frequently. This is the main idea of the improvement of Lehmer's algorithm which we will discuss in this section.

Length	Steps per digit			Divisions per step			Time (ms)		
of operands	S	D	S/D%	S	D	S/D%	S	D	S/D%
5	1.52	0.80	190.0	7.74	17.23	44.9	0.73	1.90	38.4
25	2.28	1.03	221.4	7.58	16.94	44.7	8.25	13.30	62.0
50	2.38	1.07	222.4	7.57	16.92	44.7	25.35	31.32	80.9
75	2.41	1.08	223.1	7.57	16.93	44.7	51.17	53.03	96.5
100	2.42	1.09	222.0	7.56	16.91	44.7	85.82	78.67	109.1

Fig. 1. Benchmarks of GCD computation using simple digit DPCC (columns S) vs. double digit DPCC2 (columns D).

A straight forward implementation of this idea is not very efficient (see Fig.1): the experimental speed-up ranges from 38% (5 words operands) to 109% (100 words operands), hence is rather a slow down. However, the average values in the table show that, indeed, the number of divisions which are simulated within each step increases more than twice, which leads to a decrease of the number of steps by a factor of 2.2. The lack of speed is due to the fact that the partial cosequence computation is much slower, because of the double-digits operations involved. Therefore we will concentrate in the sequel on improving the partial cosequence computation for double words (DPCC2). Few simple improvements lead to a speed-up of almost 2:

- (A1) Some double-digit operations can be reduced to simple digit operations by using
$$|u_i| \leq |u_{i+1}| \leq |v_{i+1}|, \quad |v_i| \leq |v_{i+1}|,$$
which are a consequence of (5) and (6). Therefore, only v_{i+1} has to be computed with two digits. As long as its higher digit is null, the other cofactors are also small. The speed-up ranges now from 43% to 117%.
- (A2) Digit partial cosequence computation for simple words (DPCC) can be used at the beginning of DPCC2. The speed-up grows to 58% – 140%.
- (A3) Since only the second cosequence (v_i) is needed for testing the exit condition (9), the cofactors u_k, u_{k+1} can be computed at the end, using

(10). However, our experiments show no improvement by this method: the speed-up decreases to 53% – 133%.

- (A4) Because the quotients are usually small (see [8]), simulating division by repeated subtractions is also a source of speed-up. The idea is to repeat q_{small} times the subtraction cycle, and if the quotient is not found by then, to use division. Our experiments show that $q_{small} = 32$ is a good threshold value. The increase of the speed-up is surprisingly high: we get 101% to 180%. Note that simulating division by subtraction does not give a speed-up in the original DPCC, because the single precision division is fast enough.

The experimental results for the above variants of the algorithm are presented in Fig. 2, and we keep the best variant as "version A", which will be further improved in the sequel.

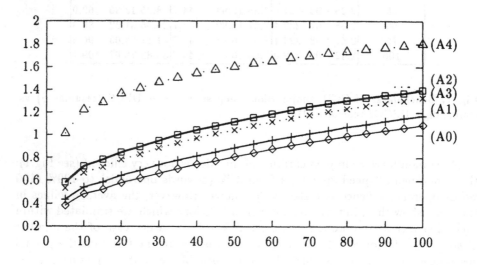

Fig. 2. Speed-up for successive improvements of version A (length of inputs from 5 to 100 32-bit words).

3   On the condition needed in digit partial cosequence computation

We develop here an "exact" condition for (7), that is, a condition which precisely indicates which is the highest k for which (7) is satisfied. Also we develop an "approximative" condition, which is weaker, but easier to test. We show then how to combine these two conditions, such that both efficiency and exactness are preserved.

3.1 An exact condition

Theorem 1. Let $a > b > 0$ be integers. Then

> the k-length quotient sequences of (a, b) and $(2^h b + A', 2^h a + B')$ match
> for any $h > 0$ and any $A', B' < 2^h$

if and only if

$$a_{i+1} \geq -u_{i+1} \quad \text{and} \quad a_i - a_{i+1} \geq v_{i+1} - v_i \quad \text{and} \quad i \text{ even}$$

or

$$a_{i+1} \geq -v_{i+1} \quad \text{and} \quad a_i - a_{i+1} \geq u_{i+1} - u_i \quad \text{and} \quad i \text{ odd}$$
$$\text{for all } i \leq k.$$

Proof. We keep all the notations as previously introduced. Let us also denote by $(A_0, A_1, \ldots, A_{k+1})$ the elements of the remainder sequence of

$$(A, B) = (2^h a + A', \ 2^h b + B').$$

Then

$$A_i = u_i(2^h a + A') + v_i(2^h b + B') = 2^h(u_i a + v_i b) + u_i A' + v_i B' = 2^h a_i + u_i A' + v_i B'.$$

Applying the relation

$$q = \lfloor M/N \rfloor \quad \text{iff} \quad 0 \leq M - qN < N \quad (M, N, q \text{ positive integers})$$

to

$$q_i = \left\lfloor \frac{A_{i-1}}{A_i} \right\rfloor = \left\lfloor \frac{2^h a_{i-1} + u_{i-1} A' + v_{i-1} B'}{2^h a_i + u_i A' + v_i B'} \right\rfloor = \left\lfloor \frac{a_{i-1} + u_{i-1}\frac{A'}{2^h} + v_{i-1}\frac{B'}{2^h}}{a_i + u_i\frac{A'}{2^h} + v_i\frac{B'}{2^h}} \right\rfloor,$$

one gets

$$0 \leq a_{i-1} + u_{i-1}\frac{A'}{2^h} + v_{i-1}\frac{B'}{2^h} - q_i \left(a_i + u_i\frac{A'}{2^h} + v_i\frac{B'}{2^h} \right) < \\ < a_i + u_i\frac{A'}{2^h} + v_i\frac{B'}{2^h}, \quad \forall h, A', B'. \tag{11}$$

Using

$$u_{i+1} = u_{i-1} - q_i u_i, \quad v_{i+1} = v_{i-1} - q_i v_i,$$

one notes that (11) is equivalent to the conjunction of

$$0 \leq \min\left\{ a_{i+1} + u_{i+1}\frac{A'}{2^h} + v_{i+1}\frac{B'}{2^h} \ \Big| \ 0 < h, \ 0 \leq A', B' < 2^h \right\},$$

$$\max\left\{ a_{i+1} - a_i + (u_{i+1} - u_i)\frac{A'}{2^h} + (v_{i+1} - v_i)\frac{B'}{2^h} \Big| 0 < h, \ 0 \leq A', B' < 2^h \right\} \leq 0,$$

The reason why the second inequality $<$ can be replaced by $\leq$ will become clear in the sequel.

Suppose i is even. Then by (4):

$$u_{i+1}, v_i \leq 0, \quad u_i, v_{i+1} \geq 0,$$

hence

$$\min \left\{ a_{i+1} + u_{i+1}\frac{A'}{2^h} + v_{i+1}\frac{B'}{2^h} \right\} = a_{i+1} + u_{i+1}\max\frac{A'}{2^h} + v_{i+1}\min\frac{B'}{2^h}$$

$$= a_{i+1} + u_{i+1} * 1 + v_{i+1} * 0 = a_{i+1} + u_{i+1}.$$

Also

$$\max \left\{ a_{i+1} - a_i + (u_{i+1} - u_i)\frac{A'}{2^h} + (v_{i+1} - v_i)\frac{B'}{2^h} \right\}$$

$$= a_{i+1} - a_i + (u_{i+1} - u_i)\min\frac{A'}{2^h} + (v_{i+1} - v_i)\max\frac{B'}{2^h}$$

$$= a_{i+1} - a_i + (u_{i+1} - u_i) * 0 + (v_{i+1} - v_i) * 1 = a_{i+1} - a_i + v_{i+1} - v_i.$$

Note that $B'/2^h < 1$ for any $h > 0$ and $B' < 2^h$. This is the reason why the second inequality $<$ in (11) can be replaced by $\leq$.

Now suppose i is odd. Similarly, one obtains

$$0 \leq a_{i+1} + v_{i+1}, \quad a_{i+1} - a_i + u_{i+1} - u_i \leq 0. \quad \square$$

The condition (9) can be obtained as a consequence, using (6).

The practical improvement of the algorithm by using this exact condition is very small. We found that the average number of divisions which are simulated in one step by using (9) is 16.92, vs. 17.12 by using the exact condition, hence an improvement of only 1.2% (see Fig. 3). However, the "exact condition" approach shows which is the maximum improvement which can be achieved by trying to increase the number of divisions per step. The speed-up over the original Lehmer's algorithm becomes 104% – 185%. We will try now to simplify the computations required for testing the condition.

Length	Steps per digit			Divisions per step			Time (ms)		
of operands	E	C	E/C%	E	C	E/C%	E	C	E/C%
5	0.80	0.80	100.0	17.42	17.23	101.1	0.70	0.72	97.2
25	1.02	1.03	99.0	17.14	16.94	101.2	5.72	5.85	97.8
50	1.06	1.07	99.1	17.13	16.92	101.2	15.48	15.83	97.8
75	1.07	1.08	99.1	17.13	16.93	101.2	29.08	29.78	97.6
100	1.07	1.09	98.2	17.11	16.91	101.2	46.50	47.63	97.6

Fig. 3. Experiments with multiprecision GCD algorithm using "exact" condition (columns E) vs. Collins' condition (columns C).

3.2 An approximative condition

Note that in the context of double-digit partial cosequence computation, the condition given by Theorem 1 is not sufficient. The cofactors $u_k, v_k, u_{k+1}, v_{k+1}$ must also be smaller than a computer word (2^{32} in our case) in order to be useful. We investigate in the sequel the size of the cofactors.

For this, we use the *continuant polynomials* (see also [8]) defined by

$$\begin{cases} P_0() = 1, \\ P_1(x_1) = x_1, \\ P_{i+2}(x_1, \ldots, x_{i+2}) = P_i(x_1, \ldots, x_i) + x_{i+2} * P_{i+1}(x_1, \ldots, x_{i+1}). \end{cases} \tag{12}$$

which are known to enjoy the symmetry

$$P_k(x_1, \ldots, x_i) = P_k(x_i, \ldots, x_1). \tag{13}$$

By comparing the recurrence relations (5) and (12) one notes

$$|u_i| = P_{i-2}(q_2, \ldots, q_{i-1}), \quad |v_i| = P_{i-1}(q_1, \ldots, q_{i-1}). \tag{14}$$

Also, by transforming (2) into

$$a_i = a_{i+2} + q_{i+1} * a_{i+1},$$

and using $a_i > a_{i+1}$, one can prove

$$a \geq a_i * P_i(q_i, \ldots, q_1), \quad b \geq a_i * P_{i-1}(q_i, \ldots, q_2).$$

Hence by (13) and (14) one has

$$|v_{i+1}| \leq a/a_i, \quad |u_{i+1}| \leq b/a_i.$$

Since a_0, a_1 are double digits, we have

$$2^{2n-1} \leq a = a_0 < 2^{2n}, \quad b = a_1 < 2^{2n},$$

where n is the bit-length of the computer word (in our implementation $n = 32$). Suppose

$$a_{i+1} > a_{i+2} \geq 2^n.$$

Then

$$|v_{i+1}| \leq a_0/a_i < a_0/2^n < 2^n < a_{i+1},$$

(note that $a_{i+1} \geq 2^n$ is sufficient here).

Also

$$a_i - a_{i+1} \geq a_i - q_{i+1} * a_{i+1} = a_{i+2} \geq 2^n,$$

$$|v_i| + |v_{i+1}| \leq |v_i| + q_{i+1} * |v_{i+1}| = |v_{i+2}| \leq a/a_{i+1} < a/2^n < 2^n.$$

The previous relations allow us to develop an alternative to (9), which is

$$a_{i+2} \geq 2^n. \tag{15}$$

For the practical implementation, this means that the high order digit of a_{i+2} is not zero, which is computationally inexpensive to test.

Also, note that when (15) holds, then

$$|v_i| \leq |v_{i+1}| \leq |v_{i+2}| \leq |v_{i+3}| \leq a/a_{i+2} < 2^n,$$

$$|u_i| \leq |u_{i+1}| \leq |u_{i+2}| \leq |u_{i+3}| \leq b/a_{i+2} < 2^n,$$

hence the sizes of the cofactors are "good" for $k = i$, $k = i + 1$, $k = i + 2$. Using this, we combine the "exact" and the "approximative" condition in order to obtain maximum of efficiency:

- In the main loop of DPCC2, the maximum i for which (15) holds is determined, and then we know $k = i$ is "good".
- After exiting the loop, we test the only second part of the condition (because $a_{i+1} \geq 2^n$ implies the first part), and if this test passes then we know $k = i+1$ is "good".
- Full condition must be tested in order to insure $k = i + 2$ is "good".

According to our experiments, if $a_{i+2} \geq 2^n$ and $a_{i+3} < 2^n$, then the maximum "good" value of k is $k = i$ in about 10% of the cases, $k = i+1$ in 20% of the cases, and $k = i + 2$ in 70% of the cases. The incidence of situations when $k = i + 3$ is "good" is insignificant.

In Fig.4 we present the comparative benchmarks for the algorithm using exact condition (columns B0) and approximative combined with exact (columns B1). One can see that the number of simulated divisions per step decreases insignificantly, but the running time improves: the speed-up is now 116% – 191% (see Fig.5). We keep the best variant as "version B", which will be further improved in the sequel.

Length	Steps per digit			Divisions per step			Time (ms)		
of operands	B0	B1	B0/B1%	B0	B1	B0/B1%	B0	B1	B0/B1%
5	0.80	0.80	100.0	17.42	17.41	100.1	0.70	0.63	111.1
25	1.02	1.02	100.0	17.14	17.13	100.1	5.72	5.30	107.9
50	1.06	1.06	100.0	17.13	17.11	100.1	15.48	14.58	106.2
75	1.07	1.07	100.0	17.13	17.12	100.1	29.08	27.80	104.6
100	1.07	1.08	99.1	17.11	17.09	100.1	46.50	44.92	103.5

Fig. 4. Benchmarks of version B (improved condition in DPCC): "exact" condition (columns B0) vs. "combined" condition (columns B1).

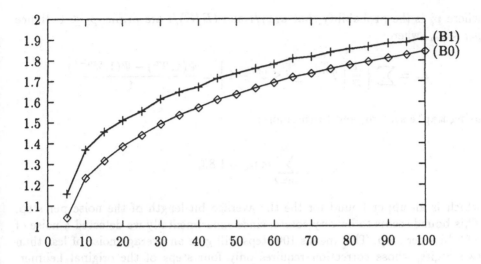

Fig. 5. Speed-up for successive improvements of version B (length of inputs from 5 to 100 32-bit words).

4 Approximative GCD computation

The final improvement is based on the following idea: instead of computation (8), which requires 4 long multiplications, compute only A_{k+1} and continue the algorithm with B and A_{k+1}. Then, since A_{k+1} is about one word shorter than B, a division will be performed at the next step, which is computationally equivalent to one long multiplication. Hence, one step of Lehmer's scheme (4 long multiplications) will be replaced by one "half–step" and one division (3 long multiplications), reducing the number of digit multiplications by 25%.

However, this new scheme lacks in correctness, since $GCD(A, B)$ divides $GCD(B, A_{k+1})$, but they might be different. Hence the final result G' of the algorithm will be, in general, bigger than the real GCD G, but still divisible by G. Then G' can be used in finding G by:

$$GCD(A, B) = GCD(A, B, G') = GCD(A \bmod G', GCD(B \bmod G', G')).$$

If the "approximative" GCD G' is "near" the real GCD G, then these two GCD computations will be inexpensive.

One can see that the "noise" $GCD(B, A_{k+1})/GCD(A, B)$ introduced at each step equals $g = GCD(u_{k+1}, B/G)$, because u_{k+1}, v_{k+1} are relatively prime. If one assumes u_{k+1} and B/G independently random, then the probability that the noise g at one step is m bits long is:

$$p_m = \sum \left\{ \frac{p_i'}{i^2} \,\middle|\, 2^{m-1} \le i \le 2^m - 1 \right\},$$

where p'_i is the probability that u_{k+1}/i and $(B/G)/i$ are prime: $p'_i \leq 1$. Hence $p_m \leq s_m$, where

$$s_m = \sum \left\{ \frac{1}{i^2} \mid 2^{m-1} \leq i \leq 2^m - 1 \right\} = \frac{\Psi(1, 2^m) - \Psi(1, 2^{m-1})}{4}$$

using **maple** system, and furthermore:

$$\sum_{m=2}^{32} m s_m \approx 1.83,$$

which is an upper bound for the the average bit-length of the noise per step. This bound seems to be very coarse, since experimentally we detected a noise of 0.52 bits per step. This means 100 steps will give an average noise of less than two digits, whose correction requires only four steps of the original Lehmer-Euclid scheme. Fig.6 presents the speed-up obtained for the "approximative" GCD computation.

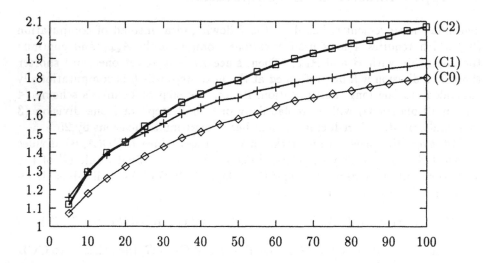

Fig. 6. Speed-up for successive improvements of version C (length of inputs from 5 to 100 32-bit words).

(C0) The straight forward implementation of the idea does not give a satisfactory result (the speed-up decreases to 107% – 180%), because in the original program a division is performed only when the length-difference of A, B exceeds one word. The experiments show that this happens only in about 7% of the cases, hence there is no alternation between Lehmer–type and division steps.

(C1) By setting the threshold for testing the length–difference at 24 bits, the percentage of the division steps becomes 49%, hence the desired alternation is realized. The speed–up improves a little (116% – 189%), but it is still not satisfactory, probably because of the additional time consumed for calling the division routine.

(C2) The final variant is obtained by replacing the calls to the division routine $A \bmod B$ (in cases when the length–difference is at most one word), with an explicit computation of $A - q * B$, where q is the quotient of the most-significant double-digits of A and B, which has the property

$$q - 1 \leq \lfloor A/B \rfloor \leq q,$$

(see [8]). Hence $|A - q * B|$ equals either $(A \bmod B)$ or $(A - A \bmod B)$, which are both good for continuing the GCD computation. This subvariant performs a little better than the previous one (speed-up 112% – 207%), it gives some improvement over version B, and also, for large lengths (80 words), overcomes the "psychological barrier" of two times speed–up in comparison with the original Lehmer algorithm. This last variant is kept as "version C" of the algorithm.

Fig.7 presents the speed–up for the 3 main versions. The absolute timings (on DECstation 5000/200) range from 0.65 ms to 41.43 ms for the improved algorithm, versus 0.73 to 85.82 ms for the original Lehmer-Euclid algorithm.

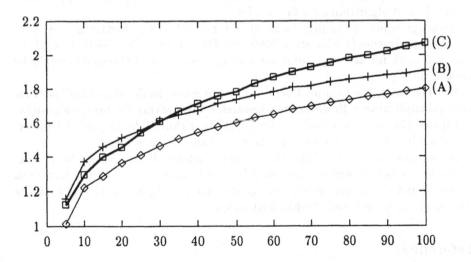

Fig. 7. Speed-up for the three main versions, depending on the length of inputs (in 32-bit words).

In fact, the figure shows an increase of the speed-up for bigger inputs. We obtained 230% speed-up for 300 word operands on a DECstation 5000/240 (see Fig.8).

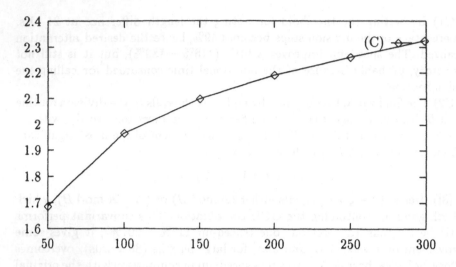

Fig. 8. Speed-up for version (C) on very long inputs (50 to 300 words of 32 bits).

Conclusions

Starting from the idea of using double digits in digit partial cosequence computation, and applying various improvements, one can speed-up the multiprecision Lehmer-Euclid algorithm by a factor of 2.

Although some of the improvements are related to the particular hardware used for experiments (DECstation 5000/200, DECstation 5000/240), most of the ideas will work for any sequential machine (a correction of thresholds might be necessary).

Until recently, the Lehmer-Euclid algorithm was considered most efficient for practical applications. However, new research on improving the binary algorithm of [14] (see [13, 16, 7, 6]) found faster GCD algorithms. The improved Lehmer-Euclid method presented here is in the same range as these ones.

An natural extension of this double-digit Lehmer-Euclid would be the extended Euclidean algorithm. Currently [4] is improving the extended Euclidean algorithm and the multiprecision integer-to-rational algorithm of [15] by using Lehmer-like approach and double-digit DPCC.

References

1. B. Buchberger. Gröbner Bases: An Algorithmic Method in Polynomial Ideal Theory. In Bose and Reidel, editors, *Recent trends in Multidimensional Systems*, pages 184–232, Dordrecht-Boston-Lancaster, 1985. D. Reidel Publishing Company.
2. B. Buchberger and T. Jebelean. Parallel rational arithmetic for Computer Algebra Systems: Motivating experiments. In *3rd Scientific Workshop of the Austrian Cen-*

ter for Parallel Computation, Weinberg, Austria, March 1992. Report ACPC/TR 93-3, February 1993.

3. G. E. Collins. Lecture notes on arithmetic algorithms, 1980. Univ. of Wisconsin.
4. M. Encarnacion. A Lehmer-type improvement of the modular residue to rational conversion. Technical report, RISC–Linz, 1993. To appear.
5. T. Granlund. GNU MP: The GNU multiple precision arithmetic library, 1991.
6. T. Jebelean. A Generalization of the Binary GCD Algorithm. In *ISSAC'93: International Symposium on Symbolic and Algebraic Computation*, Kiev, Ukraine, July 1993.
7. T. Jebelean. Comparing Several GCD Algorithms. In *ARITH-11: IEEE Symposium on Computer Arithmetic*, Windsor, Canada, June 1993.
8. D. E. Knuth. *The art of computer programming*, volume 2. Addison-Wesley, 2 edition, 1981.
9. D. H. Lehmer. Euclid's algorithm for large numbers. *Am. Math. Mon.*, 45:227–233, 1938.
10. R. T. Moenck. Fast computation of GCDs. In *ACM V^{th} Symp. Theory of Computing*, pages 142–151. ACM, 1973.
11. W. Neun and H. Melenk. Very large Gröbner basis calculations. In Zippel, editor, *Computer algebra and parallelism. Proceedings of the second International Workshop on Parallel Algebraic Computation*, pages 89–100, Ithaca, May 1990. LNCS 584, Springer Verlag.
12. A. Schönhage. Schnelle Berechung von Kettenbruchentwicklugen. *Acta Informatica*, 1:139–144, 1971.
13. J. Sorenson. Two fast GCD algorithms. Submitted to *J. of Algorithms*, 1993.
14. J. Stein. Computational problems associated with Racah algebra. *J. Comp. Phys.*, 1:397–405, 1967.
15. P. S. Wang, M. J. T. Guy, and J. H. Davenport. P-adic reconstruction of rational numbers. *ACM SIGSAM Bulletin*, 16(2):2–3, May 1982.
16. Ken Weber. The accelerated integer gcd algorithm. Technical report, Kent State University, 1993. To appear.

Storage Allocation for the Karatsuba Integer Multiplication Algorithm

Roman E. Maeder

Theoretical Computer Science, ETH Zürich, Switzerland

Abstract. The Karatsuba (or 3-for-4) integer multiplication algorithm needs some amount of auxiliary memory (in addition to the memory needed to store the result). We give sharp upper and lower bounds for the amount of auxiliary storage required and discuss implementation issues. Since we must never allocate too few memory cells, an asymptotic formula is not sufficient. The bounds are $2(n - 4 + 3\lfloor\log_2(n - 3)\rfloor)$ and $2(n - 6 + 2\lceil\log_2(n - 2)\rceil)$, respectively. The formula for the amount of auxiliary storage is obtained by an exact closed-form solution of a recursion formula involving the floor and ceiling functions. Comparing the upper and lower bounds shows that this simple allocation formula does not waste much memory for any input size.

1 Introduction

The Karatsuba algorithm is a practical integer multiplication algorithm with a low overhead for small input sizes and a good asymptotic complexity. It is the method of choice in general-purpose symbolic and computer algebra programs where arithmetic with arbitrary size numbers is an important part of the system.

The paper is organized as follows: First, we recall the Karatsuba algorithm for fast integer multiplication and give an implementation that allocates its temporary storage from a pool of auxiliary memory. Next, we derive recursion formulas for the amount of auxiliary memory required and solve them exactly for certain input sizes. Then we show that the amount used for any input size lies always between the minimum and maximum so obtained.

Note that a formula that gives only the *order* of storage required (it is $O(n)$) is not sufficient for our purpose, since we need an upper bound valid for *all* input sizes to actually allocate memory. By deriving a lower bound as well we can see how much memory we waste in the worst case. The difficulty in deriving exact bounds is caused by the floor and ceiling functions present in the recursion formulas. These can be ignored for asymptotic estimates, but not for finding sharp bounds.

2 The Karatsuba Integer Multiplication Algorithm

We use the standard representation of arbitrary size integers. An unsigned arbitrary-size integer is stored as an array of digits in a positional number system with base B. The array $a_0, a_1, \ldots a_{n-1}$ of length n represents the number

$$a = \sum_{i=0}^{n-1} a_i B^i. \tag{1}$$

The digits are, of course, in the range $0 \le a_i < B$.

The naive multiplication algorithm [3] needs $O(n^2)$ digit multiplications. The Karatsuba algorithm [1] (English translation in [2]) divides the input numbers a and b of size n into two halves. The product can be formed by only three recursive multiplications of size $n/2$ and some additions and subtractions. This gives a complexity of $O(n^{\log_2 3})$ digit multiplications. Let $m = \lceil n/2 \rceil$. We can write

$$a = a_l + a_h B^m, b = b_l + b_h B^m, \tag{2}$$

where

$$a_l = \sum_{i=0}^{m-1} a_i B^i, a_h = \sum_{i=0}^{m-1} a_{i+m} B^i. \tag{3}$$

(If n is odd, we assume $a_n = 0$.) The product $c = ab$ can be expressed as

$$c = a_l b_l + ((a_l + a_h)(b_l + b_h) - a_l b_l - a_h b_h) B^m + a_h b_h B^{2m}. \tag{4}$$

The case where the number b has fewer digits than a will be treated in Sect. 4.2. During the computation some auxiliary storage is required to hold the sums $a_l + a_h$ and $b_l + b_h$ and the products $a_l b_l$ and $a_h b_h$ (the same space can be used for $a_l b_l$ and $a_h b_h$). The third product $(a_l + a_h)(b_l + b_h)$ can go directly into the result array. Allocation and deallocation of this auxiliary storage inside each recursive level would lead to a large overhead and possible memory fragmentation. We will see that it would not even save storage space. Instead we will allocate all required memory in one large array. Each invocation of the routine will take some of the work array for storing its intermediate results and will pass the unused rest on to the recursive invocations.

In the following algorithm the notation $a+i$ denotes the part of array a starting with element a_i (as in the programming language C). Within the procedure the parameters are treated as arrays with a lower index of 0. We will use the unspecified data type digit for the array elements (usually one uses short or int). An array is therefore of type (digit *).

```
1 void
  karatsuba(digit *a, digit *b, digit *c, digit *w, int la, int lb)
  // add the product of a and b to c.
  // we assume la >= lb > [la/2]. c must be la + lb + 1 in size
  // the array w is used as work array (temporary storage)
2 {
3   if (la <= 4) { // use naive method
4          long_multiplication(a, b, c, la, lb);
5          return;
6   }
7   m = (la+1)/2;                    // [la/2]
```

```
 8    copyto(w + 0, a + 0, m);              // a_0,...,a_{m-1} into w_0,...,w_{m-1}
 9    w[m] = 0;                             // clear carry digit
10    addto(w + 0, a + m, la - m);          // form a_l + a_h into w_0,...,w_m
11    copyto(w + (m+1), b + 0, m);          // b_0,...,b_{m-1} into w_{m+1},...,w_{2m}
12    w[m+1+m] = 0;                         // clear carry digit
13    addto(w + (m+1), b + m, lb - m);      // form b_l + b_h into w_{m+1},...,w_{2m+1}
      // compute (a_l + a_h)(b_l + b_h) into c_m,...,c_{3m+1}
14    karatsuba(w + 0, w + (m+1), c + m, w + 2*(m+1), m+1, m+1);
15    lt = (la - m) + (lb - m) + 1;         // space needed for a_h b_h
16    clear(w + 0, lt);                     // clear result array
      // compute a_h b_h into w_0,...,w_{la+lb-2m-1}
17    karatsuba(a + m, b + m, w + 0, w + lt, la - m, lb - m);
18    addto(c + 2*m, w, (la - m) + (lb - m)); // add a_h b_h B^{2m}
19    subfrom(c + m, w, (la - m) + (lb - m)); // subtract a_h b_h B^m
20    lt = m + m + 1;                       // space needed for a_l b_l
21    clear(w + 0, lt);                     // clear result array
      // compute a_l b_l into w_0,...,w_{2m-1}
22    karatsuba(a, b, w + 0, w + lt, m, m);
23    addto(c + 0, w, m + m);               // add a_l b_l
24    subfrom(c + m, w, m + m);             // subtract a_l b_l B^m
25    return;
26 }
```

A few explanations of the code above:

- karatsuba() adds the product of a and b to the old contents of c. The output array c is assumed cleared. Recursively this precondition is established in lines 16 and 21.

- The procedure clear(digit *r, int l) sets l Elements of the array r to zero.

- The procedure copyto(digit *r, digit *a, int l) copies the digits a_0, ..., a_{l-1} into r_0, ..., r_{l-1}.

- The procedure addto(digit *r, digit *a, int l) adds the digits a_0, ..., a_{l-1} to the old contents of r_0, ..., r_{l-1}, propagating the carry as usual. Any carry out of the last place r_{l-1} is propagated further into r_l, r_{l+1}, ... This is the reason for setting the next higher digit to zero in lines 9 and 12.

- The procedure subfrom(digit *r, digit *a, int l) subtracts a from the old contents of r, propagating the borrow as usual. Any borrow out of the last place r_{l-1} is propagated further. The way this is called in line 19 there will never be a negative result since line 18 guarantees that at least one higher element of c is nonzero and can therefore absorb the borrow.

- This formulation of Karatsuba's algorithm avoids negative intermediate results. Such results can occur in the formulation in [3], where the sums $a_h - a_l$ and $b_l - b_h$ are formed instead of $a_h + a_l$ and $b_l + b_h$.

- The final result has, of course, at most $l(a) + l(b)$ digits. The intermediate result after line 18 could cause a carry into the next digit. Therefore we must allocate and clear one digit more. This is recursively done in lines 15 and 20.

3 Amount of Auxiliary Storage Needed

Let $w(n)$ be the amount of working memory needed for input size n. At line 14 we have used $2(\lceil n/2 \rceil + 1)$ elements of the work array to hold the two sums $a_l + a_h$ and $b_l + b_h$. The recursive call costs another $w(\lceil n/2 \rceil + 1)$ elements to hold its result. The other two recursive calls need strictly less auxiliary memory. No storage is needed for $n \leq 4$. This gives the following recursion:

$$w(n) = \begin{cases} 0, & n \leq 4 \\ w(\lceil n/2 \rceil + 1) + 2(\lceil n/2 \rceil + 1) & \text{otherwise} \end{cases} \tag{5}$$

We need a tight upper bound $t_1(n)$ on $w(n)$ for actually allocating the storage. We will derive such a bound. Comparison with a tight lower bound $t_0(n)$ will show that we will not waste more than $2\log_2 n$ memory cells for any n. For the remainder of this section we will assume $n > 4$, since the formulas hold trivially for smaller n.

The worst case (using the largest number of recursive steps) occurs if the numbers in the sequence $n_0 = n$, $n_{i+1} = \lceil n_i/2 \rceil + 1$ down to 4 are all *odd*. This happens for $n = 2^k + 3$, for some $k > 0$, since

$$\left\lceil \frac{2^k + 3}{2} \right\rceil + 1 = 2^{k-1} + 3. \tag{6}$$

For these values of n we can solve the recursion exactly (see [4] for solution methods for such equations):

$$w(2^k + 3) = 2(n - 4 + 3k). \tag{7}$$

Verification is by an easy induction. We note that $k = \lfloor \log_2(n - 3) \rfloor$. This leads to

$$t_1(n) = 2(n - 4 + 3\lfloor \log_2(n - 3) \rfloor) \tag{8}$$

Theorem 1. *For all n we have $w(n) \leq t_1(n)$.*

Proof. By induction.

- assume $n = 5$. Then $w(5) = 8 = t_1(5)$.
- assume $n > 5$ odd. Then $\lceil n/2 \rceil + 1 = (n + 3)/2 > 4$. Therefore the second line of (5) applies. We have

$$\begin{aligned} w(n) &= w((n + 3)/2) + 2(n + 3)/2 \\ &\leq 2((n + 3)/2 - 4 + 3\lfloor \log_2((n + 3)/2 - 3) \rfloor) + 2(n + 3)/2 \\ &= 2(n - 4 + 3(\lfloor \log_2((n + 3)/2 - 3) \rfloor + 1)) \\ &= 2(n - 4 + 3\lfloor \log_2((n - 3)/2) + 1 \rfloor) \\ &= 2(n - 4 + 3\lfloor \log_2((n - 3)/2) + \log_2 2 \rfloor) \\ &= 2(n - 4 + 3\lfloor \log_2(n - 3) \rfloor) \\ &= t_1(n). \end{aligned} \tag{9}$$

(We may take integers into the floor function.)

- assume $n > 4$ even. Then $n - 1 > 4$ and $\lceil n/2 \rceil = \lceil (n-1)/2 \rceil$. We have

$$w(n) = w(n-1) \leq t_1(n-1) \leq t_1(n) \tag{10}$$

since t_1 is monotonous.

□

t_1 is therefore a possible choice for the amount of storage to be allocated. Now we will show that we do not waste too much for other values of n.

The best case (using the smallest number of recursive steps) occurs if the numbers in the sequence $n_0 = n$, $n_{i+1} = \lceil n_i/2 \rceil + 1$ down to 4 are all *even*. This happens for $n = 2^k + 2$, for some $k > 0$, since

$$\left\lceil \frac{2^k + 2}{2} \right\rceil + 1 = 2^{k-1} + 2. \tag{11}$$

For these values of n we can also solve the recursion exactly:

$$w(2^k + 2) = 2(n - 6 + 2k). \tag{12}$$

We note that $k = \lceil \log_2(n-2) \rceil$. This leads to

$$t_0(n) = 2(n - 6 + 2\lceil \log_2(n-2) \rceil) \tag{13}$$

Theorem 2. *For all n we have $w(n) \geq t_0(n)$.*

Proof. The proof is analogous to the proof of Theorem 1.

- assume $n = 6$. Then $w(6) = 8 = t_0(6)$.
- assume $n > 6$ even. Then $\lceil n/2 \rceil + 1 = n/2 + 1 > 4$. Therefore the second line of (5) applies. We have

$$\begin{aligned}
w(n) &= w(n/2 + 1) + 2(n/2 + 1) \\
&\geq 2(n/2 + 1 - 6 + 2\lceil \log_2(n/2 + 1 - 2) \rceil) + 2(n/2 + 1) \\
&= 2(n - 6 + 2(\lceil \log_2(n/2 - 1) \rceil + 1)) \\
&= 2(n - 6 + 2\lceil \log_2((n-2)/2) + 1 \rceil) \\
&= 2(n - 6 + 2\lceil \log_2((n-2)/2) + \log_2 2 \rceil) \\
&= 2(n - 6 + 2\lceil \log_2(n-2) \rceil) \\
&= t_0(n).
\end{aligned} \tag{14}$$

- assume $n > 4$ odd. Then $\lceil n/2 \rceil = \lceil (n+1)/2 \rceil$. We have

$$w(n) = w(n+1) \geq t_0(n+1) \geq t_0(n) \tag{15}$$

since t_0 is monotonous.

□

From Theorems 1 and 2 we conclude that $w(n)$ is always between $t_0(n)$ and $t_1(n)$. The two differ only by $2\log_2 n$. Therefore we will never waste much space if we use t_1 for storage allocation.

4 Other Considerations

4.1 Run-time Computation of Storage Required

Instead of using (8) for sizing the work array we could also evaluate the recursion (5) at run time. It takes only $O(\log n)$ steps to do so.

4.2 Non-square Multiplication

If the problem is not square, i.e. $lb < la$, we still need this amount of work space for $lb > \lceil la/2 \rceil$. This is so because the size of the first recursive call (program line 17) is still of maximum size. If $lb \leq \lceil la/2 \rceil$, this recursive subdivision should not be used. We should simply compute $a_l b$ and $a_h b$ recursively and then add them together to give $ab = a_l b + a_h b B^m$.

4.3 When to Switch to the Trivial Algorithm

For any implementation the threshold for switching to the trivial multiplication method should be chosen experimentally. It may be higher than 4 if fast digit multiplication hardware is available [5]. For many machines (including SPARC and M680x0) a value of 20 is optimal (the minimum is rather flat). For a threshold n_0 (5) becomes

$$w(n) = \begin{cases} 0, & n \leq n_0 \\ w(\lceil n/2 \rceil + 1) + 2(\lceil n/2 \rceil + 1) & \text{otherwise} \end{cases} \quad (16)$$

The exact solutions to the recursion equation become

$$t_0(n) = \begin{cases} 0, & n \leq n_0 \\ 2(n - \alpha_0 + 2\lceil \log_2(n - 2) \rceil) & \text{otherwise} \end{cases} \quad (17)$$

$$t_1(n) = \begin{cases} 0, & n \leq n_0 \\ 2(n - \alpha_1 + 3\lfloor \log_2(n - 3) \rfloor) & \text{otherwise} \end{cases} \quad (18)$$

The constants α_0 and α_1 can be determined as follows:

- Let $k = \lceil \log_2(n_0 - 2) \rceil$. Then $2^{k-1} + 2 \leq n_0 < 2^k + 2$. We then get $w(2^k + 2) = 2(2^{k-1} + 2)$ since the recursion stops after one step. Equating coefficients with $t_0(2^k + 2)$ gives

$$\alpha_0 = 2^{k-1} + 2k. \quad (19)$$

- Let $k = \lceil \log_2(n_0 - 3) \rceil$. Then $2^{k-1} + 3 \leq n_0 < 2^k + 3$. We then get $w(2^k + 3) = 2(2^{k-1} + 3)$ since the recursion stops after one step. Equating coefficients with $t_1(2^k + 3)$ gives

$$\alpha_1 = 2^{k-1} + 3k. \quad (20)$$

Acknowledgements

We wish to thank the anonymous referees for their helpful suggestions for improving the clarity of the presentation.

References

1. A. Karatsuba and Yu. Ofman. Multiplication of many-digital numbers by automatic computers. *Doklady Akademii nauk SSSR*, 145(2), 1962. (in Russian).
2. A. Karatsuba and Yu. Ofman. Multiplication of multidigit numbers on automata. *Soviet physics doklady*, 7(7), 1963.
3. D. E. Knuth. *Seminumerical Algorithms*, volume 2 of *The Art of Computer Programming*. Addison-Wesley, second edition, 1981.
4. Donald E. Knuth, Ronald L. Graham, and Oren Patashnik. *Concrete Mathematics*. Addison-Wesley, 1989.
5. Wolfgang W. Kuechlin, David Lutz, and Nicholas Nevin. Integer multiplication in PARSAC-2 on stock microprocessors. In *Proc. AAECC-9, New Orleans*. Lecture Notes in Computer Science 539. Berlin: Springer 1991.

Process Scheduling in DSC and the Large Sparse Linear Systems Challenge*

A. Diaz, M. Hitz, E. Kaltofen, A. Lobo and T. Valente**

Department of Computer Science, Rensselaer Polytechnic Institute
Troy, NY 12189-3590, USA
Internet: diaza, hitzm, kaltofen, loboa@cs.rpi.edu; t_valent@colby.edu

Abstract. New features of our DSC system for distributing a symbolic computation task over a network of processors are described. A new scheduler sends parallel subtasks to those compute nodes that are best suited in handling the added load of CPU usage and memory. Furthermore, a subtask can communicate back to the process that spawned it by a co-routine style calling mechanism. Two large experiments are described in this improved setting. We have implemented an algorithm that can prove a number of more than 1,000 decimal digits prime in about 2 months elapsed time on some 20 computers. A parallel version of a sparse linear system solver is used to compute the solution of sparse linear systems over finite fields. We are able to find the solution of a 100,000 by 100,000 linear system with about 10.3 million non-zero entries over the Galois field with 2 elements using 3 computers in about 54 hours CPU time.

1 Introduction

In Diaz et al. (1991) we introduced our DSC system for distributing large scale symbolic computations over a network of UNIX computers. There we discuss in detail the following features:

— The distribution of so-called parallel subtasks is performed in the application program by a DSC user library call. A daemon process, which has established IP/TCP/UDP connections to equivalent daemon processes on the participating compute nodes, handles the call and sends the subtask to one of them. Similarly, the control flow of the application program is synchronized by library calls that wait for the completion of one or all subtasks.

— DSC distributes not only remote procedure calls to precompiled programs, but also programs that are first compiled on the machine serving the subtask. This enables the distribution of dynamically generated "black-box"

*This material is based on work supported in part by the National Science Foundation under Grant No. CCR-90-06077, CDA-91-21465, and under Grant No. CDA-88-05910.

**Valente's current address is Department of Mathematics and Computer Science, Colby College, Waterville, Maine 04901.

functions (c.f. Kaltofen and Trager 1990) and easy use of computers of different architecture.

— DSC can be invoked from both Common Lisp and C programs. It can distribute within a local area network (LAN) and across the Internet.

— The interface to the application program consists of seven library functions. Processor allocation and interprocess communication is completely hidden from the user.

— The progress of a distributed computation can be monitored by an independently run controller program. This controller also initializes the DSC environment by establishing server daemons on the participating computers.

— We document experiments with DSC on a parallel version of the Cantor/Zassenhaus polynomial factorization algorithm and the Goldwasser-Kilian/Atkin (GKA) integer primality test.

New experiments with the GKA primality test that run on so-called "titanic" integers, i.e., integers with more than 1000 decimal digits, and experiments with a parallel sparse linear system solver, namely, Coppersmith's block Wiedemann algorithm (Coppersmith 1992), have lead to several key modifications to DSC. In this article we describe these changes, as well as the results obtained by applying the improved environment to both titanic primality testing and sparse linear system solving.

Unlike on a massively parallel computer, where each processor has the same computing power and internal memory, a network of workstations and machines is a diverse computing environment. At the time the application program distributes a subtask, the DSC server has to determine which machine will receive this subtask. Our original design used a round-robin schedule, which resulted in quite bad subtask-to-processor allocation. The new scheduler continuously receives the CPU load and memory usage of all participating machines, which are probed by resident daemon processes at 10 minute intervals. In addition, the application program supplies an estimate of the amount of memory and a rough measure of CPU usage. The scheduler then makes a sophisticated selection of which processor is to handle the subtask. If certain threshold values are not met, the subtask gets queued for later distribution under hopefully better load conditions on the network. The details of the scheduling algorithm are described in §2.1. Without this very fine tuned distribution scheduler, neither the primality tester nor the sparse linear system solver could have been run on as large inputs as the ones we had. Note that DSC's ability to account for the heterogeneity of the compute nodes is one distinguishing mark to other parallel computer algebra systems such as Maple/Linda (Char 1990), PARSAC-2 (Collins et al. 1990), the distributed SAC-2 of Seitz (1992), or PACLIB (Hong and Schreiner 1993).

DSC supports a very coarse grain parallelism. This was quite successful for the primality tester, where each parallel subtask is extremely compute intensive but uses a moderate amount of memory. However, the Wiedemann sparse linear system solver can be implemented by slicing the coefficient matrix and storing each slice on a different processor. These slices will repeatedly be multiplied by a sequence of vectors. We have implemented a mechanism whereby the subtasks remain loaded in memory (or swap space) on first return, and can be continued at the point following the previous return with different data supplied by the calling program, much like co-routines (see §2.3). This introduces a finer grain parallelism and allows two-way communication between the subtask and the parent process. This co-routine mechanism tends to make use of the distributed memory more than the parallel compute power.

DSC has also been modified internally in two important ways. First, the environment can now be initialized on a user supplied UDP port number. Several users can thus set up individual DSC servers without interfering with one another. We note that the inter-process communication does not take place on the system level, where a single port number could have been reserved for DSC. Hence no system modifications are necessary to run DSC, which is often desired when linking to off-site computers. Nonetheless, the port number is public and servers could be started, perhaps maliciously, to communicate with an existing environment. We guard against such mishap by tagging each message with a key set by the individual user. More details on these enhancements are found in §2.2.

Our first test problem has been the GKA primality test applied to numbers with more the 1000 decimal digits. We are successful in proving the primality of a 1111 digit number on a LAN of some 20 computers in about 2 months turnaround time. The details and observations of this experiment are described in §2.4. Our second test problem is a distributed version of the Coppersmith block Wiedemann algorithm. This algorithm for solving unstructured sparse linear systems has very coarse grain size, unlike classical methods such as conjugate gradient, which makes it very suitable for the DSC environment. We have implemented two variants, one for entries being from prime finite fields whose elements fit into 16 bits, and one for entries from $GF(2)$, the field with two elements. In the latter case, we not only realize Coppersmith's processor internal parallelism by performing the bit operations simultaneously on 32 elements stored in a single computer word, but we further "doubly" block the method and distribute across the network. The details of our experiments are described in §3; we are successful in speeding up the solution of $100,000 \times 100,000$ linear systems with 10.3 million entries over $GF(2)$ by factors of 3 and more. Such large runs would very likely not have completed using our old round-robin scheduling, since only the selection of compute nodes with large memory makes our programs feasible.

2. New DSC Features

2.1 The DSC Scheduler

The goal of process scheduling in DSC is to locate available resources in the network and to distribute subtasks without creating peak loads on any node. Selection of a suitable computer is based on three factors:

Load on nodes	*Rating of resources*	*Requirements for subtask*
– Long term	– MIPS of processors	– crude estimates by user
– Most recent	– Installed memory	

The hardware resources are defined in three fields of the node list: the core memory size in MByte, the CPU power in MIPS and the number of processors. In the application program the user has to specify an estimate for the expected memory needs in MByte and a "fuzzy" value (LOW, MEDIUM, HIGH) for CPU usage.

In order to provide the scheduler with information about the current and the expected load at each compute node, a method of data collection had to be developed. The ideal load-meter would provide exact data in real time with some corrections derived from trend predictions. This would require the monitoring program being tied to the operating system on a low level. Unfortunately this would place excessive burden on the user (request for higher privileges) and it would make the system less portable. However, most of the time it is not desirable to have measurements with high resolution. The readings should reflect trends for longer time periods rather than being just snapshots. As a first solution the UNIX ps command was chosen to measure CPU and memory usage about 8 to 10 times an hour. Due to the latency (up to one minute) involved with ps, and for better modularity, a separate process "DSC_ps" was added in the current version of DSC.

Once a DSC server is running, it spawns off the DSC_ps process for its node. DSC_ps maintains a table of statistics, which is saved to disk after each update. At the end of the hour, the mean of CPU and memory usage is averaged with the previous value for this hour of the day. At the end of the day (or week) the values for this day of the week (or week of the year) are adjusted by the latest readings. From all four levels (current reading, hour, day and week) a weighted average is computed to include long term effects. The resulting two values (CPU, memory) are sent to the local DSC server which in turn will communicate the update to all other servers on the network. The backup file allows the initialization of load parameters according to anticipated patterns of usage. DSC_ps will then adapt those guessed values with respect to the new readings.

The scheduler in the DSC server uses the values received from DSC_ps, the ratings from the node database, and the estimated needs of the next task to select the target machine. For this purpose a sorted list of compute nodes which satisfy a minimum requirement of available memory is maintained. Based on the memory estimate of the application, all nodes which would stay above a certain threshold (allowing for some moderate paging) are preselected. Among them the one with the lowest CPU usage is finally chosen for distribution. If none of the nodes can satisfy the requirements, the job is put back into a queue until the load on one of the computers decreases to a sufficient level.

After distributing the new task the server adjusts for the expected change in the load parameters of the selected node. Because of the long latency period, it cannot wait for the next readings of the actual load when it has to distribute many tasks in a short period of time. In time it can replace the estimates by the actual values whenever new load readings are received from the other server. This has the convenient side effect that the distributing server does not have to rely on transient measurements resulting from the startup phase of the new task (which can involve compilation of source code). Most of the time it will receive the steady-state readings because the server of the selected compute nodes will send the update of the load parameters with low priority.

2.2 Interprocess Communication and Message Validation

DSC uses the User Datagram Protocol (UDP) for most of its communication and Transmission Control Protocol (TCP) stream sockets for file transfer. For the sake of portability, all inter-process communication adheres to the DARPA Internet standard TCP/IP/UDP as implemented in UNIX 4.2/4.3bsd (see Diaz 1992). This low level approach avoids the high latency present in the UNIX rsh and ftp commands, and it provides real time information on subtasks and compute nodes for possible control actions such as subtask rescheduling. Before a user starts an application program that distributes parallel subtasks over the network, the DSC server daemons must be started. These daemon programs execute in the background and monitor a single UDP datagram address for new external stimuli from other DSC servers, the DSC controller program, the resource and work load monitor daemon program DSC_ps, and application programs. In order for a client process to contact a DSC server, the client must have a way of identifying the DSC server it wants. This can be accomplished by knowing the 32-bit Internet address of the host on which the server resides and the 16-bit port number which identifies the destination of the datagram on the host machine. Each DSC server must be using the same UDP port number in order to communicate with the others. UDP port numbers are not reserved and can be allocated by any process. The run time port number allocation option allows the user to automatically poll the machines in the configured DSC network to find a suitable port number for the initiation of a set of DSC servers. This is done via the DSC control program and consequently all DSC servers can be started using the determined available port number for their UDP communication.

If the control program could not establish a connection to a DSC server via the port number specified in the configuration file, the control program will assume that no active DSC server is monitoring this port. Consequently, the control program searches for an available port number which is not used on any machines in the "farm" of compute nodes in the DSC network. Optionally, the user may specify its own port number thereby bypassing the runtime port number allocation mechanism.

Once a port number has been determined the control program will start up the remote DSC servers via a rsh command supplying the executable and the port number to the remote or local compute node. Once all DSC servers are active communication takes place only via the IP/TCP/UDP protocols.

The primary function of the DSC server daemon program is to monitor a single UDP datagram address for incoming messages. Each message is a request for the DSC server to perform some action. However, in order to act only on messages received from the user that started the server, all messages contain a message validation tag which is specified by the user. If for any reason the message validation tag received by the DSC server does not match the server's message validation tag, the message is ignored and the invalid action request is logged. This avoids the inadvertent message passing that could occur when multiple DSC systems execute concurrently in the same open network computing environment using the same datagram port number.

2.3 Co-Routines

The C and Lisp DSC application programmer can take advantage of the resources found in the DSC network by utilizing 5 base functions callable from a user's program. The function **dscpr_sub** is used for the activation of parallel subtasks and designating their respective resource usage specifications. The calls to **dscpr_wait**, **dscpr_next**, and **dscpr_kill** are used to wait on a specific parallel subtask or on the completion of all parallel subtasks, to wait for the next completed parallel subtask and to kill a specific parallel subtask, respectively. Finally, the function **dscdbg_start** can be used to track a task and is useful when one wishes to debug tasks using interactive debuggers such as Unix's **dbx**.

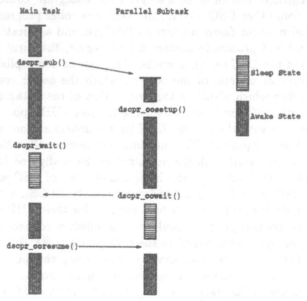

Figure 1: Co-routine flow of control.

In order to meet the sparse linear system challenge (see §4), where there is a need to maintain large amounts of data within parallel subtasks, the C User Library has been extended to allow the user to implement co-routines (Kogge 1991, §9.6.3). The function **dscpr_cosetup** must be called at the beginning of any parallel subtask that is to be treated as a co-routine. This initialization is necessary so that the wake up signal received by a parallel subtask from the DSC server can be interpreted as a command to resume execution of the subproblem. Specifically, the **dscpr_cosetup** function specifies how the subtask process will handle an asynchronous software interrupt by providing the address of an internal DSC function that wakes the process from a sleep state when the corresponding interrupt signal is detected. When the subtask calls the **dscpr_cowait** function, it enters a sleep state and optionally transmits a data file back to its parent. Once a co-routine parallel subtask or a set of co-routine parallel subtasks has been spawned by a call to dscpr_sub, the returned indices have been recorded, and a successful wait has completed, the parent task can send a wake up call to a sleeping parallel subtask via the dscpr_coresume function. Arguments to this function are an integer which uniquely identifies a parallel subtask (returned from the spawning call to dscpr_sub) and a string which identifies which input file if any should be sent to the co-routine before the parallel subtask is to be resumed. This call essentially generates the software interrupt needed for the waking of the sleeping parallel subtask. Figure 1 denotes the relationship that could exist between DSC utility function calls in a main task and its co-routine parallel subtask child.

2.4 The GKA Primality Test

In this section, we describe new experimental results with our distributed implementation of the Goldwasser-Kilian/Atkin (GKA) primality test, which uses elliptic curves to prove an integer p prime; for earlier results, see Kaltofen et al. 1989, Diaz et al. 1991. In particular, we discuss here our success in proving "titanic" integers, i.e., integers with more than 1000 decimal digits, prime (see also Valente 1992, Morain 1991).

Let us briefly summarize the algorithm. The test has two phases: in the first phase, a sequence $\{p_i\}$ of probable primes is constructed, such that $p = p_0 > p_1 > p_2 > \cdots > p_n$. Each p_{i+1} is obtained from p_i by first finding a discriminant d such that p_i splits as $\pi\bar{\pi}$ in the ring of integers of the field $\mathbf{Q}(\sqrt{d})$. If $(1 - \pi)(1 - \bar{\pi})$ is divisible by a sufficiently large (probable) prime q, we set p_{i+1} to q, thus "descending" from p_i to p_{i+1}. We then repeat the process, seeking a descent from p_{i+1}. The first phase terminates with p_n having fewer than 10 digits. In the second phase, it is necessary to construct, for each p_i from the first phase, an appropriate elliptic curve over $\mathrm{GF}(p_i)$ which is used to prove p_i prime, provided p_{i+1} is prime. This results in a chain of implications

$$p_n \text{ prime} \implies p_{n-1} \text{ prime} \implies \cdots \implies p_0 = p \text{ prime}.$$

In our experiment, we started with a probable prime number of 1111 decimal digits. Our code is written in the C programming language calling the Pari library functions (Batut et al. 1991) for arbitrary precision integer arithmetic. Each time a descent is required in the first phase, a list of nearly 10,000 discriminants is examined. In fact, we chose to search all d with $|d| \leq 100,000$, where $\mathbf{Q}(\sqrt{d})$ has class number ≤ 50. Unfortunately, when p is titanic, few if any of these discriminants will induce a descent. For our prime of 1111 digits, we distributed the search for a descent from p_i to p_{i+1} to 24 subtasks, each of which is given approximately 400 discriminants to examine. The first subtask to find a descent reports it to the main task which then redistributes in order to find a descent from p_{i+1}. We required 204 descents before a prime of 10 digits was obtained. Our first phase run with the 1111 digit number as input began on January 12, 1992, and ended on February 13, 1992. The total elapsed time for

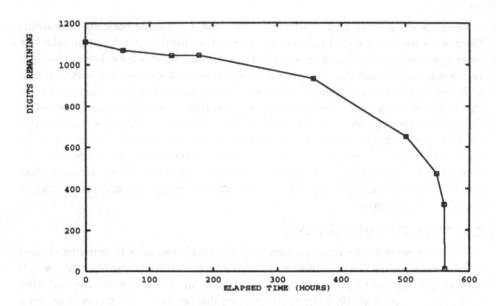

Figure 2: Graph of progress of GKA first phase.

the run was measured at about 569 hours, or approximately $3\frac{1}{2}$ weeks. Figure 2 depicts the progress of this run during this period.

Notice that after 135 elapsed hours, the 1111 digit number is "reduced" to a number having 1044 digits. After an additional 43 hours, it appears that we regress, because the number shown now has 1045 digits! In fact, what has happened is that our program failed to find a descent from the 1044 digit number, and was forced to backtrack to a larger prime and find an alternate descent. Slow but steady progress is evident, until the last day, when the 322 digit number rapidly shrinks, and the first phase suddenly ends. Interestingly, it appears that about half of the total elapsed time of this run is spent merely reducing the original number to subtitanic size.

For our second phase run with 1111 digit inputs, there are a total of 204 descents to process. Typically, each of the 20 or so workstations is given about 10 descents. For each descent, the subtask must construct a class equation for the class field over $\mathbf{Q}(\sqrt{d})$, find a root of the class equation, then use this root to construct an elliptic curve over the appropriate prime field $GF(p)$. Once this curve is found, verification proceeds by finding a point on the curve which serves as a witness to the primality of p. Difficulties arise when the root-finder must handle class equations of degree 30 or more. Since the second phase is so sensitive to the degree of the class equation, it is critical that in the first phase we do whatever is possible to insure that only discriminants of relatively low class numbers are passed on to the second phase. Because of these factors, our phase two run for this titanic input takes approximately three weeks elapsed time. In a situation like this where a distributed subtask has a long running time it is much more difficult to schedule the task on a processor with expected low load, thus insuring high performance.

3 Distributed Sparse Linear System Solving

3.1 Introduction and Background Theory

We now discuss our experience with DSC in the solution of homogeneous systems of sparse linear equations. The problem is to find a non-zero w such that $Bw = 0$, where B is a matrix of very large order N. The classical method of Gaussian elimination is not well suited for this task as the memory and time complexity are bounded by functions that are quadratic and cubic, respectively, in N.

Wiedemann (1986) gives a Las Vegas randomized algorithm with time complexity governed by $O(N)$ applications of B to a vector plus $O(N^2)$ arithmetic operations in the field of entries. The algorithm does not alter the sparsity of the coefficient matrix and requires only a linear amount of extra storage. The algorithm yields a non-zero solution with high probability. Wiedemann also gives a similar method for non-singular systems, although one could easily transform the non-singular to the homogeneous case by the addition of an extra column.

Coppersmith (1992) presents a generalization of Wiedemann's algorithm which can be implemented in a distributed setting. For the purposes of this paper we shall now give a brief description of it, which is outlined in Figure 3.

(0) Select random x, z such that $x^{\mathrm{tr}} Bz$ has full rank;
$y \leftarrow Bz$;
comment: x, y, z are $N \times n$ block vectors
with random entries from the ground field.

(1) $L \leftarrow \lfloor 2N/n \rfloor + 5$;
for $i = 0, 1, \ldots, L$ do
$a^{(i)} = \left(x^{\mathrm{tr}} B^i y\right)^{\mathrm{tr}}$;
comment: the $m \times n$ matrices $a^{(i)}$ are coefficients
of a polynomial $A(\lambda)$ of degree L.

(2) $\Lambda \leftarrow$ find_recurrence(A);
comment: $\Lambda(\lambda)$ is obtained by the generalized,
block Berlekamp/Massey algorithm.

(3) for $l = 1, \ldots, n$ do
$w_l \leftarrow$ evaluate(Λ, B, z, l);
comment: This steps yields w_l such that $Bw_l = 0$
and hopefully $w_l \neq 0$.

Figure 3: Outline of the block Wiedemann algorithm.

In the initializing step (0), one generates block vectors x and z with dimensions $N \times m$ and $N \times n$ whose entries are randomly chosen from the ground field. The vectors meet the conditions that $x^{tr}Bz$ has full rank. In subsequent discussions we shall take $m = n$ and refer to n as the blocking factor of the algorithm.

Next, step (1) is the computation of $L = \lfloor 2N/n \rfloor + 5$ terms of a sequence whose i^{th} term is $a^{(i)} = (x^{tr}B^iy)^{tr}$ where $y = Bz$. It is clear that the individual columns of $a^{(i)}$ can be computed independently and in parallel. To do this we distribute the vector x and the νth column of y along with B and collect the νth columns of $a^{(i)}$. Each of the parallel subtasks involves no more than $\lfloor 2N/n \rfloor + 5$ multiplications of B by vectors (Kaltofen 1993). The original Wiedemann algorithm requires $2N$ such applications.

Step (2) is to find a polynomial $\Lambda(\lambda)$ representing a linear recurrence that generates the sequence. In the original Wiedemann Algorithm, the $a^{(i)}$ are field elements and their recurrence is obtained by the Berlekamp/Massey Algorithm. In the present context, the $a^{(i)}$ are $n \times n$ matrices and we have to use Coppersmith's generalization. The individual $a^{(i)}$ are treated as the coefficients of the polynomial $A(\lambda) = \sum_{j=0}^{L} a^{(j)}\lambda^j$. We iteratively compute the matrix polynomial

$$F_j(\lambda) = \begin{bmatrix} \Lambda_j(\lambda) \\ \Psi_j(\lambda) \end{bmatrix}.$$

During each iteration $1 \leq j \leq L$ we also compute a discrepancy matrix Δ_j, which is the coefficient of λ^j in $F_j(\lambda)A(\lambda)$.

The first n rows of F_j are analogous to a bank of "current" shift registers and the last n rows represent a "previous" bank, as in the original Berlekamp/Massey algorithm. The task at hand is to compute a non-singular linear transformation T_j that will zero out the first n rows of Δ_j. Initially, the rows of F are assigned nominal degrees and in determining T_j, one selects a pivot row in Δ_j corresponding to a row of lowest nominal degree in F_j and uses it to zero out the appropriate columns of Δ_j. Completing the iterative step, F_j is updated by setting $F_{j+1} = DT_jF_j$, where $D = \text{diag}[1, \ldots, 1, \lambda, \ldots, \lambda]$ increments the nominal degrees of the last n rows by 1. Overall, the whole step requires $O(nN^2)$ field operations. At present this step is done sequentially.

In step (3) one obtains a vector w_l which is non-zero with high probability, and which satisfies $Bw_l = 0$. It involves a Horner-type evaluation at B of a polynomial derived from Λ with coefficients that are N-dimensional vectors and, additionally, $O(N^2)$ arithmetic operations. This step requires no more than $N/n + 2$ multiplications of B by vectors. There are n options of finding vectors in the kernel of B, one for each $1 \leq l \leq n$. With high probability the solutions found are nontrivial, at least, if the field of entries is sufficiently large. For $GF(2)$, hopefully, some of the solutions will be nontrivial. That these different solutions sample the entire null space can so far only be argued heuristically. For the details the reader is referred to (Coppersmith 1992 and Kaltofen 1993).

Our implementation is in the C programming language on computers running UNIX. The programs are written generically, meaning that the underlying field of entries can be changed with very little difficulty. We have successfully used DSC to distribute the task outlined in step (1). The DSC scheduler was given a list of approximately 30 machines of diverse processing power and memory capacity and the processing and storage requirements of the subtasks. From this information it selected suitable target hosts.

3.2 The Generic Arithmetic Package

In our implementation we sought the ability to program the arithmetic operations generically. This was accomplished by writing all the field arithmetic operations as macros and implementing the macros in a way specific to the underlying field. At the moment, arithmetic can be done in the fields $GF(2^k)$ and $GF(p)$ where k can be from 1 to 31 and p is a prime requiring at most 15 bits for its binary representation. These restrictions come from the maximum size of a word in the target machines. The macro implementations and corresponding basic datatypes are selected by setting a single software switch at compile time. The actual programs require no changes.

For the $GF(2^k)$ case the binary operations are implemented using the bit operations of *exclusive-or*, *and*, and *shift* available in the C language. Field division is done by computing the inverse of an element with the extended Euclidean algorithm inside of which division is done using bit operations. In $GF(2)$, a single bit is sufficient to represent an element and hence the operations of addition and multiplication are exactly the bit-operations *exclusive-or* and *and*, respectively. In Coppersmith's implementation, 32 bit vectors are packed into a single vector of machine words. Thus, on a single processor, 32 vectors can simultaneously be multiplied into a matrix with bit entries. We have used the same approach in a special implementation for $GF(2)$. In addition, we do "double blocking" in step (1), where we distribute several packed vectors to different compute nodes.

For the field $GF(p)$ the binary operations are implemented with built-in integer arithmetic operations. The limit of 15 bits keeps any intermediate product of two members of the field from overflowing the bounds of an unsigned 31-bit integer. The maximum permitted value of p is thus $2^{15} - 19 = 32749$.

3.3 Experimental Results and Observations

We conducted tests in the field $GF(32749)$ using sparse square matrices with row dimensions of 10,000 and 20,000, and in the field $GF(2)$ with dimensions 10,000, 20,000, 52,250, and 100,000 respectively. In the case of dimension 10,000 there are between 23 and 47 non-zero elements per row and approximately 350,000 non-zero entries in total; in the case of dimension 20,000, 57 to 73 non-zero elements per row and 1.3 million non-zero entries in total. In the case of dimension 52,250 there are between 9 and 34 non-zero elements per row and altogether 1.1 million non-zero entries. The largest matrix contained 89 to 117 entries per row and a total of 10.3 million entries. The matrix of dimension 52,250, arising from integer factoring by the MPQS method, was supplied to us by A. M. Odlyzko, while the other matrices were generated with randomly placed random non-zero entries. The tests were done with blocking factors of 2, 4, and 8 in the $GF(32749)$ case and 32, 64, and 96 in $GF(2)$. The calculation of the sequence $(a^{(i)})$ was done in a distributed fashion using DSC. Pointers to the matrix and the vectors x and y_ν were sent out to separate machines and the corresponding components of A were returned. The computation of the generator polynomial and the evaluation of the solution were done on a single SUN-4 machine rated at 28.5 MIPS. Compilation was done with the optimizer flag activated. Figures 4 and 5 below give the actual CPU time taken for each task. The evaluation time is the time to find the first non-zero solution.

N	Task		Blocking Factor	
		2	4	8
10,000	(1) $\langle a^{(i)} \rangle$	$7^h 29'$	$3^h 54'$	$2^h 09'$
	(2) b-massey	$2^h 25'$	$4^h 08'$	$8^h 00'$
	(3) evaluation	$3^h 47'$	$1^h 59'$	$1^h 05'$
	total	$13^h 41'$	$10^h 06'$	$11^h 14'$
20,000	(1) $\langle a^{(i)} \rangle$	$57^h 17'$	$28^h 43'$	$15^h 21'$
	(2) b-massey	$9^h 48'$	$16^h 36'$	$33^h 39'$
	(3) evaluation	$29^h 42'$	$14^h 44'$	$7^h 53'$
	total	$96^h 47'$	$60^h 02'$	$56^h 53'$

Figure 4: CPU Time (hours^{h}minutes$'$) for different blocking factors with all arithmetic in GF(32749). Each processor is rated at 28.5 MIPS.

It can be seen in both tables that the time for computing the sequence $\langle a^{(i)} \rangle$ decreases as the blocking factor increases. In the field GF(32749), the cost drops approximately in half each time the blocking factor doubles. This is as expected because the length of the sequence and hence the number of $B^i y$ computations is $O(N/n)$. In the GF(2) case the overall trend is still visible but the rate of decrease is less because more work has to be done in unpacking the doubly blocked bit-vectors x and y_ν. We note that it took us about 114 hours CPU time to solve the system with $N = 52,250$ by the original Wiedemann method (blocking factor $= 1 \times 1$). The memory requirement per task is the quantity needed to store B plus $O(nN)$ field elements for the vectors and intermediate results.

For the computation of the linear recurrence in step (2), the complexity is $O(nN^2)$ and we thus expect the CPU time to increase with blocking factor. This is borne out very well by the results in the table. The memory requirement of this step is $O(nN)$. It is also clear that step (2) dominates when n is large and this is a potential bottleneck. In the evaluation step we report only the time taken to find the first non-zero solution. As stated above, other non-zero solutions may be derived at a similar cost. Note that in Figure 5 the time of 28 minutes includes the time of computing one additional solution that was zero.

As a final note, we observed that the scheduler met the minimum desired goal of sending one subtask at a time to a target machine. It could also recognize when a host had surplus capacity and in such a case would send more than one task there if conditions permitted. It distinguished and eliminated from the

N	Task		Blocking Factor		
			1×32	2×32	3×32
10,000	(1)	$\langle a^{(i)} \rangle$	$0^h 10'$	$0^h 06'$	$0^h 05'$
	(2)	b-massey	$0^h 06'$	$0^h 08'$	$0^h 10'$
	(3)	evaluation	$0^h 06'$	$0^h 02'$	$0^h 02'$
		total	$0^h 22'$	$0^h 16'$	$0^h 17'$
20,000	(1)	$\langle a^{(i)} \rangle$	$1^h 12'$	$0^h 40'$	$0^h 30'$
	(2)	b-massey	$0^h 25'$	$0^h 31'$	$0^h 39'$
	(3)	evaluation	$0^h 29'$	$0^h 28'$	$0^h 10'$
		total	$2^h 06'$	$1^h 39'$	$1^h 19'$
52,250	(1)	$\langle a^{(i)} \rangle$	$3^h 53'$	$2^h 11'$	$1^h 37'$
	(2)	b-massey	$2^h 30'$	$3^h 09'$	$3^h 54'$
	(3)	evaluation	$1^h 15'$	$0^h 33'$	$0^h 22'$
		total	$7^h 38'$	$5^h 53'$	$5^h 53'$
100,000	(1)	$\langle a^{(i)} \rangle$	$77^h 37'$	$44^h 05'$	$27^h 28'$
	(2)	b-massey	$10^h 03'$	$12^h 28'$	$15^h 42'$
	(3)	evaluation	$74^h 37'$	$27^h 48'$	$11^h 09'$
		total	$162^h 17'$	$84^h 31'$	$54^h 19'$

Figure 5: CPU Time (hours^{h}minutes$'$) for different blocking factors with all arithmetic in GF(2). Each processor is rated at 28.5 MIPS.

schedule machines with high power and memory capacity that were under high load conditions at the time of distribution. We experienced an exceptional case in which two nodes were rendered inoperative by external causes. The scheduler diagnosed the condition, identified the subtasks and successfully restarted them on two other active machines.

4 Conclusions

Using intelligently scheduled parallel subtasks in DSC we have been able to prove titanic integers prime. We have also been able to solve sparse linear systems with over 10.3 million entries over finite fields. Both tasks have been accomplished on a network of common computers. We have solved linear systems with over 100,000 equations, over 100,000 unknowns, and over 10 million non-zero entries over GF(2). The challenge we propose is to solve such systems over GF(p) for word-sized primes p, and ultimately over the rational numbers. In order to meet our challenge, we will explore several improvements to our current approach, by which we hope to overcome certain algorithmic bottlenecks in the block Wiedemann algorithm. As Figure 4 shows, higher parallelization of step (1) slows step (2). One way to speed step (2) with high blocking factor is to use a blocked Toeplitz linear system solver (Gohberg et al. 1986) instead of the generalized Berlekamp/Massey algorithm. The latter method can be further improved to carry out step (2) in $O(n^2N(\log N)^2 \log\log N)$ arithmetic operations using FFT-based polynomial arithmetic and doubling (see Bitmead and Anderson (1980), Morf (1980), and a May 1993 addendum to Kaltofen (1993)).

Another way to speed step (2) is to set up a pipeline between the subtasks that generate the components of $a^{(i)}$ and the program that computes the linear recurrence. Each subtask would compute a segment of $M \leq 2N/n$ sequence elements at a time, and pass it on to the Berlekamp-Massey program which could begin working on these terms of A. Meanwhile the subtasks would compute the next M terms of the sequence. We plan to use co-routines to implement this pipeline.

Literature Cited

Batut, C., Bernardi, D., Cohen, H., and Olivier, M., "User's Guide to PARI-GP," *Manual*, February 1991.

Bitmead, R. R. and Anderson, B. D. O., "Asymtotically fast solution of Toeplitz and related systems of linear equations," *Linear Algebra Applic.* **34**, pp. 103–116 (1980).

Char, B. W., "Progress report on a system for general-purpose parallel symbolic algebraic computation," in *Proc. 1990 Internat. Symp. Symbolic Algebraic Comput.*, edited by S. Watanabe and M. Nagata; ACM Press, pp. 96–103, 1990.

Collins, G. E., Johnson, J. R., and Küchlin, W., "PARSAC-2: A multi-threaded system for symbolic and algebraic computation," *Tech. Report* **TR38**, Comput. and Information Sci. Research Center, Ohio State University, December 1990.

Coppersmith, D., "Solving linear equations over GF(2) via block Wiedemann algorithm," *Math. Comput.*, p. to appear (1992).

Diaz, A., "DSC," *Users Manual (2nd ed.)*, Dept. Comput. Sci., Rensselaer Polytech. Inst., Troy, New York, 1992.

Diaz, A., Kaltofen, E., Schmitz, K., and Valente, T., "DSC A System for Distributed Symbolic Computation," in *Proc. 1991 Internat. Symp. Symbolic Algebraic Comput.*, edited by S. M. Watt; ACM Press, pp. 323–332, 1991.

Gohberg, I., Kailath, T., and Koltracht, I., "Efficient solution of linear systems of equations with recursive structure," *Linear Algebra Applic.* **80**, pp. 81–113 (1986).

Hong, H. and Schreiner, W., "Programming in PACLIB," *SIGSAM Bull.* **26/4**, pp. 1–6 (1992).

Kaltofen, E., "Analysis of Coppersmith's block Wiedemann algorithm for the parallel solution of sparse linear systems," in *Proc. AAECC-10*, Springer Lect. Notes Comput. Sci. **673**, edited by G. Cohen, T. Mora, and O. Moreno; pp. 195–212, 1993.

Kaltofen, E. and Trager, B., "Computing with polynomials given by black boxes for their evaluations: Greatest common divisors, factorization, separation of numerators and denominators," *J. Symbolic Comput.* **9/3**, pp. 301–320 (1990).

Kaltofen, E., Valente, T., and Yui, N., "An improved Las Vegas primality test," *Proc. ACM-SIGSAM 1989 Internat. Symp. Symbolic Algebraic Comput.*, pp. 26–33 (1989).

Kogge, P. M., *The Architecture of Symbolic Computers*; McGraw-Hill, Inc., New York, N.Y., 1991.

Morain, F., "Distributed primality proving and the primality of $(2^{3539} + 1)/3$," in *Advances in Cryptology—EUROCRYPT '90*, Springer Lect. Notes Comput. Sci. **473**, edited by I. B. Damgård; pp. 110–123, 1991.

Morf, M., "Doubling algorithms for Toeplitz and related equations," in *Proc. 1980 IEEE Internat. Conf. Acoust. Speech Signal Process.*; IEEE, pp. 954–959, 1980.

Seitz, S., "Algebraic computing on a local net," in *Computer Algebra and Parallelism*, Springer Lect. Notes Math. **584**; Springer Verlag, New York, N. Y., pp. 19–31, 1992.

Valente, T., "A distributed approach to proving large numbers prime," *Ph.D. Thesis*, Dept. Comput. Sci., Rensselaer Polytech. Instit., Troy, New York, December 1992.

Wiedemann, D., "Solving sparse linear equations over finite fields," *IEEE Trans. Inf. Theory* **IT-32**, pp. 54–62 (1986).

Gauss: a Parameterized Domain of Computation System with Support for Signature Functions

Michael B. Monagan

Institut für Wissenschaftliches Rechnen
ETH Zentrum, CH 8092 Zürich, Switzerland
monagan@inf.ethz.ch

Abstract. The fastest known algorithms in classical algebra make use of signature functions. That is, reducing computation with formulae to computing with the integers modulo p, by substituting random numbers for variables, and mapping constants modulo p. This idea is exploited in specific algorithms in computer algebra systems, e.g. algorithms for polynomial greatest common divisors. It is also used as a heuristic to speed up other calculations. But none exploit it in a systematic manner. The goal of this work was twofold. First, to design an AXIOM like system in which these signature functions can be constructed automatically, hence better exploited, and secondly, to exploit them in new ways. In this paper we report on the design of such a system, Gauss.

1 Introduction

In 1980 Schwarz [11] proposed the following *probabilitistic* method for testing if a matrix of polynomials in $x_1, x_2, ..., x_n$ over the integers is singular or not. This method formalized an idea that was already being used in an ad-hoc way for speeding up various calculations in computer algebra systems, and other places at the time.

Procedure TESTZERO

Input: an n by n matrix A over $\mathbf{Z}[x_1, ..., x_n]$, a failure tolerance ϵ, a degree bound d on $det(A)$

Output: false implies $det(A) \neq 0$, true implies $det(A) = 0$ with probability $\geq 1 - \epsilon$

error $\leftarrow 1$
while error $> \epsilon$ do
    choose prime $p > d$
    $A' \leftarrow A$ modulo p
    for $i = 1..n$ choose α_i at random from $\mathbf{Z}_p$
    $A' \leftarrow A$ evaluated at $x_i = \alpha_i$ modulo p
    $D \leftarrow det(A')$ modulo p
    if $D = 0$ then error $\leftarrow$ error $\times d/p$ else output false
output true

If this procedure outputs "false" then it has proven that the matrix A is non-singular. If the procedure outputs "true" then the procedure may make an error. The idea behind this approach is it will make errors *controllably* low probability ϵ where typically, one would arrange to have $\epsilon < 10^{-50}$. Schwarz's contribution is firstly, a theorem that says that the error bound is met provided the prime p is larger than the degree the determinant, a polynomial in $\mathbb{Z}[x_1, ..., x_n]$, which can be bounded easily in advance. Secondly, if the primes p are chosen to be at least $> 2d$ so that the probability of error is decreased by at least a factor of 2 at each step, then the complexity of the method will be satisfactory from a theoretical viewpoint. In practice, p will be much larger than d, so that only a few iterations of the loop are required before a small probability of error is achieved. Another well known application of such a *probabilistic* algorithm is primality testing. We refer the reader to Rabin [7] and Solovay & Strassen [8] for two different probabilistic algorithms for primality testing.

The idea of using modular mappings in this way is not used much in computer algebra systems. That is a pity because for many specific problems, such as testing whether a linear system has a solution or not, this approach is often computationally the only hope for an answer. In [3], and [4], Gonnet extended the class of expressions for which signature functions were available from rational expressions in $x_1, x_2, ..., x_n$ over $\mathbb{Q}$ to include, roughly speaking, firstly, unnested exponentials, logarithms, and trigonometric functions, and secondly, simple roots. This work led to the routine **testeq** in Maple which tests to see if an expression is zero or not. However, the ideas are not used by the rest of Maple. There is no facility for testing if a matrix is singular or not. In [9] we have looked at signatures for algebraic numbers and algebraic functions. One of the methods for algebraic integers, is based on the ideas in [10]. We will sketch another method later in this paper.

In our experiments, we have attempted to build such signature functions in a more systematic manner, so they can be exploited more generally. To do this, we have designed an AXIOM like system [1] which runs on top of Maple. We have called our system *Gauss*. This paper is organized as follows. In Section 2 we present the design of Gauss. In Section 3 we show how signature functions are built into Gauss. In a subsequent paper, we will report in more detail on the actual signature functions that we use in [9]. In Section 4 we give some novel applications of signature functions in Gauss. We end with some conclusions and comments about Gauss.

2 Gauss

We have designed a system which supports parameterized types and parameterized abstract types. In AXIOM terminology, these are called *domains* and *categories* respectively. The principle motivation is to allow one to implement an algorithm in a *generic* setting. Let us motivate the reason for doing this and illustrate how this is done in Gauss with an example. Consider computing the determinant of a square Matrix over a field using the algorithm Gaussian elimination. In Gauss the code would look like this:

```
GaussianEliminationDeterminant := proc(A,n,F) local d,i,j,k,m;
    d := F[1];
    for k to n do
        for i from k to n while F['=']( A[i,k], 0 ) do od;
        if i>n then RETURN(0) fi;
        if i<>k then
            ... # interchange row i with row k
            d := F['-'](d)
        fi;
        d := F['*'](d,A[k,k]);
        for i from k+1 to n do
            m := F['/'](A[i,k],A[k,k]);
            for j from k+1 to n do
                A[i,j] := F['-'](A[i,j],F['*'](m,A[k,j]))
            od
        od
    od;
    RETURN( d )
end;
```

The key difference between this code and other coding styles is the parameter
F, the field. It is a collection of all the operations for computing over the field
called a *domain*. In the Maple [6] code above, we see that we are using a Maple
table, a hash table, for representing a domain. Thus a domain in Gauss consists
essentially of a table of Maple procedures.

2.1 Using Gauss

The above example gives the reader an idea of how one programs in Gauss.
Before we give more details, we show first how one uses Gauss interactively, and
make some initial comments about the functionality which all domains in Gauss
provide.
The Gauss package is available in Maple V Release 2 after executing the command with(Gauss).

```
> with(Gauss);
---------------------- Gauss version 1.0 ----------------------
Initially defined domains are Z and Q the integers and rationals

                            [init]
```

Initially, the two domains Z, the integers $\mathbb{Z}$, and Q, the rationals $\mathbb{Q}$, have been
defined. Let's do some operations with them:

```
> Z[Gcd](8,12);
```

4

Principle 1: All named operations in Gauss begin with an upper case letter and must be "package-called", i.e. the domain subscript must be explicitly used.

```
> Q['+'](1/2,1/3,1/4);
```

$$\frac{13}{12}$$

```
> Z[Gcd];
```

igcd

Principle 2: Gauss uses the Maple representation for integers and rationals, and the builtin Maple functions where appropriate, e.g. the igcd function. Since Gauss is written in Maple, i.e. Gauss code is Maple code which is interpreted, this starting point is rather crucial for efficiency.

Next we want to perform some operations in $\mathbb{Q}[x]$. First we have to create the domain of computation for objects in $\mathbb{Q}[x]$. This is done by calling a domain constructor, in this case the constructor DenseUnivariatePolynomial (or DUP for short). A domain constructor in Gauss is a Maple procedure (with zero or more parameters), which returns a Gauss domain. The domain constructor DUP takes two parameters, the coefficient ring and the name of the variable.

```
> D := DUP(Z,x):
> a := D[Random]();
```

a := [-37, -35, -55]

Principle 3: All domains in Gauss have a Random function which creates a "random" example from the domain.

```
> d := D[Diff](a);
```

d := [-35, -110]

```
> D[Output](d);
```

$$- 110 \, x - 35$$

Principle 4: All domains in Gauss support an Output function which converts from the domains internal data representation to the Maple form.

```
> m := D[Input](y^4-10*x^2-y);
```

m := FAIL

```
> m := D[Input](x^4-10*x^2-1);
```

m := [-1, 0, -10, 0, 1]

Principle 5: All domains in Gauss support an Input function which tries to convert a Maple expression into the Gauss representation. We do require that D[Input](D[Output](x)) always succeeds as this is also used as a general mechanism for converting from one data type to another.

```
> D[Type](m);
```

true

Principle 6: Since Gauss is implemented on top of Maple, there is no means for static type checking. Type checking is therefore done at run-time in a Maple style manner. Note, we are not advertising this particular feature as a "good" feature of Gauss.

In the remaining parts of this section on Gauss, we will assume that the reader is somewhat familiar with the AXIOM approach. We will present Gauss from AXIOM's point of view, due to lack of space. As we go, we will point out differences.

First, we wish to give some reasons for why we have designed another system instead of using AXIOM. AXIOM suffers from being big and inflexible. If you want to make major changes to fundamental domains, then you will have a lot of work to do. Especially because you would have to recompile dependencies, and the compiler is very slow. If you want to experiment with this, it will be too painful to bear. What we really needed was a small flexible system. That is essentially what Gauss is.

2.2 Domains and Categories in Gauss

Domains (parameterized types) are created by executing Maple functions. The parameters to a domain can be values and other domains. We have seen how domains are used in the determinant example.

Categories (parameterized abstract types) are just domains with missing functions. They are also parameterized by values and domains. Their purpose is to provide code which does not depend on the data representation, and hence can be shared amongst more than one domain.

These ideas, how it is done in Gauss, and the differences with AXIOM are best shown by looking at carefully chosen examples. In Gauss, here is the definition for the category Set, the category to which all other categories belong. The call Set() creates a domain which belongs to the category Set. It defines what operations a domain which is a Set must have, and implements one of them. The call Set(S) extends a domain by adding these operations to the domain set.

```
Set := proc() local S;
    if nargs = 0 then S := newCategory() else S := args[1] fi;
    if hasCategory(S,Set) then RETURN(op(S)) fi;
    addCategory(S,Set);
```

Specifies that S, the domain being constructed, belongs to the category Set.

```
    defOperations( {'=','<>'}, [S,S] &-> Boolean, S );
```

Adds the definitions for the two operations equality and inequality which have the signature indicated to the domain being constructed S.

```
    defOperation( Random, [] &-> S, S );
    defOperation( Input, Expression &-> Union(S,FAIL), S );
    defOperation( Output, S &-> Expression, S );
    defOperation( Type, Expression &-> Boolean, S );
    S['<>'] := subs(D = S,proc(x,y) not D['='](x,y) end);
```

Adds a default definition for the operation inequality in terms of equality.

```
S[Output] := x -> x; # use Maple by default
op(S)
end:
```

The Euclidean Domain in Gauss is

```
EuclideanDomain := proc() local E;
    if nargs = 0 then E := newCategory() else E := args[1] fi;
    E := UniqueFactorizationDomain(E);
```

Inherits (adds to E) all the operations defined and implemented from the category UniqueFactorizationDomain.

```
    addCategory(E,EuclideanDomain);
    defOperation( EuclideanNorm, E &-> Integer, E );
    defOperation( SmallerEuclideanNorm, [E,E] &-> Boolean, E );
    defOperations( {Rem,Quo}, {[E,E] &-> E,[E,E,Name] &-> E}, E );
    defOperation( Gcdex, {[E,E,Name,Name] &-> E,[E,E,Name] &-> E}, E );
    defOperation( Powmod, [E,Integer,E] &-> E, E );
    E[Quo] := subs('D' = E, proc(x,y,r) local t,q;
        t := D[Rem](x,y,q); if nargs = 3 then r := t fi; q
      end);
    E[Div] := subs('D' = E, proc(x,y) local q;
        if D[Rem](x,y,q) <> D[0] then FAIL else q fi
      end);
    E[SmallerEuclideanNorm] := subs('D' = E, proc(x,y)
        evalb( D[EuclideanNorm](x) < D[EuclideanNorm](y) )
      end);
    E[Powmod] := subs('D' = E, proc() PowerRemainder(D,args) end);
    E[Gcd] := subs('D' = E, proc() EuclideanAlgorithm(D,args) end);
```

The default algorithm for computing GCD's is the Euclidean Algorithm. It is implemented in the same manner as the GaussianElimination algorithm.

```
    E[Gcdex] := subs('D' = E, proc() PrincipalIdeal(D,args) end);
    op(E)
end:
```

Notice that of the new operations defined in the EuclideanDomain category, only the Euclidean norm and remainder operations are not defined. The other operations are defined either in-line, or as calls to out-of-line procedures. We include here the code for the EuclideanAlgorithm, because this code illustrates the way one implements algorithms in Gauss, and it is probably the best generic implementation possible of the Euclidean algorithm. Note that the algorithm is n-ary, i.e. it computes the GCD of zero or more arguments.

```
EuclideanAlgorithm := proc(E) local a,b,r,s;

    s := {args[2..nargs]} minus {E[0]};
```

A property of Gauss inherited from Maple is that for Gauss domains which have canonical forms, equal values are represented uniquely, and hence here, duplicates can be removed very efficiently.

```
    if s = {} then RETURN(E[0]) fi;
    sort( [op(s)], E[SmallerEuclideanNorm] );
```

The idea here is to start with the smallest values, because the size of GCD's always get smaller in the Euclidean norm sense, and usually, they this carries over to the computational cost too.

```
a := E[Normal](s[1]);
for b in subsop(1=NULL,s) while a <> E[1] do
     b := E[Normal](b);
     while b <> E[0] do
     r := E[Normal](E[Rem](a,b));
```

This is the monic Euclidean algorithm. That is, the Normal function here is making the remainder unit normal. This is a generic attempt to control intermediate expression swell.

```
          a := b;
          b := r
          od;
     od;
     a
end:
```

We continue with an outline of the category for univariate polynomials.

```
UnivariatePolynomial := proc() local P,R,env,U,'?';

     R := args[1];
     if not hasCategory(R,Ring) then
          ERROR('1st argument must be a ring') fi;
```

This is typical error checking. In the new version of Maple, it could have been done in the procedure declaration as proc(R:Ring).

```
     if nargs = 1 then P := newCategory(); else P := args[2] fi;
     addCategory(P,UnivariatePolynomial);

     if hasCategory(R,Field)
     then P := EuclideanDomain(P)
     ...
     else P := Ring(P)
     fi;

     if hasCategory(R,OrderedSet) then P := OrderedSet(P); fi;
```

Here we have included the usual theorems about univariate polynomials and an example of multiple inheritance. Multiple inheritance in Gauss means that more than one set of operations is added to the domain.

```
     defOperation( CoefficientRing, P &-> Ring, P );
     P[CoefficientRing] := R;
     ...
```

There are lots of operations defined for polynomials of course. One difference between AXIOM and Gauss is that programs in Gauss can get their hands on parts of domains, their parameters e.g. here the coefficient ring.

```
     env := ['D' = P, 'C' = R];
     ...
     P[EuclideanNorm] := subs(env, proc(a) D[Degree](a) end);
```

This is a Maple problem. We are in the process of building the domain P. Here we are inserting the Maple procedure for the EuclideanNorm operation. It needs to call the operation Degree from the domain P. The reason for the reference to D[Degree] instead of P[Degree] is because Maple does not support nested scopes, and the substitution is a just hack to make this work. Nested scopes will be supported in a future version of Maple. We mention this because apart from having to package call every operation, which is not really serious, this is the only dissatisfying feature of the readability of the code.

```
    ...
    op(P);
end;
```

Finally, let's look at a domain which inherits the definitions and code from the category UnivariatePolynomial.

```
    DenseUnivariatePolynomial := proc() local x,P,R,env;

    R := args[1];
    P := UnivariatePolynomial(R);
```

Inherit the definitions and code from the category

```
    ...
    if hasProperty(R,CanonicalForm) then
        addProperty(P,CanonicalForm);
        if hasProperty(R,UniquelyRepresented) then
                addProperty(P,UniquelyRepresented);
                P['='] := <evalb(x=y)>;
        fi;
    fi;
```

This shows that Gauss keeps some other kind of information around called properties. For categories, we have properties associated with the mathematical properties of the category like "commutativity". For domains, we have properties associated with the data representation, like "CanonicalForm" and "UniquelyRepresented". Here we are exploiting the fact that if we have a canonical form for the coefficients, then because we are using a dense expanded representation, we have a canonical form for the polynomial ring we are constructing. Secondly, if the coefficients are also uniquely represented in memory, then because we are using a Maple list, the polynomials will be uniquely represented. And why is this information useful? Because now we can use a faster implementation of equality, based on machine address – which is what this Maple code is doing.

```
    # Representation is a Maple list of coefficients
    P[0] := [R[0]];
    P[1] := [R[1]];
    ...
    op(P)
end:
```

3 Signature Functions in Gauss

The signature functions that we use, in general, map values into finite rings, namely the integers mod n where n is not necessarily prime, and $\mathbb{Z}_p[x]/(m(x))$ where $m(x)$ is not necessarily irreducible, and p is a prime integer. These finite rings are chosen so that there are a low percentage of un-invertible elements. To simplify the presentation of this section however, we assume finite fields.

Our first idea is that domains for which we can define signature functions should support the operation

$$\text{ModularMapping: } (N, N) \rightarrow \text{UNION}((F:\text{FiniteField}, D \rightarrow F), \text{FAIL})$$

where N is the set of natural numbers, D is the domain for which this operation is being defined. Thus ModularMapping outputs a pair, a finite field F, and a mapping from D to F. Since for probabilistic algorithms, we will need more than one such mapping, ModularMapping takes an index as the first argument and hence can be used to generate a sequence of mappings. The second argument is a lower bound for the cardinality of the finite field. If D is itself a finite field, then the only non-trivial mapping available is the identity mapping. For this reason, **ModularMapping** is allowed to fail.

The operation ModularMapping has been explicitly coded for the Integer domain and the FiniteField category. In all other cases, it is automatically constructed by the system when a domain is constructed. The constructions for polynomials and quotient fields are simple. We include the code for polynomials here. Note, we have removed the magic substitutions (which are used for simulating nested scopes) from the Maple code to make it easier to read. I.e. the reader should assume that Maple has nested scopes.

```
PolynomialHomomorphism := proc(n,p) local R,F,f,alpha,D,x;
    R := P[CoefficientRing];
    F := R[ModularHomomorphism](n,p);
    if F = FAIL then RETURN(FAIL) fi;
    f := F[1]; F := F[2];
    beta := F[Random]();
    D := DUP(F,x);
    proc(x) local b;
            b := map(f,P[ListCoeffs](x));
            b := D[Polynom](b);
            D[Eval](b,beta);
    end, F
end:
```

In [9] we have looked at several possible constructions for algebraic number fields. Here we sketch a general method for algebraic extensions. Given the algebraic extension $\mathbb{K}[x]/(a)$, we first obtain a modular mapping f into a finite field $\mathbb{F}_q$ for the ground field $\mathbb{K}$ such that the leading coefficient of a does not vanish under f. Then we map the defining polynomial a into $\mathbb{F}_q[x]$ yielding a'. Since $\mathbb{F}_q$ is a finite field, we factor a' and choose to work with the smallest factor b and hence construct the finite field $\mathbb{F}_q[x]/(b)$ in which to compute signatures.

We want to show how we can implement Schwarz's algorithm using this ModularMapping function. Before we can do this, we need some way to determine a degree bound for the degree of the determinant. One way to do this is to implement a special routine. But this would mean coding a new version of the determinant routine. Another way of computing a bound is to simply count the number of multiplications m done in the coefficient field and note that 2^m is a bound on the degree. This is not very good bound but it can be easily implemented. A better method is given in the next section. Assuming we have a degree bound D we can sketch the implementation.

```
# Input: an m by n matrix A over R, a degree bound D,
#        and an error tolerance E
# Output: rank(A) with probability > 1-E of being correct
ProbabilisticMatrixRank := proc(M,A,D,E)
local R,m,n,i,F,N,f,B,rank,error;
    R := M[CoefficientRing]; m := M[Rows](A); n := M[Cols](A);
    error := 1;
    rank := 0;

    for i while error > E do
        F := R[ModularHomomorphism](i,D);
        if F = FAIL then RETURN(FAIL) fi;
        f := F[1]; F := F[2];
        N := Matrix(m,n,F); # Create an m by n matrix domain over F
        B := N[Input](N[Output](N[Map](f,A)));
        rank := max(rank,N[Rank](B));
        error := error * D/F[Size];
    od;
    rank
end;
```

Other applications of ModularMapping include testing if a polynomial divides another, computing the degree of the GCD of two univariate polynomials, hence testing whether two polynomials are relatively prime or not. A good application of the latter is in determining whether a univariate polynomial is square free or not, i.e. whether $GCD(a, a') = 1$. The coding of these applications is straightforward.

4 Applications

We give several other applications of signatures in this section.

4.1 Signature domains and forms of expressions

The **Signature** domain computes with signatures for rational functions in a set of variables over a constant field in which we can compute signatures. The **Signature** domain simply replaces input expressions by signatures; the ensuing computation computes over the finite ring defined by the signature function. This eliminates intermediate expression swell and provides us with probabilistic zero equivalence but restricts us to the field operations $+, -, \times$, and $/$.

When given as a parameter to a domain constructor, such as DenseUnivariatePolynomial or SquareMatrix, we get a computation in which part of the computation is being done with signatures, and another part is being done with variables. This allows one to look at the *structure* or *form* of the result. This is best understood by looking at an example.

```
> S := Signature(a,b,c):
> R := RationalFunction(S,[x,y,z]):
```

We are computing with rational expressions in 6 symbols. But the symbols a, b, c will be mapped to a finite field on input, so the actual computation will be done with x, y, z only.

```
> M := SM(3,R):  # Create a 3 by 3 matrix domain over R
> A := [[a,b,c],[a*x,a*y,a*z],[a*x^2,a*y^2,a*z^2]]:
> M[Output](M[Inv](M[Input](A)));
```

$$
\left[\left[\frac{*yz^2 + *y^2z}{\%1}, \frac{*z^2 + *y^2}{\%1}, \frac{*z + *y}{\%1}\right],\right.
$$

$$
\left[\frac{*xz^2 + *x^2z}{\%1}, \frac{*z^2 + *x^2}{\%1}, \frac{*z + *x}{\%1}\right],
$$

$$
\left.\left[\frac{*xy^2 + *x^2y}{\%1}, \frac{*y^2 + *x^2}{\%1}, \frac{*y + *x}{\%1}\right]\right]
$$

$$
\%1 := (y + *x)z^2 + (*y^2 + *x^2)z + *xy^2 + *x^2y
$$

The *'s appearing in the output are signatures which are neither 0 nor 1. Thus the output tells us what the inverse of the matrix looks like as a function of x, y, z by telling us which coefficients are 0 and also which are 1. Note, one could also output the coefficients which are small integers, e.g. -1 might also be of interest.

4.2 Computation Sequences and Automatic Code Generation

The ComputationSequence domain creates a "computation sequence" i.e. a sequence of all the field operations done during a calculation, but without doing the operations symbolically. This gives us a third possibility for automatic code generation. For example, consider the problem of generating efficient numerical code for inverting a given matrix A where the entries of A are a function of $x_1, x_2, \ldots, x_n$. There are three possible approaches:

1. The numeric approach: Here the values of the parameters are bound before the matrix is constructed. The resulting matrix is purely numerical, and a numerical solver is used.

2. The symbolic approach. Here, the parameters are bound after the inverse is computed symbolically. The problem with this approach, is that the symbolic inverse may have no compact representation.

3. The computation sequence approach. Here, the sequence of operations that would be performed in the numeric approach is created in advance. Signatures are used to determine intermediate zeroes so that non-zero pivots are selected.

We illustrate the idea with an example.

```
> S := ComputationSequence(a,b,c):
> M := SM(3,S):
> A := [[a,b,c],[a,b,a],[a,c,b]]:
> B := M[Input](A):
> M[Det](B);
t1 = 1/a
t2 = a-c
t3 = c-b
t4 = b-c
t5 = -a
t6 = t5*t3
t7 = t6*t2
```

We have output the computation sequence as a sequence of assignment statements on the fly, that is, as each arithmetic operation is executed. The t variables are temporary variables. The last variable t7 is the determinant.

4.3 A Probabilistic Method for Computing the Degree of a Polynomial

We give an algorithm that given a polynomial represented by a computation sequence, such as the result of the above determinant calculation, determines probabilistically the degree of each variable. The method was also used in the DAGWOOD system [5]. This is quite an important utility because it means that we can now remove the need to compute degree bounds from many calculations. For example, this can be used to improve the performance of the sparse polynomial interpolation by determining the actual degrees of the polynomials rather than using a bound.

The idea is to create a computation sequence f for the algorithm, then to compute the degree d of it by trying to interpolate the polynomial. After having evaluated f at i points, and interpolated those i points, we evaluate the interpolated polynomial at a next point x_{i+1} yielding y_{i+1} and compare with $f(x_{i+1})$. If the values agree, then we output i as the "probable" degree of the polynomial. If not, we iterate.

To do this efficiently, one needs an incremental version of a polynomial interpolation algorithm that interpolates the polynomial for each successive point in a linear number of operations in F. We have adapted algorithm 5.2 "Newton Interpolation Algorithm" from [2] for this purpose.

Procedure Probabilistic Polynomial Degree Bound

Input: f a computation sequence for a polynomial in n variables over F a finite ring, and k the index of the variable for which the degree is sought

Output: the degree of the kth variable of f

for $i = 1..n$ do choose a_i at random from F
choose x_0 at random from F
$y_0 \leftarrow f(a_1, \ldots, a_{k-1}, x_0, a_{k+1}, \ldots a_n)$
for $i = 1..\infty$ do
    $g \leftarrow$ interpolate $x_1, \ldots, x_i$ and $y_1, \ldots, y_i$ over F incrementally
    choose t at random from F such that t is not in $x_1, \ldots, x_i'$
    $y_{i+1} \leftarrow f(a_1, \ldots, a_{k-1}, t, a_{k+1}, \ldots, a_n)$
    if $g(a_1, \ldots, a_{k-1}, t, a_{k+1}, \ldots, a_n) = y_{i+1}$ then output i
    $x_{i+1} \leftarrow t$

Note this assumes that the cardinality of F is greater than the degree, and that the computation sequence f is polynomial in x_k, otherwise, the algorithm will not terminate. Also, if the computation sequence includes divisions, the algorithm may fail due to an unlucky division by zero. A practical implementation must allow for these cases.

5 Conclusion

We have designed a system that supports parameterized domains, a la AX-IOM, in which signature functions are automatically created for many integral domains. This makes it possible to use probabilistic methods to do various calculations much faster than is otherwise possible. We have shown some examples of how signature functions can be used to solve various kinds of problems, in addition to the standard applications.

We expect that the reader will have questions about Gauss so the remainder of this conclusion gives some general information about Gauss. The main advantage of Gauss in Maple is that it allows us to implement generic algorithms, in an AXIOM-like manner. The primary advantages are that it is very simple, very flexible, and very small. We have found that programmers have had no difficulty in writing code.

Is Gauss efficient? Yes and no. Gauss code is written in Maple, thus interpreted. The overhead of Gauss, in comparison with Maple code, consists of a Maple table subscript and procedure call for every operation. The subscript is cheap, the procedure calling mechanism in Maple is relatively expensive. But in many cases, we can directly make use of builtin Maple functions. We find that Gauss runs typically not much slower than the Maple interpreter. For polynomial arithmetic over $\mathbb{Z}, \mathbb{Q}, \mathbb{Z}_p$, Gauss is much slower than Maple because $+, -, \times$ and division are builtin. What we have done to get back the efficiency, is to make Maple polynomial domains in Gauss which use the Maple representation for polynomials, and hence also Maple's builtin operations. On the other hand, in some cases, Gauss has turned out to be much faster than Maple because no time is wasted analyzing what kind of expression was input.

Is Gauss code aesthetically pleasing to read and easy to write? An obvious disadvantage of Gauss is that one must explicitly "package call" each operation. There is no compiler, so no type analysis to avoid having to do this. However, it has been our experience that this is of little or no hindrance to programming in Gauss. The drawback is in interactive usage. The other main deficiency is that since Maple does not support nested scopes, this has to be simulated which makes the code look somewhat ugly. We intend to resolve this deficiency by adding nested scoping to Maple.

References

1. Jenks R., Sutor R.: *axiom – The Scientific Computation System*, Springer, 1992.
2. Geddes K.O., Labahn G., Czapor S.R.: *Algorithms for Computer Algebra* Kluwer, 1991.
3. Gonnet G.H.: Determining Equivalence of Expressions in Random Polynomial Time. it Proceedings of the 16th ACM Symposium on the Theory of Computing (1984) 334–341
4. Gonnet G.H.: New Results for Random Determination of Equivalence of Expressions. *Proceedings of the 1986 Symposium on Symbolic and Algebraic Computation* (1986) 127–131
5. Freeman T., Imirzian G., Kaltofen, E.: DAGWOOD: A System for Manipulating Polynomials Given by Straight-Line Programs. *Proceedings of the 1986 Symposium on Symbolic and Algebraic Computation* (1986) 169–175
6. Char B.W., Geddes K.O., Gonnet G.H., Leong B.L., Monagan M.B., and Watt S.M.: *Maple V Language Reference Manual.* Springer-Verlag, New York, 1991.
7. Rabin, M.O.: Probabilistic Algorithm for Testing Primality. *J. of Number Theory* **12** (1980) 128–138
8. Solovay, R. Strassen, V.: A fast Monte-Carlo Test for Primality. *SIAM J. of Computing* **6** (1977) 84–85
9. Monagan M.B.: Signatures + Abstract Types = Computer Algebra - Intermediate Expression Swell. Ph.D. Thesis, University of Waterloo, 1989.
10. Monagan M.B.: A Heuristic Irreducibility Test for Univariate Polynomials *J. Symbolic Comp.* **13** No. 1 (1992) 47–57
11. Schwartz J.T.: Fast probabilistic algorithms for verification of polynomial identities. *J. ACM.* **27** (1980) 701–717

On Coherence
in Computer Algebra

Andreas Weber

Wilhelm-Schickard-Institut für Informatik
Universität Tübingen
72076 Tübingen, Germany
E-mail: weber@informatik.uni-tuebingen.de
Fax: +49 70 71 29 59 58

Abstract. Modern computer algebra systems (e. g. AXIOM) support a rich type system including parameterized data types and the possibility of implicit coercions between types. In such a type system it will be frequently the case that there are different ways of building coercions between types. An important requirement is that all coercions between two types coincide, a property which is called *coherence*.

We will prove a coherence theorem for a formal type system having several possibilities of coercions covering many important examples. Moreover, we will give some informal reasoning why the formally defined restrictions can be satisfied by an actual system.

1 Introduction

Modern computer algebra systems (e. g. AXIOM [12]) support many different types which correspond to mathematical structures. By the use of parameterized data types (also called *type constructors*) and of generic operators they can handle a variety of structures in a uniform way.

If such a system is internally strongly typed — which is desirable from a software engineering point of view — then a user interface should have the possibility to *insert coercion functions* in appropriate places. So if a user types in $3 + x$ the system should coerce the integer 3 into the constant integral polynomial 3.

A more complex example will exemplify the power of implicit coercions, but also a possible problem with them, which will be the main topic of this paper.

Consider the expression

$$t - \begin{pmatrix} 1 & 0 \\ 3 & \frac{1}{2} \end{pmatrix}$$

which — as a mathematician would conclude — denotes a 2×2-matrix over $\mathbb{Q}[t]$ where t is the usual shorthand for t times the identity matrix. In an AXIOM like type system, this expression involves the following types and type constructors: The integral domain I of integers, the unary type constructor FF which forms the quotient field of an integral domain, the binary type constructor UP which forms the ring of univariate polynomials over some ring in a specified indeterminate, and the type constructor $M_{2,2}$ building the 2×2-matrices over a commutative ring.

In order to type this expression correctly several of the following coercions have to be used.

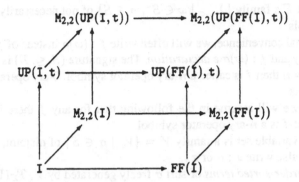

There are different ways to coerce I to $M_{2,2}(UP(FF(I), t))$. Of course one wants the embedding of I in $M_{2,2}(UP(FF(I), t))$ to be independent of the particular choice of the coercion functions.

In this example this independence seems to be the case, but how can we *prove* it? Moreover, not all coercions which would be desirable for a user share this property. Consider for example the binary type constructor "direct sum" $\oplus$ defined for Abelian groups. One could coerce A into $A \oplus B$ via a coercion ϕ_1 and B into $A \oplus B$ via a coercion ϕ_2. But then the image of A in $A \oplus A$ depends on the choice of the coercion function!

Following [3] and [18] we will call a type system *coherent*, if the coercions are independent of the way they are deduced in the system.[1]

In the following we will look at different kinds of coercions which occur and we will state some conditions which will yield the coherence of the system. Besides the technical proof of the coherence theorem we will give some informal discussions about the significance of these conditions.

We will describe our type system in terms of an order-sorted algebra as suggested by Rector [16] and Comon, et al. [7] and we will develop the coherence theorem in a category theoretic framework as suggested by Reynolds [17].

2 Preliminaries

2.1 Category Theory

For the notions of category theory we refer the reader to the literature. A comprehensive source is [14].

We will call a category C a *preorder*, if for any two objects A, B of the category and any arrows $f, g : A \longrightarrow B$ we will have $f = g$.

---

[1] Notice that the terminus "coherence" is used similarly in category theory (see e. g. [14]) but is used quite differently in connection with order-sorted algebras (e. g. in [20], [10], [16]).

2.2 Order-Sorted Signatures

An *order-sorted signature* is a triple $(S, \leq, \Sigma)$, where S is a set of sorts, $\leq$ a partial order on S, and Σ a family $\{\Sigma_{\omega,\sigma} \mid \omega \in S^*, \sigma \in S\}$ of not necessarily disjoint sets of operator symbols.

For notational convenience, we will often write $f : (\omega)\sigma$ instead of $f \in \Sigma_{\omega,\sigma}$; $(\omega)\sigma$ is called an *arity* and $f : (\omega)\sigma$ a *declaration*. The signature $(S, \leq, \Sigma)$ is often identified with Σ. If $|\omega| = n$ then f is called a *n-ary operator symbol*. 0-ary operator symbols are *constant symbols*.

As in [19] we will assume in the following that for any f there is only a single $n \in \mathbb{N}$ such that f is a n-ary operator symbol.

A σ-sorted variable set is a family $V = \{V_\sigma \mid \sigma \in S\}$ of disjoint, nonempty sets. For $x \in V_\sigma$ we also write $x : \sigma$ or x_σ.

The set of *order-sorted terms* of sort σ freely generated by V, $T_\Sigma(V)_\sigma$, is the least set satisfying

- if $x \in V_{\sigma'}$ and $\sigma' \leq \sigma$, then $x \in T_\Sigma(V)_\sigma$
- if $f \in \Sigma_{\omega,\sigma'}$, $\omega = \sigma_1 \cdots \sigma_n$, $\sigma' \leq \sigma$ and $t_i \in T_\Sigma(V)_{\sigma_i}$, then $f(t_1, \ldots, t_n) \in T_\Sigma(V)_\sigma$.

The set of all order-sorted terms over Σ freely generated by V will be denoted by

$$T_\Sigma(V) := \bigcup_{\sigma \in S} T_\Sigma(V)_\sigma.$$

The set of all *ground terms* over Σ is $T_\Sigma := T_\Sigma(\{\})$.

If $t \in T_\Sigma(V)_\sigma$ we will also write $t : \sigma$.

A signature is *regular*, if each term $t \in T_\Sigma(V)$ has a least sort.

The *complexity* of a term $t \in T_\Sigma(V)$, $\text{com}(t)$ is inductively defined as follows:

- $\text{com}(t) = 1$, if $t \in V_\sigma$ or $t \in \Sigma_{\epsilon,\sigma}$ for some $\sigma \in S$,
- if $f \in \Sigma_{\omega,\sigma'}$, $\omega = \sigma_1 \cdots \sigma_n$, and $t_i \in T_\Sigma(V)_\sigma$, then

$$\text{com}(f(t_1, \ldots, t_n)) = \max(\text{com}(t_1), \ldots, \text{com}(t_n)) + 1.$$

3 The Language of Types

As in the paper of Comon, et al. [7], on a type system for a computer algebra system, a *type* will just be an element of the set of all order-sorted terms over a regular signature $(S, \leq, \Sigma)$ freely generated by some family of infinite sets $V = \{V_\sigma \mid \sigma \in S\}$.

The sets V_σ are the sets of *type variables*, the sorts are called *properties* and have, e. g. field, ring, ... as typical elements. The ordering of the sorts reflects the strength of the properties. For instance, field $\leq$ ring.

A type denoted be a ground term is called a *ground type*, a non-ground type is called a *polymorphic type*.

A type denoted by a constant symbol will be called a *base type*. Typical examples will be integer, boolean, ...

The non-constant operator symbols are called *type constructors*, and have

> list : (any)any
> list : (ordered_set)ordered_set
> UP : (commutative_ring symbol)commutative_ring
> UP : (integral_domain symbol)integral_domain
> FF : (integral_domain)field

as typical examples.

This formalism is well suited to express the subset of the type system of AXIOM [12], in which only non-parameterized categories[2] are considered and the sorts correspond to the non-parameterized categories.

In the following we will assume that we have an actual programming language in which the types can be described by our framework.

Moreover, we will assume that we have a *semantics* for the ground types which satisfies the following conditions:

- The ground types correspond to mathematical objects in the sense of universal algebra or model theory.
- Functions between ground types are set theoretical functions. If we say that two functions $f, g : t_1 \longrightarrow t_2$ are equal ($f = g$) we mean equality between them as set theoretic objects.

Remark. Since we only need a set theoretic semantics for *ground types* and functions between ground types, the obvious interpretations of the types as set theoretic objects will do.

Of course, equality between two functions will be in general an undecidable property, but this will not matter in the following discussion, since we will always give some particular reasoning for the equality of two functions between two types.

4 Coercions

We will assume that we have a mechanism to declare some functions between types to be *implicit coercions* between these types (or simply *coercions*). If there is a coercion $\phi : t_1 \longrightarrow t_2$ we will write $t_1 \hookrightarrow t_2$.

Remark. The requirement of set theoretic ground types and coercion functions excludes some constructions — if we gave all types the "obvious" set theoretic interpretation —, as the one used in in [15, Lemma 2], which assumes a coercion from the space of functions $FS(D, D)$ over some domain D into this domain. Such coercions which correspond to certain constructions of models of the λ-calculus (see e. g. [2]) seem to be of theoretical interest only. At least for the purpose of a computer algebra system the requirement of set theoretic coercion functions does not seem to be a restriction at all!

Relying on the set theoretic semantics for our types and coercion functions we can give the following definition of coherence.

---

[2] Category in the sense of the AXIOM language, not in the sense of category theory!

Definition 1 (Coherence). A type system is *coherent* if the following condition is satisfied:

> For any ground types t_1 and t_2 of the type system, if $\phi, \psi : t_1 \longrightarrow t_2$ are coercions then $\phi = \psi$.

4.1 General Assumptions

It will be convenient to declare each identity function on a type to be an implicit coercion.

Assumption 1. *For any ground type t the identity on t will be a coercion. If $\phi : t_1 \longrightarrow t_2$ and $\psi : t_2 \longrightarrow t_3$ are coercions, then the composition $\phi \circ \psi : t_1 \longrightarrow t_3$ of ϕ and ψ is a coercion.*

Lemma 2. *If assumption 1 holds, then the set of ground types as objects together with the coercion functions as arrows form a category.*

Proof. Since composition of functions is associative and the identity function is a coercion, all axioms of a category are fulfilled. □

In the following we will always assume that assumption 1 holds even if we do not mention it explicitly.

4.2 Base Types

For this presentation[3] it is possible to define only a few types as base types, but to construct them by type constructors whenever possible.[4] Since there are only very few coercions between base types, the following assumption seems to be easily satisfiable.

Assumption 2 (Base Types). *The subcategory of base types and coercions between base types forms a preorder, i.e. if t_1 and t_2 are base types and $\phi, \psi : t_1 \longrightarrow t_2$ are coercions then $\phi = \psi$.*

4.3 Structural Coercions

Definition 3 (Structural Coercions). The n-ary type constructor ($n \geq 1$) f induces a *structural coercion*, if there are sets $A_f \subseteq \{1, \ldots, n\}$ and $M_f \subseteq \{1, \ldots, n\}$ such that the following condition is satisfied:

Whenever there are declarations $f : (\sigma_1 \cdots \sigma_n)\sigma$ and $f : (\sigma'_1 \cdots \sigma'_n)\sigma'$ and ground types $t_1 : \sigma_1, \ldots, t_n : \sigma_n$ and $t'_1 : \sigma'_1, \ldots, t'_n : \sigma'_n$ such that $t_i = t'_i$ if $i \notin A_f \cup M_f$ and there are coercions

$$\phi_i : t_i \longrightarrow t'_i, \quad \text{if } i \in M_f,$$
$$\phi_i : t'_i \longrightarrow t_i, \quad \text{if } i \in A_f,$$
$$\phi_i = \mathrm{id}_{t_i} = \mathrm{id}_{t'_i}, \text{if } i \notin A_f \cup M_f,$$

---

[3] Efficiency of computation does not play the central rôle in this investigation!

[4] As an example consider the field of rational numbers, which can be constructed as the quotient field of the integers.

then there is a *uniquely defined* coercion

$$\mathcal{F}_f(t_1,\ldots,t_n,t'_1,\ldots,t'_n,\phi_1,\ldots,\phi_n) : f(t_1,\ldots,t_n) \longrightarrow f(t'_1,\ldots,t'_n).$$

The type constructor f is *covariant in its i-th argument*, if $i \in \mathcal{M}_f$. It is *contravariant in its i-th argument*, if $i \in \mathcal{A}_f$.

As can be seen by the examples given below the family of coercion functions

$$\{\mathcal{F}_f(t_1,\ldots,t_n,t'_1,\ldots,t'_n,\phi_1,\ldots,\phi_n) \mid t_i \in T_\Sigma(\{\})_{\sigma_i}, t'_i \in T_\Sigma(\{\})_{\sigma'_i}, \phi_i : t_i \longrightarrow t'_i\}$$

will very often be just a single (*polymorphic*) function. Such a system will have only finitely many coercion functions which define the induced structural coercion, although in our framework there might be infinitely many.

Assumption 3 (Structural Coercions). *Let f be n-ary type constructor which induces a structural coercion and let $f(t_1,\ldots,t_n)$, $f(t'_1,\ldots,t'_n)$, and $f(t''_1,\ldots,t''_n)$ be ground types. Assume that*

$$t_i \hookrightarrow t'_i \hookrightarrow t''_i, \text{ if } i \in \mathcal{M}_f,$$
$$t''_i \hookrightarrow t'_i \hookrightarrow t_i, \text{ if } i \in \mathcal{A}_f,$$
$$t_i = t'_i = t''_i, \quad \text{if } i \notin \mathcal{A}_f \cup \mathcal{M}_f.$$

and let $\phi_i : t_i \longrightarrow t'_i$, $\phi'_i : t'_i \longrightarrow t''_i$ (if $i \in \mathcal{M}_f$), and $\phi'_i : t''_i \longrightarrow t'_i$, $\phi_i : t'_i \longrightarrow t_i$ (if $i \in \mathcal{A}_f$) be coercion functions. For $i \notin \mathcal{A}_f \cup \mathcal{M}_f$ let ϕ and ϕ' be the appropriate identities.

Then the following conditions are satisfied:

1. *$\mathcal{F}_f(t_1,\ldots,t_n,t_1,\ldots,t_n,\phi_1,\ldots,\phi_n)$ is the identity on $f(t_1,\ldots,t_n)$,*
2. *$\mathcal{F}_f(t_1,\ldots,t_n,t''_1,\ldots,t''_n,\phi_1 \circ \phi'_1,\ldots,\phi_n \circ \phi'_n) =$*
   $\mathcal{F}_f(t_1,\ldots,t_n,t'_1,\ldots,t'_n,\phi_1,\ldots,\phi_n) \circ \mathcal{F}_f(t'_1,\ldots,t'_n,t''_1,\ldots,t''_n,\phi'_1,\ldots,\phi'_n).$

Typical examples of type constructors which induce a structural coercion are `list`, UP, $M_{n,n}$ or the "function space" type constructor $FS(A, B)$.

The former examples give rise to structural coercions, because the constructed type is in some way a sequence[5] built of the type parameter. The coercions between the constructed types are then obtained by *mapping* the coercions between the type parameter into the sequence. Since a mapping of functions distributes with function composition, assumption 3 will be satisfied by these examples.

Another mechanism gives rise to the structural coercion in the case of the "function space" type constructor, as is well known (see e. g. [5]). It is contravariant in its first argument and covariant in its second argument, as the following considerations show: Let A and B be two types where there is an implicit coercion ϕ from A to B. If f is a function from B into a type C, then $f \circ \phi$ is a function from A into C. Thus any function from B into C can be coerced into a function from A into C. Thus an implicit coercion from $FS(B, C)$ into $FS(A, C)$ can be defined, i. e. $FS(B, C) \hookrightarrow FS(A, C)$. If

---

[5] Using parameterized categories in the terminology of AXIOM which correspond to parameterized type classes in the terminology of HASKELL [11] it is possible to define "sequences" formally, see e. g. [6].

$C \hookrightarrow D$ by an implicit coercion ψ, then $\psi \circ f$ is a function from A into D, i.e. an implicit coercion from $\mathrm{FS}(A, C)$ into $\mathrm{FS}(A, D)$ can be defined.

In this case assumption 3 is satisfied because of the associativity of function-composition.

4.4 Direct Embeddings in Type Constructors

Definition 4 (Direct Embeddings). Let $f : (\sigma_1 \cdots \sigma_n)\sigma$ be a n-ary type constructor. If for some ground types $t_1 : \sigma_1, \ldots, t_n : \sigma_n$ there is a coercion function $\Phi^i_{f,t_1,\ldots,t_n} : t_i \longrightarrow f(t_1, \ldots, t_n)$, then we say that f *has a direct embedding at its i-th position*.

Moreover, let $\mathcal{D}_f = \{i \mid f$ has a direct embedding at its i-th position$\}$ be the *set of direct embedding positions of f*.

Remark. In a system, a type constructor represents a parameterized abstract data type which is usually built uniformly from its parameters. So the family of coercion functions

$$\{\Phi^i_{f,t_1,\ldots,t_n} \mid t_i \in T_\Sigma(\{\})_{\sigma_i}\}$$

will very often be just one (*polymorphic*) function. In this respect the situation is similar to the one in Sec. 4.3.

Assumption 4 (Direct Embeddings). *Let $f : (\sigma_1 \cdots \sigma_n)\sigma$ be a n-ary type constructor. Then the following conditions hold:*

1. *$|\mathcal{D}_f| = 1$.*
2. *The coercion functions which give rise to the direct embedding are unique, i.e. if $\Phi^i_{f,t_1,\ldots,t_n} : t_i \longrightarrow f(t_1, \ldots, t_n)$ and $\Psi^i_{f,t_1,\ldots,t_n} : t_i \longrightarrow f(t_1, \ldots, t_n)$, then $\Phi^i_{f,t_1,\ldots,t_n} = \Psi^i_{f,t_1,\ldots,t_n}$.*

Many important type constructors such as $\mathtt{list}$, $\mathtt{M}_{n,n}$, $\mathtt{FF}$, and in general the ones describing a "closure" or a "completion" of a structure — such as the p-adic completions or an algebraic closure of a field — are unary. Since for unary type constructors the condition $|\mathcal{D}_f| = 1$ is trivial and the second condition in assumption 4 should be always fulfilled, the assumption holds in these cases.

For n-ary type constructors ($n \geq 2$) the requirement $|\mathcal{D}_f| = 1$ might restrict the possible coercions. Consider the "direct sum" type constructor for Abelian groups which we have already seen that it could lead to a type system that is not coherent if we do not restrict the possible coercions. For a type constructor

$$\oplus : (\mathsf{Abelian\_group}\ \mathsf{Abelian\_group})\mathsf{Abelian\_group}$$

the requirement $|\mathcal{D}_f| = 1$ means that it is only possible to have either an embedding at the first position or at the second position.

In our framework the types $A \oplus B$ and $B \oplus A$ will be different. However, the corresponding mathematical objects are *isomorphic*. Having a mechanism in a language that represents certain isomorphic mathematical objects by the same type[6] the declaration

---

[6] In the example of the "directs sums" which are commutative and associative this could be done using the techniques of representing associative-commutative operators in certain term-rewriting systems (see e. g. [4]).

of both natural embeddings to be coercions would not lead to an incoherent type system. Notice that such an additional mechanism, which corresponds to factoring the free term-algebra of types we regard by some congruence relation, will be a conservative extension for a coherent type system. If a type system was coherent, it will remain coherent. It is only possible that a type system being incoherent otherwise becomes coherent.

Let $f : (\sigma\sigma')\sigma$ be a binary type constructor with σ and σ' incomparable having direct embeddings at the first and second position, and let $t : \sigma$ and $t' : \sigma'$ be ground types such that $t' \hookrightarrow f(f(t, t'), t')$. Then there are two possibilities to coerce t' into $f(f(t, t'), t')$ which might be different in general. In the case of types R : c_ring and x : symbol the coercions of x into $UP(UP(R, x), x)$ are unambiguous, if $UP(UP(R, x), x)$ and $UP(R, x)$ are the same type. However, it does not seem to be generally possible to avoid the condition $|\mathcal{D}_f| = 1$ even in cases where a type constructor is defined for types having incomparable properties.

The naturally occurring direct embeddings for types built by the type constructors FF and UP show that in the context of computer algebra there are cases in which a coercion is defined into a type having an incomparable property, into a type having a stronger property, into a type having a weaker property, or into a type having the same property. Since the order on the properties correspond to an "inheritance hierarchy", this behavior of coercions shows an important difference between coercions occurring in computer algebra and the "subtypes" occurring in object oriented programming.

The next assumption will guarantee that structural coercions and direct embeddings will interchange nicely.

Assumption 5 (Structural Coercions and Embeddings). *Let f be a n-ary type constructor which induces a structural coercion and has a direct embedding at its i-th position. Assume that $f : (\sigma_1 \cdots \sigma_n)\sigma$ and $f : (\sigma_1' \cdots \sigma_n')\sigma$, $t_1 : \sigma_1, \ldots, t_n : \sigma_n$, and $t_1' : \sigma_1', \ldots, t_n' : \sigma_n$. If there are coercions $\psi_i : t_i \longrightarrow t_i'$, if the coercions $\Phi^i_{f,t_1,\ldots,t_n}$ and $\Phi^i_{f,t_1',\ldots,t_n'}$ are defined, and if f is covariant at its i-th argument, then the following diagram is commutative:*

$$
\begin{array}{ccc}
t_i & \xrightarrow{\quad\psi_i\quad} & t_i' \\[2mm]
\Big\downarrow{\scriptstyle\Phi^i_{f,t_1,\ldots,t_n}} & & \Big\downarrow{\scriptstyle\Phi^i_{f,t_1',\ldots,t_n'}} \\[2mm]
f(t_1,\ldots,t_n) & \xrightarrow[\mathcal{F}_f(t_1,\ldots,t_n,t_1',\ldots,t_n',\psi_1,\ldots,\psi_n)]{} & f(t_1',\ldots,t_n')
\end{array}
$$

If f is contravariant at its i-th argument, then the following diagram is commutative:

$$
\begin{array}{ccc}
t_i & \xrightarrow{\quad\psi_i\quad} & t_i' \\[2mm]
\Big\downarrow{\scriptstyle\Phi^i_{f,t_1,\ldots,t_n}} & & \Big\downarrow{\scriptstyle\Phi^i_{f,t_1',\ldots,t_n'}} \\[2mm]
f(t_1,\ldots,t_n) & \xleftarrow[\mathcal{F}_f(t_1,\ldots,t_n,t_1',\ldots,t_n',\psi_1,\ldots,\psi_n)]{} & f(t_1',\ldots,t_n')
\end{array}
$$

The type constructors list, UP, $M_{n,n}$ may serve as examples of constructors which induce structural coercions and can also have direct embeddings: It might be useful to have coercions from elements into one element lists, from elements of a ring into a constant polynomial or to identify a scalar with its multiple with the identity matrix.

As was already discussed in Sec. 4.3, in all these examples the parameterized data types can be seen as sequences and the structural coercions — i.e. $\mathcal{F}_{UP}(I, x, FF(I), x, \psi, id_x)$ — can be seen as a kind of "mapping" operators.

The direct embeddings are "inclusions" of elements in these sequences. Since applying a coercion function to such an element and then "including" the result in a sequence will yield the same result as first including the element in the sequence and then "mapping" the coercion function into the sequence, assumption 5 will be satisfied by these examples. For example, $\mathcal{F}_{UP}(I, x, FF(I), x, \Phi^1_{FF,I}, id_x)$ is the function which maps the coercion function $\Phi^1_{FF,I}$ to the sequence of elements of I in UP(I, x) which represents the polynomial.

So the three "squares" having the lower left corner in common in the "cube" of coercions in our motivating example are instances of the commutative diagrams in assumption 5.

If the mathematical structure represented by a type t_i in assumption 5 has non-trivial automorphisms, then it is possible to construct the structural coercion

$$\mathcal{F}_f(t_1, \ldots, t_n, t'_1, \ldots, t'_n, \psi_1, \ldots, \psi_n)$$

in a way such that the assumption is violated: just apply a non-trivial automorphism to t_i! However, such a construction seems to be artificial. Moreover, the argument shows that a possible violation of assumption 5 "up to an automorphism" can be avoided by an appropriate definition of $\mathcal{F}_f(t_1, \ldots, t_n, t'_1, \ldots, t'_n, \psi_1, \ldots, \psi_n)$.

4.5 The Main Result

The assumptions 1, 2, 3, 4, and 5 are "local" coherence conditions imposed on the coercions of the type system. In the following theorem we will prove that the type system is "globally" coherent, if these local conditions are satisfied.

Theorem 5 (Coherence). *Assume that all coercions between ground types are only built by one of the following mechanisms:*

1. *coercions between base types;*
2. *coercions induced by structural coercions;*
3. *direct embeddings in a type constructor;*
4. *composition of coercions;*
5. *identity function on ground types as coercions.*

If the assumptions 1, 2, 3, 4, and 5 are satisfied, then the set of ground types as objects and the coercions between them as arrows form a category which is a preorder.

Proof. By assumption 1 and lemma 2 the set of ground types as objects and the coercions between them as arrows form a category.

For any two ground types t and t' we will prove by induction on the maximum of the complexities of t and t' that if $\phi, \psi : t \longrightarrow t'$ are coercions then $\phi = \psi$.

If $\max(\operatorname{com}(t), \operatorname{com}(t')) = 1$ then the claim follows from assumption 2. Now assume that the induction hypothesis holds for k, and let $\max(\operatorname{com}(t), \operatorname{com}(t')) = k+1$. Assume w. l. o. g. that $t \hookrightarrow t'$ and that $\phi, \psi : t \longrightarrow t'$ are coercions.

First notice that $t \hookrightarrow t'$ implies $\operatorname{com}(t) \leq \operatorname{com}(t')$. Thus we can assume that $t' = f(u_1, \dots, u_n)$ for some n-ary type constructor f.

The coercions ϕ and ψ are compositions of coercions between base types, direct embeddings in type constructors and structural coercions. Because of assumption 3 and the induction hypothesis we can assume that there are ground types s_1 and s_2 and unique coercions $\psi_1 : t \longrightarrow s_1$ and $\psi_2 : t \longrightarrow s_2$ such that

$$\phi = \mathcal{F}_f(\dots, t, \dots, s_1, \dots, \psi_1, \dots) \tag{1}$$

or

$$\phi = \psi_1 \circ \Phi^i_{f, \dots, s_1, \dots} \tag{2}$$

Similarly,

$$\psi = \mathcal{F}_f(\dots, t, \dots, s_2, \dots, \psi_2, \dots) \tag{3}$$

or

$$\psi = \psi_2 \circ \Phi^j_{f, \dots, s_2, \dots} \tag{4}$$

If ϕ is of form 1 and ψ is of form 3, then $\phi = \psi$ because of assumption 3 and the uniqueness of $\mathcal{F}_f$. If ϕ is of form 2 and ψ is of form 3, then $\phi = \psi$ because of assumption 5. Analogously for ϕ of form 1 and ψ of form 4.

If ϕ is of form 2 and ψ is of form 3 then assumption 4 implies that $i = j$ and $s_1 = s_2$. Because of the induction hypothesis we have $\psi_1 = \psi_2$ and hence $\phi = \psi$ again by assumption 4. □

5 Conclusion and Comparison to Related Work

We have shown that the type system of a language allowing structural coercions and direct embeddings into constructors will be coherent, if some "local" conditions are satisfied. By these classes of coercions we have covered many important cases occurring in computer algebra. Since coercions of this form which violate our stated assumptions will frequently lead to a type system which is not coherent and can thus have an unexpected behavior, a check of the assumptions by a programmer who wants to add a coercion of this form to a system will help to improve the reliability of a system.

Using some informal reasoning — up to now no formal semantics for a larger computer algebra system has been given — we have shown that assumptions 1, 2, 3, and 5 will hold in many examples given by the present computer algebra systems and we did not find any natural counter-examples. The assumption 4 is also satisfied in many cases. However, natural counter-examples exist. A possible solution, which requires further research, seems to be that not a type system which can be described by a free term algebra of types is used but one which can be described by a quotient of a free term algebra.

Extensions of our work to other classes of coercions seem to be desirable. Using a type system which can be described by a free term algebra there will be many important cases in which two types are isomorphic. For instance, for any integral domain I the types FF(FF(I)) and FF(I) are isomorphic; another example is given by the isomorphism between the direct sums $A \oplus B$ and $B \oplus A$ of two Abelian groups. Using coercions from one type into the other and vice versa these type isomorphisms can be modeled using the technical framework of coercions. However, in many cases it seems to be more appropriate to use other frameworks for modeling type isomorphisms (cf. Sec. 4.4).

There are some other cases of coercions we did not cover, e. g. the ones between types built by different type constructors, such as the one which is needed in order to express that polynomials over some ring R are also functions, i. e. UP(R, x) $\hookrightarrow$ FS(R, R).

Nevertheless, the coercion rules we have considered are more general than many found in the literature.

Rector [16] and Fortenbacher [8] examine type systems for computer algebra programs which include coercions. Without using our terminology both simply require that their systems have a coherent coercion system but they do not make further investigations. Interestingly, Fortenbacher gives a concrete instance of the commutative diagram in assumption 5 as an example.

In the paper of Comon, et al. [7] about a type system for a computer algebra system it is required that all type constructors are covariant in all arguments.

In the papers of Mitchell [15], Fuh and Mishra [9], and Kaes [13] about subtyping (in the sense of implicit coercions) structural coercions are the only coercions which are present besides coercions between base types. Moreover, all type constructors have to be covariant or contravariant in all arguments in [15] and [9].

However, in all these papers the question of coherence is not addressed, or only in form of a hint to the literature [17]. In [17] Reynolds gives some general category theoretic conditions for a first-order language which imply a coherent system. Moreover, he also addresses questions whether generic operators are well defined in the presence of implicit coercions. However, in his system all types are unstructured. So there is no possibility to state some conditions on simple base types and the type constructors which would yield the "nice" properties of the entire system.

Acknowledgments. The present paper is part of the author's forthcoming PhD-thesis written under the supervision of Prof. R. Loos at the University of Tübingen. I would like to thank Prof. Loos for initiating and supervising my research. I also want to express my gratitude to F. Haug for several helpful discussions. The suggestions of the referees helped to improve the paper. The macros of M. Barr were used to produce the diagrams.

References

1. Association for Computing Machinery. *Proceedings of the 1992 ACM Conference on Lisp and Functional Programming*, San Francisco, CA, June 1992.

2. H. P. Barendregt. *The Lambda Calculus — Its Syntax and Semantics*, volume 103 of *Studies in Logic and the Foundations of Mathematics*. North-Holland, Amsterdam, second edition, 1984.

3. V. Breazu-Tannen, T. Coquand, C. A. Gunter, and A. Scedrov. Inheritance as implicit coercion. *Information and Computation*, 93(1):172–222, July 1991.

4. R. Bündgen. Reduce the redex ⟶ ReDuX. In C. Kirchner, editor, *Fifth International Conference on Rewriting Techniques and Applications (RTA '93)*, Lecture Notes in Computer Science, Montréal, Canada, June 1993. Springer-Verlag. To appear.

5. L. Cardelli. A semantics of multiple inheritance. *Information and Computation*, 76:138–164, 1988.

6. K. Chen, P. Hudak, and M. Odersky. Parametric type classes. In LFP '92 [1], pages 170–181.

7. H. Comon, D. Lugiez, and P. Schnoebelen. A rewrite-based type discipline for a subset of computer algebra. *Journal of Symbolic Computation*, 11:349–368, 1991.

8. A. Fortenbacher. Efficient type inference and coercion in computer algebra. In A. Miola, editor, *Design and Implementation of Symbolic Computation Systems (DISCO '90)*, volume 429 of *Lecture Notes in Computer Science*, pages 56–60, Capri, Italy, Apr. 1990. Springer-Verlag.

9. Y.-C. Fuh and P. Mishra. Type inference with subtypes. *Theoretical Computer Science*, 73:155–175, 1990.

10. J. A. Goguen and J. Meseguer. Order-sorted algebra I: Equational deduction for multiple inheritance, polymorphism, and partial operations. *Theoretical Computer Science*, 105(2):217–273, Nov. 1992.

11. P. Hudak, S. Peyton Jones, P. Wadler, et al. Report on the programming language Haskell — a non-strict, purely functional language, version 1.2. *ACM SIGPLAN Notices*, 27(5), May 1992.

12. R. D. Jenks and R. S. Sutor. *AXIOM: The Scientific Computation System*. Springer-Verlag, New York, 1992.

13. S. Kaes. Type inference in the presence of overloading, subtyping, and recursive types. In LFP '92 [1], pages 193–205.

14. S. Mac Lane. *Categories for the Working Mathematician*, volume 5 of *Graduate Texts in Mathematics*. Springer-Verlag, New York, 1971.

15. J. C. Mitchell. Type inference with simple subtypes. *Journal of Functional Programming*, 1(3):245–285, July 1991.

16. D. L. Rector. Semantics in algebraic computation. In E. Kaltofen and S. M. Watt, editors, *Computers and Mathematics*, pages 299–307, Massachusetts Institute of Technology, June 1989. Springer-Verlag.

17. J. C. Reynolds. Using category theory to design implicit conversions and generic operators. In N. D. Jones, editor, *Semantics-Directed Compiler Generation, Workshop*, volume 94 of *Lecture Notes in Computer Science*, pages 211–258, Aarhus, Denmark, Jan. 1980. Springer-Verlag.

18. J. C. Reynolds. The coherence of languages with intersection types. In T. Ito and A. R. Meyer, editors, *Theoretical Aspects of Computer Software — International Conference TACS '91*, volume 526 of *Lecture Notes in Computer Science*, pages 675–700, Sendai, Japan, Sept. 1991. Springer-Verlag.

19. G. Smolka, W. Nutt, J. A. Goguen, and J. Meseguer. Order-sorted equational computation. In H. Aït-Kaci and M. Nivat, editors, *Resolution of Equations in Algebraic Structures, Volume 2*, chapter 10, pages 297–367. Academic Press, 1989.

20. U. Waldmann. Semantics of order-sorted specifications. *Theoretical Computer Science*, 94(1):1–35, Mar. 1992.

Subtyping Inheritance in Languages for Symbolic Computation Systems *

Paolo Di Blasio and Marco Temperini

Dipartimento Informatica e Sistemistica - Università "La Sapienza"
Via Salaria 113, I-00198 Roma, Italy
E-mail: {diblasio,marte}@disco1.ing.uniroma1.it

Abstract. Object-oriented programming techniques can be fruitfully applied to design languages for symbolic computation systems. Unfortunately, basic correctness problems still exist in object-oriented languages, due to the interaction between polymorphism and method redefinition. Here a mechanism of subtyping inheritance is presented, in order to propose a solution of these problems. A subtyping inheritance mechanism (*Enhanced Strict Inheritance*) is defined by deriving from the characteristics of a presented model of subtyping. As the base of the subtyping rule, the monotonic (or covariant) rule is chosen. Once it is supported by the programming language of a symbolic computation system, our mechanism allows for a safe treatment of polymorphism induced by inheritance.

1 Introduction

In this paper we present the *Enhanced Strict Inheritance (ESI) mechanism*, together with the theoretical features of its underlying subtyping system. Our aim is to define an object-oriented programming (OOP) language, suitable as the software development tool of a symbolic computation system. The subtyping model is defined by a covariant rule, and is based on the concept of *replacement* [CaW85]. The subclassing rule is shown to be coherent with this model of subtyping. We also present a technique for establishing, at compile-time, the run-time correctness of polymorphic statements. If such a technique is used, the amount of work needed by the run-time system of the language decreases, making the execution of programs much more efficient. This aspect should be of considerable importance, since very strict efficiency needs arise in the fields of symbolic computation.

The development of new software systems for Symbolic Computation, and for Computer Algebra in particular, has surged ahead in recent years. New systems have been developed from scratch, while others have been renewed and increased in power [Wan85] [Hea87] [ChG91] [Jen92] [Sac92]. The new generation of symbolic computation systems were developed keeping in mind the evolution of powerful interfaces, allowing for the application of defined algorithms over defined

---

* This work has been partially supported by Progetto Finalizzato "Sistemi Informatici e Calcolo Parallelo" of CNR under grant n. 92.01604.69.

structures [Wol91] [ChG91]. However, Symbolic Computation and, in particular, Computer Algebra, presents particular necessities that need to be accomplished from a software engineering point of view. In some of these systems, the use of built in programming languages has been considered as a tool for expressing new data structures and algorithms. Moreover, a large common software basis can be exploited as a library, in both developing algorithms and data structures, and enriching the library itself introducing new pieces of software. At present, the idea of constructing and maintaining libraries of mathematical software is being taken into consideration in the development of the SACLIB system [Sac92], which is still in progress.

However, one aspect in the development of new systems apparently left out, is the adoption of a powerful software development methodology. It should be supported by a system language, in order to flank the specification of data stuctures and of the related algorithms. Leaving out this aspect is responsible for the complex maintainance needed by systems and, thus, for the difficulties found in the expansion of systems by reusing or modifying what is already defined in them. In fact these difficulties increase their importance with the dimension of the whole system. Even if the existing code is organized in library items, that methodological deficiency may result in serious difficulties in establishing the intercommunications among the library modules used in a program. Let us think, for example, of the relationships that should hold – and, in some way, be tested – among data structures defined separately and then collected in the same algorithm.

In order to give a solution to the described problem, different aspects of programming languages have been investigated. Some researches have been utilizing concepts of OOP paradigm [LiM91]. The main needs that have been detected are the use of an *abstract data types* (ADT henceforth) *language*, the *viewing* feature, and the *hierarchycal* organization of data structures and algorithms. The possibility of splitting the specification and implementation hierarchies has been recognized as a means for exploiting some general peculiarities, which are common to sets of different structures [Bau90] [LiT92].

AXIOM [Jen92] implements the concept of Category and Domain in order to express the duality of the specification and the implementation of the structures. Views [Abd86] features the viewing operation, in order to enhance the expressive power of hierarchies: by viewing, it is possible to consider a concrete structure as instance of several abstract types, as is needed. In [LiT92] the data structure to be classified are distinguished into abstract, parametric and ground. They are implemented by means of the respective class constructs of an OOP language. Moreover, in [Tem92], features of OOP methodology have been studied and selected in order to define a suitable mechanism of inheritance, supported by a language, for the design and implementation of symbolic computation systems. The need for polymorphism and redefinition of attributes in subclassing have been shown to be source of potential problems, mainly because of the possibility of incorrectness at run-time of statements that were checked and declared correct at compile time. Let us not misunderstand what is meant by should be generated about the term "correctness": a program is correct if its statements which satisfied the rules of the compile-time type checking, never result in run-time type errors, due to the polymorphic behaviour of some of their elements.

A first approach to a subtyping model of inheritance is in [Car84], where method redefinition is constrained by a contravariant (or anti-monotonic) subtyping rule. This approach leads to a well-founded type theory in which the above cited problems are solved at the cost of redefining the method by the unnatural contravariant rule [DaT88].

On the contrary, a covariant redefinition rule is more natural and expressive from a software engineering point of view. Such a redefinition rule can be safely supported by a programming language , provided that a suitable mechanism of dynamic binding is defined in an environment of strict inheritance with redefinition [ReT90]. The criticism to the contravariant redefinition rule in favour of a covariant one has been presented in [ChM92] too, where a function subtyping rule based on the equivalence between subset and subtype is described. In [Ghe91], the contravariant subtyping rule is preserved, over the attributes of record types, while a static type checking system is presented, for the treatment of overloaded functions defined by a covariant rule.

The paper is organized as follows. In section 2, the subtyping rule, modelling the idea of *replacement*, is presented. In section 3, the subtyping system is enclosed in a programming language framework, by defining the subtyping inheritance mechanism. The problems in dynamic binding, due to the interaction of polymorphism and redefinition, are solved by the definition of a suitable mechanism for method lookup. Section 4 concludes this paper with some remarks about related works and future developments.

2 A Model of Subtyping

In modelling the subtyping system we adopt the most generally accepted idea based on the concept of *replacement*. This concept expresses the fact that objects of a given type can be safely replaced by objects of any related subtype. The aim of this section is to provide the definition of a covariant subtyping rule which is able to model the idea of *replacement*. The subtyping relation inferred by such a definition is stated over the following type structure. Such a type structure is denoted in terms of its typed data attributes (VARIABLES) and functional attibutes (FUNCTIONS).

TYPE C: VARIABLES $v_1: T_1; \ldots; v_m: T_m$

FUNCTIONS $m_1: P_1 \longrightarrow R_1; \ldots; m_n: P_n \longrightarrow R_n$.

The declaration $v_i: T_i$ denotes a variable v_i of type T_i, and $m_i: P_i \longrightarrow R_i$ denotes a function which takes an argument of type P_i and returns a result of type R_i.

We adopt such a structure in order to state the distinction between the structural and functional component of a type. Any concrete instance of type C is also called an *object of type* C.

Since each function is defined in a stated type specification, the arguments of its signature do not indicate the type in which the function itself is defined. In order to make our notations during the following discussion simpler, without any loss of generality, the functional attributes have only one argument.

Furthemore, let us stress that the complete specification of a function, in a given programming language, should include both the signature and the implementation. Nevertheless, in the following we consider a function as completely specified based only on its signature, in order to state function compatibility without considering implementation compatibility. This choice is made because in general the latter problem is unsolvable. Now we can define the covariant subtyping rule $\leq$.

Definition 1 (subtyping rule $\leq$).
1. given the type C, C is a subtype of C ($C \leq C$)
2. let C and C' be two types given as

TYPE C: VARIABLES $v_1: T_1; \ldots; v_m: T_m$

FUNCTIONS $m_1: P_1 \longrightarrow R_1; \ldots; m_n: P_n \longrightarrow R_n$

TYPE C': VARIABLES $v_1: T'_1; \ldots; v_m: T'_m; \ldots; v_{m+q}: T'_{m+q}$

FUNCTIONS $m_1: P'_1 \longrightarrow R'_1; \ldots; m_n: P'_n \longrightarrow R'_n; \ldots; m_{n+p}: P'_{n+p} \longrightarrow R'_{n+p}$

we say that C' is a *subtype* of C ($C' \leq C$) iff
- $T'_1 \leq T_1; \ldots; T'_m \leq T_m$
- $P'_1 \leq P_1; \ldots; P'_n \leq P_n$
- $R'_1 \leq R_1; \ldots; R'_n \leq R_n$

Given $C' \leq C$, the type C can also be called a *supertype* of C'. We can also use $C' < C$ to mean *strict subtyping*, i.e.: $(C' \leq C \wedge C' \neq C)$.

A graphical interpretation of a set of types defined according to the subtyping rule $\leq$ is a direct acyclic graph (DAG), in which each arc starts from a supertype and leads to a subtype. So we can speak in terms of subtyping hierarchy.

In the following, the signatures $o.v$ and $o.m(q)$ stand respectively for the access to the variable attribute v in the object o and for the invocation of the function m on the object o, with argument q.

It is simple to see that the variable access and function invocation make sense (we say "they are legal") only if certain conditions do hold. Namely, assuming the object o is of type T ($o : T$), $o.v$ can be considered as legal only if a variable v is defined in T, and $o.m(q)$ can be considered as legal only if a function m is defined in T, or in some supertype of T, such that the actual parameter q is compatible with the formal one declared for m. In order to better express the conditions roughly stated above, and verify that Def. 1 gives rise to a subtyping system satisfying the concept of *replacement*, we state the following additional definitions.

Definition 2 (function applicability). Given a function invocation $o.m(q)$ with $o : O$ and $q : Q$, a function $m : P \longrightarrow R$, defined in a type C, is said to be *applicable* to the objects o, q iff $O \leq C$ and $Q \leq P$.

Let us note that this definition uses the subsumption rule implicitly. This rule states that if o is of type T, then it is of any supertype of T. This means that, since o and q are of types O and Q respectively, they are also of types C and P.

Moreover, note that if the function $m : P \longrightarrow R$ is applicable to the objects o and q then the function invocation $o.m(q)$ is meaningful and can be performed, giving rise to a result of type R.

A function applicable to the objects o and q gives rise to a so-called *legal invocation* $o.m(q)$. Analogously one can state a definition of *legal access* to a variable attribute.

Definition 3 (legal access and legal invocation).
1. o being an object of type O, the variable access $o.v$ is *legal* iff the variable v is defined within the type O;
2. the function invocation $o.m(q)$, with $o : O$ and $q : Q$, is legal iff in O or in some supertype of O, say $\overline{O}$, a function m is defined such that the condition of function applicability of m to o and q does hold.

We say that an attribute defined in a type C is "redefined" in a subtype C', if it has different type, in case of data attributes, or it has different domain or codomain, in case of functional attributes. In such a case we say that the attribute definition in C' *covers* the one in C.

Definition 4 (attribute covering). Given the types C and C', with $C' \leq C$, we say that

1. the variable $v_j : T'_j$ in C' covers the variable $v_j : T_j$ in C iff $T'_j < T_j$;
2. the function $m_k : P'_k \longrightarrow R'_k$ in C' *covers* the function $m_k : P_k \longrightarrow R_k$ in C iff either $P'_k < P_k$ or $R'_k < R_k$ or both.

We can assert that a covering attribute is a specialization of the covered one. With respect to point 2. in Def. 4, we also note that since a function m, defined into the type C, could be covered by different redefinitions in subtypes of C, it follows that starting from any legal function invocation, several different applicable functions could exist in different but related types. It is simple to realize that among all such applicable functions, one has to be applied and this, as specialized as possible.

Below we set up the criterion for determining such a "selected function". In order to show that, given a legal invocation, the selected function does exist and can be found, we identify as $App_{o.m(q)}$ the set of all types in which a method m is defined and is applicable to the objects o, q. Let us note that $App_{o.m(q)}$ is always non-empty, by definition of legal invocation.

It is simple to note that, once a set of classes has been defined following the subtyping rule of Def.1, the resulting hierarchy is a partially ordered set under the subtyping relation $\leq$. Obviously $App_{o.m(q)}$ is a subset of such a *poset*. This property should be carefully considered with respect to function invocations. Actually a legal invocation $o.m(q)$ could give rise to two dramatically different configurations for the related $App_{o.m(q)}$, depending on the existence of the *bottom* element B in it (i.e. an element $B \in App_{o.m(q)}$ such that $\forall B' \in App_{o.m(q)}, B \leq B'$):

a) if $App_{o.m(q)}$ has a *bottom* element B then the function m can be correctly selected from B itself; such m is, among the applicable functions, the most specialized;

b) if $App_{o.m(q)}$ admits more than one minimal element w.r.t. the $\leq$ ordering, then such elements contain several *equally most specialized* functions m among the applicable ones. All minimal elements of $App_{o.m(q)}$ are types whose functional attribute m are equally applicable. Normally the problem of choosing one single m from among all the applicable ones is solved by adding further selection criteria which do not act on semantic characteristics of the functions. Here we do not treat these criteria, see [Agr91]. In the following, S denotes the application of such a criterion.

Let MIN be the set of minimal elements of $App_{o.m(q)}$ w.r.t. $\leq$, then the selected function m is defined in one of the types of MIN. In case b) it should be chosen by the mentioned criterion. In case a) MIN has only one element (the bottom of $App_{o.m(q)}$) so the selected function is uniquely determined.

In order to prove that the rule of Def.1 models the concept of subtyping as replacement, let us stress that the *replacement* of any object o of type O by an object o' of type $O' \leq O$ means that:

1. given a legal variable access $o.v$, then $o'.v$ is legal too, and its application doesn't violate the context of the $o.v$ occurrence;
2. given a legal function invocation $o.m(q)$, with $q : Q$, then $o'.m(q)$ is legal too, and its application doesn't violate the context of the function invocation occurrence.

We identify, here, the context of both a variable access and a function invocation with the type expected for the result of their execution.

In the context of an assignment – like $x := o.v$, or $x := o.m(q)$ – the expected type for $o.v$ and $o.m(q)$ is a subtype of the type of x.

In the context of a function invocation – like $\overline{o}.\overline{m}(o.v)$ or $\overline{o}.\overline{m}(o.m(q))$ – the expected type for $o.v$ and $o.m(q)$ is a subtype of the type of the correspondent formal argument of $\overline{m}$.

The previous notes about the coherence of an expression with the related context can be summarized by the following statements. In case 1. the result type of $o'.v$ must be a subtype of the type expected for $o.v$. In case 2. the result type of $o'.m(q)$ must be a subtype of the type expected for $o.m(q)$.

A replacement in a variable access, as in 1., is always safe if $O' \leq O$, due to the covariant redefinition stated by the subtyping rule. The same safeness is not guaranteed in case 2.: one must be more careful when searching for the function that answers an invocation $o.m(q)$, once a replacement $o'.m(q)$ has occurred. In this case we have to think of the context in which such an invocation appears. Below we examine this case.

Previously we called S the application that selects a type from the set MIN (resp. MIN') of minimal types in $App_{o.m(q)}$ (resp. $App_{o'.m(q)}$).

Let us note that once o is replaced by o' in the function invocation, at least the function m in $S(MIN)$ could be executed without violating the context. Actually this might not be the most specialized function which can be applied. Neither we can choose the function to be executed directly in MIN', since in this minimal set, the case Fig.1 could occur, in which a function returns a result of the wrong type. In this example the function invocation $o.m(q)$ is in a context of type of at least R_0 (we can suppose here, without any loss of generality, that $S(MIN)$ is exactly O). This context could be violated, after the replacement of o by o', because of the possible call of the function m defined in the type S – suppose $S \in S(MIN')$ and $R_S \not\leq R_0$.

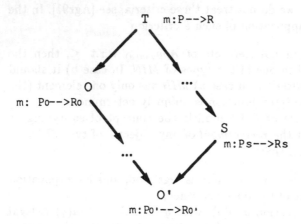

Fig. 1. Context of a function invocation

The following Lemma establishes the conditions for ensuring a safe replacement in case of function invocations.

Lemma 5. *Given a legal invocation $o.m(q)$, with $o : O$ and $q : Q$, then for all $o' : O'$, with $O' \leq O$, also $o'.m(q)$ is legal. Moreover, provided that the selected function m (in $o'.m(q)$) is the one defined in $C'=S(MIN_C)$ where MIN_C is the set of minimal elements of $(App_{o'.m(q)} \cap \{S \mid S \leq C\})$, with $C=S(MIN)$, then the result type of $o'.m(q)$ is a subtype of the result type of $o.m(q)$.*

Proof. $o'.m(q)$ is legal because there exists at least the function m in C that is applicable to o' and q.

Moreover the result type of $o'.m(q)$, with m defined in C', is a subtype of the result type of $o.m(q)$, with m defined in C ($C' \leq C$), due to the covariant subtyping rule stated on functional attributes. □

Now we can establish formally that the Def.1 rule models the concept of subtyping as replacement.

Theorem 6. *The Def.1 subtype rule models the concept of replacement.*

Proof. If $o.v$ with $(o : O)$ is legal, then the variable v is defined in O and hence in O' with $O' \leq O$. So $o'.v$ with $(o' : O')$ is legal. Moreover, since o' is of type O', the result of accessing any variable in o' is of a subtype of the one expected for $o : O$. This is due to the covariant relation induced through the subtyping rule. If $o.m(q)$, with $(o : O)$, is legal, then the Lemma 5 gives us the conditions so that $o'.m(q)$, with $(o' : O')$, is legal too, and the results are compatible. $\square$

Concluding this section we give the following example in order to show the main features of the subtyping model described above.

Assume that the types $WINDOW \leq RECTANGLE \leq POLYGON$ are defined. In each one of them let a function *Intersection* be defined:

TYPE POLYGON: VARIABLES ...

METHODS ...Intersection(p: POLYGON): POLYGON ...

TYPE RECTANGLE: VARIABLES ...

METHODS ...Intersection(r: RECTANGLE): POLYGON ...

TYPE WINDOW: VARIABLES ...

METHODS ...Intersection(w: WINDOW): WINDOW ...

Let be $p_1.Intersection(r_1)$ a legal invocation with $p_1 : POLYGON$ and $r_1 : RECTANGLE$. Once p_1 is replaced by an object w_1 of type $WINDOW$ the following can be stated:

1. $App_{w_1.Intersection(r_1)} = \{POLYGON, RECTANGLE\}$;
   $WINDOW \notin App_{w_1.Intersection(r_1)}$ because the argument r_1 makes the function *Intersection* of $WINDOW$ not applicable;
2. $RECTANGLE$ is the bottom element of $App_{w_1.Intersection(r_1)}$;

So, by choosing the most specialized applicable function *Intersection* in $RECTANGLE$ it is possible to replace an instance of the supertype $POLYGON$ with one of the subtype $WINDOW$. Moreover the compatibility of the result is preserved after the replacing, because the expected result of this legal invocation, on object p_1, and the effective one, on object w_1, are objects of the same type $POLYGON$.

3 Subtyping Inheritance

In order to define the management of a subtyping inheritance mechanism in an object-oriented system, three main topics must be set up: the definition of the subclassing rule, the treatment of polymorphic assignment and the management of method invocation. So this section is dedicated to showing the effective connection between the concepts of:

a) Subtype and Subclass
b) Replacement and Polymorphic Assignment
c) Set of minimal elements of $App_{o.m(q)}$ and Method Lookup

First we state a class structure. A class is an implementation of a data type expressed possibly by inheriting attributes defined in other classes.

Definition 7 (class structure).
class *name* {inheritance list}
$v_1 : C_1; \cdots; v_m : C_m;$
$m_1(p_1 : P_1) : R_1$ is $\{m_1$ implementation$\}$;
...
$m_n(p_n : P_n) : R_n$ is $\{m_n$ implementation$\}$;
endclass

In the class structure the *inheritance list* specifies the set of its direct super-classes; the "list of attributes" specifies a set of typed instance variables, used to denote the state of an object, and a set of methods, according to the choices made in the previous section.

The subclassing relation is established through an inheritance mechanism, that we call *Enhanced Strict Inheritance* (*ESI* henceforth). *ESI* is a strict inheritance mechanism in which the covariant redefinition of inherited attributes is supported. The concept of covariant redefinition depends on the *ESI* relation it helps to define strictly, so the following two definitions are mutually recursive. In Def.8 the notation C' *is a subclass* of C is used to mean that either $C'=C$ or C' is defined by inheriting C through *ESI*.

Definition 8 (covariant redefinition). If C' is a subclass of C:

1. the redefinition $v : C'_v$ in C' of the variable $v : C_v$ defined in C is a covariant redefinition iff C'_v is a subclass of C_v.
2. the redefinition $m(p : P') : R'$ in C' of the method $m(p : P) : R$ defined in C is a covariant redefinition iff, either P' is a subclass of P or R' is a subclass of R or both.

Definition 8 corresponds to the concept of *attribute covering* introduced in the previous section.

Definition 9 (Enhanced Strict Inheritance). Given a class C' and a set of classes $\{C_1, \ldots, C_n\}$, with $n \geq 1$, the *ESI* is defined by the following three cases:

1. $C' \leq_{ESI} C'$;
2. if $n=1$ and $C' \neq C_1$, then $C' \leq_{ESI} C_1$ iff C' inherits every attribute of C_1, possibly redefining some of them by covariant redefinition;
3. if $n>1$, then $C' \leq_{ESI} \{C_1, \ldots, C_n\}$ iff C' inherits every method of $C_1, \ldots, C_n$ possibly redefining some of them by covariant redefinition and C' inherits all the variable attributes defined in the classes $C_1, \ldots, C_n$ ensuring the following constraint: if v is an attribute belonging to more than one class $\{C_i\}$

with $i \in I$, then either v is redefined in C' ensuring the covariant redefinition w.r.t. the classes C_i, for each $i \in I$, or a variable $v : V_j (j \in I)$ is inherited such that $\forall i \in I \; V_j \leq V_i$.

A straightforward comparison between $\leq_{ESI}$ and the subtyping relation $\leq$ defined in Sec.2 shows how $\leq_{ESI}$ implements $\leq$, in terms of object-oriented programming.

Now we pass onto the discussion of points b) and c).

In OOP languages a *polymorphic assignment* is an instruction in which a variable c declared to be of class C is assigned by an object o of class C' which is a subclass of C. In the following we apply this definition to the subclassing relation $\leq_{ESI}$, stating what is intended by correct polymorphic assignment.

Definition 10 (correct polymorphic assignment). Let be := the assignment operator. A polymorphic assignment $c := o$, with $c : C$, is correct iff o is an expression returning an object of class $C' \leq_{ESI} C$.

The polymorphic assignment is the fundamental implementation of the concept of *replacement*. Usually OOP languages don't support a real subtyping inheritance mechanism, so the use of polymorphic assignment might prove unsafe.

Before considering an example that shows this problem, let us give the definition of syntactically correct method invocation. It is a simple derivation, in OOP terms, of the definition of *legal invocation* given in Sec.2.

Definition 11 (syntactically correct method invocation). A method invocation $o.m(q)$, whose o is declared class C and q of class Q, is syntactically correct iff a method $m(p : P) : R$, with P superclass of Q, is defined either in C or in a superclass C' of C.

On the problem of unsafeness cited above, Figure 2 provides the same example of the previous section.

In Fig.2 we polymorphically assign the object referred to by w_1 to the variable p_1, being $WINDOW \leq_{ESI} POLYGON$. Let us note that the method invocation $p_1.Intersection(r_1)$, called with argument r_1 of class $RECTANGLE$, is correct from a syntactical point of view.

As established by the usual dynamic binding rules, the code version of *Intersection* is selected in the class $WINDOW$, since p_1 refers to an object of $WINDOW$ class. But the execution isn't correct because the actual parameter r_1 should be of class $WINDOW$, while it is actually of class $RECTANGLE$. Thus a type failure occurs at run-time due to an unsatisfactory treatment of the coexistence of polymorphism and redefinition [ReT90].

Hence the compile-time correctness of an instruction might not correspond to its safeness at run-time. This is due to the fact that a variable declared at compile-time as being of a certain class could dynamically make reference at run-time to an object of any related subclass (its *actual class*).

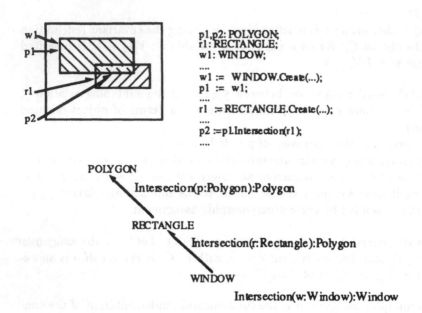

```
p1,p2: POLYGON;
r1: RECTANGLE;
w1: WINDOW;
....
w1 :=  WINDOW.Create(...);
p1 := w1;
....
r1 := RECTANGLE.Create(...);
....
p2 :=p1.Intersection(r1);
....
```

POLYGON

      Intersection(p:Polygon):Polygon

RECTANGLE

      Intersection(r:Rectangle):Polygon

WINDOW

      Intersection(w:Window):Window

Fig. 2. Unsafeness of a syntactically correct method invocation

In order to implement the idea of subtyping as *replacement*, we need to manage the polymorphic assignment under *ESI* correctly. So dynamic binding must be improved by a suitable mechanism for method lookup. Such a mechanism should return a selected executable method: the most specialized method possible or one among the most specialized methods.

It is simple to note that, in Def.9, a class may inherit more than one method, with the same name, from different superclasses. For each method invocation, depending on the actual parameter, only one of them will be chosen by the method lookup algorithm. This situation does not occur while inheriting variables.

Moreover if the method invocation is syntactically correct, in the case of *Single ESI* a unique executable method always exists. In terms of the subtyping model proposed in Sec.2, this means that the set of classes corresponding to $App_{o.m(q)}$ always has a bottom element. In the case of *Multiple ESI* more than one executable method might exist because $App_{o.m(q)}$ might not have a bottom element. Infact this means that, given a syntactically correct method invocation $o.m(q)$ over a variable o declared of class C, with argument $q : Q$, we can retrieve either a unique most specialized method (indicated by $\overline{m}$) or a set of equally most specialized methods taken from a suitable set of classes (indicated by $\overline{M}$). The method $\overline{m}$ or the set $\overline{M}$ are detected in the set of classes which are superclasses of C' (the actual class of the object o, $C' \leq_{ESI} C$) and subclasses of C_0 (the superclass of C which makes $o.m(q)$ syntactically correct). This corresponds to Lemma 5 in Sec.2.

However, in order to answer the method invocation, only one executable method must be selected. If a unique $\overline{m}$ is given as above, then it is the selected method; otherwise a further ordering criterion, say K, is needed in order to state from which class, in the set $\overline{M}$, the selected method must be taken.

The following *Method Lookup* algorithm, given $o.m(q)$ syntactically correct, performs a breadth first search on a subgraph of the class hierarchy. This subgraph is composed by the superclasses of the actual class of o (ACTUAL). The notation $DS(C)$ stands for the set of direct superclasses of the class C.

Algorithm: *Method Lookup*
Input: a syntactically correct method invocation $o.m(q)$ with $o : C$, $q : Q$;
the superclass of C, C_0, which makes $o.m(q)$ syntactically correct;
a criterion K
Output: the executable most specialized method $\overline{m}$

```
M̄:=∅
enqueue(ACTUAL, QUEUE);
while not empty-queue(QUEUE) do
  begin
    C:= front(QUEUE);
    dequeue(QUEUE);
    if ∀C̄ ∈ M̄, C̄ ≰ESI C
      then if (a method m(p : P) : R is defined in C, s.t. Q ≤ESI P)
        then cons(C, M̄)
        else for all S such that (S ∈ DS(C) and S ≤ESI C0)
          do if S ∉ QUEUE
            then enqueue(S, QUEUE)
              endif
          endif
      endif
  end
  return m̄:= K(M̄)
end algorithm
```

The algorithm searches through all the superclasses of ACTUAL analyzing only the methods m in classes which are subclasses of C_0. The elements in $\overline{M}$ are the minimal classes having a method which is correctly executable over the actual classes of o and q, and whose result is a subtype of the type we expected. From this set, the method $\overline{m}$ is selected by means of a criterion K. Discussing K is not significant in this context, since once the *Method Lookup* algorithm is stated, our purposes of run-time correctness are independent of any selecting criteria over $\overline{M}$.

The effectiveness of the proposed algorithm is shown by once again taking into consideration the example of Fig.2. Once the algorithm for method lookup receives the syntactically correct method invocation $p_1.Intersection(r_1)$, it starts searching the selected method from the actual class of p_1 ($WINDOW$). There the method *Intersection* fails to meet the condition of applicability, as $RECTANGLE \nleq_{ESI} WINDOW$. So the search continues on the direct superclass (unique in the example). This class is $RECTANGLE$ where the method intersection is applicable. This terminates the method lookup.

A significant feature of this method lookup algorithm is the selection of the abstraction level within which the given method should be selected, depending on the whole context of the method invocation (the actual class of the object within which the method is invoked and the actual class of the argument). Actually, in the example of Fig. 2, also the method *Intersection* of the class *POLYGON* could be applicable, but the one of the class *RECTANGLE* was selected because it was more specialized, i.e. defined in a superclass closer to the one actual.

The following example shows the behaviour of the algorithm in the more general case of multiple inheritance. Suppose we have the hierarchy of Fig.3, and want to perform the method invocation $o.m(q)$, where $o : C_0$ refers to an object of the class ACTUAL, and q is declared of class Q. Moreover, suppose that $Q \leq_{ESI} P_3$, P_6 and P_A, $P_2 \leq_{ESI} Q$, in order to evaluate the applicability of methods.

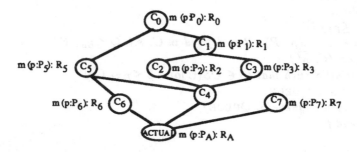

Fig. 3. Method lookup algorithm over a multiple inheritance hierarchy

First we have QUEUE={ACTUAL} and $\overline{M}=\emptyset$. During the first iteration of the while loop, ACTUAL is assigned to C and dequeued; no method m is applicable in ACTUAL, so its direct superclasses are enqueued: QUEUE={C_6, C_4}, $\overline{M}=\emptyset$. Note that C_7 is not enqueued, since it is not a subclass of C_0. At the second iteration, $C=C_6$ and the method m in C is executable, hence $\overline{M}$ becames {C_6}. Then C_4 is assigned to C and dequeued; since no executable method is detected in C, its direct superclasses are added to QUEUE. After the third iteration we have QUEUE={C_5, C_2, C_3} and $\overline{M}$={C_6}. During the fourth iteration C_5 is assigned to C and dequeued; actually it is a superclass of $C_6 \in \overline{M}$, hence it could never be a minimal class to be added in $\overline{M}$, so it is no longer considered. During the fifth iteration C_2 is assigned to C and dequeued; its method m is not executable so its direct superclass C_1 is added to QUEUE; hence we have QUEUE={C_3, C_1} and $\overline{M}$={C_6}. When C_3 is analyzed (sixth iteration) it is added to $\overline{M}$ and only C_1 remains to be inspected. Anyway C_1 will not be added to $\overline{M}$, because it is a superclass of $C_3 \in \overline{M}$. Now QUEUE is empty and a most specialized method $\overline{m}$ can be selected from the classes in $\overline{M}$ by means of the criterion K.

4 Remarks and Conclusions

We have presented a subtyping inheritance mechanism (*ESI*) for an object-oriented programming language, suitable for the specification of data structures and algorithms of a symbolic computation system.

ESI reflects the classical view of OOP: classes are ADT implementations; they embed data structures and related methods; message passing is used to invoke methods. It is based on a covariant subtyping rule, and is applied to hierarchies of classes. We have shown how the concept of *replacement* in subtyping matches polymorphic assignment in *ESI*, and allows us to exploit features of polymorphism combined with a natural mechanism of attribute redefinition. The correspondence between the static and dynamic correctness of programs developed with *ESI* is guaranteed by a suitable enhancement of the usual dynamic binding features. After an invocation, the method lookup selects the method to be executed based on the whole set of passed arguments - not only on the object receiving the method invocation.

A similar approach to method selection is used in the multimethod languages [Agr91]. Actually, we remain consistent with an ADT approach, where functions are attributes of the class and not organized in generic functions.

In particular, the proposed subtyping model is based on a covariant functional redefinition rule. A similar approach to functional overloading is presented in [Ghe91], where the algorithm for the function selection works over the whole set of arguments. There, the subtyping rule is the traditional contravariant rule of [Car84] and covariant redefinition is bound to generic funtions, which are not attributes of the data structure. Here, the covariant subtyping rule applies to the functional attributes of types.

So, we are working to embed a multimethod approach in an ADT model of covariant subtyping. The first step is to increase the expressive power of the class construct, by allowing for the encapsulation in the same class of several methods having different signatures but identical names. In the framework of a system for symbolic computation, the definition of *overloaded methods* in the same class, can permit the safe treatment of several cases in which coercion is needed.

We are also developing a denotational semantics of the *ESI* mechanism which would allow a complete set of formal tests on its consistency and its implementation.

5 References

[Abd86] S.K. Abdali, G.W. Cherry, N. Soiffer, *An Object Oriented Approach to Algebra System Design*, in B.W. Char (Ed.) Proc. of ACM SYMSAC '86, 1986.

[Agr91] R. Agrawal, L.G. DeMichiel, B.G. Lindsay, *Static Type Checking of Multi-Methods*, in Proc. of OOPSLA'91, 1991.

[Bau90] G. Baumgartner, R. Stansifer, *A Proposal to Study Type Systems for Computer Algebra*, RISC-Linz Tech. Report 90-07.0, Mar. 1990.

[Car84] L. Cardelli, *A Semantics of Multiple Inheritance*, in Proc. of Symp. on Semantics of Data Type, LNCS 173, Springer Verlag 1984.

[CaW85] L. Cardelli, P. Wegner, *On Understanding types, Data Abstraction and Polymorphism*, ACM Computing Surveys, Vol 17,4, 1985.

[ChG91] B.W. Char, K.O. Geddes, G.H. Gonnet, B.C. Leong, M.B. Monagan, S.M. Watt, *Maple V Language Reference Manual*, Springer Verlag, 1991.

[ChM92] I. Choi, M. Mannino, V. Tseng, *Graph Interpretation of Methods: a Unifying Framework for Polymorphism in Object-Oriented Programming*, OOPS Messenger, Vol 1,2, 1992.

[DaT88] S. Danforth, C. Tomlinson, *Type Theories and Object-Oriented Programming*, ACM Computing Surveys, Vol 20,1, 1988.

[Hea87] A.C. Hearn, *Reduce-3 User's Manual*, Rand Corporation, S.Monica, California, USA, 1987.

[Ghe91] G. Ghelli, *A Static Type System for Message Passing*, in Proc. of OOPSLA'91, 1991.

[Jen92] R.D. Jenks, R.S. Sutor, *Axiom, the scientific computation*, Springer Verlag, New York, 1992.

[LiM91] C. Limongelli, A. Miola, M. Temperini, *Design and Implementation of Symbolic Computation Systems*, in P.W.Gaffney, E.N.Houstis (Eds) Proc. IFIP TC2/WG2.5 Working Conference on Programming Environments for High Level Scientific Problem Solving, Elsevier Scientific Publisher, 1991.

[LiT92] C. Limongelli, M. Temperini, *Abstract Specification of Structures and Methods in Symbolic Mathematical Computation*, Theoretical Computer Science, 104, 1992.

[ReT90] M. Regio, M. Temperini, *Type Redefinition and Polymorphism in Object-Oriented Languages*, in Proc. TOOLS PACIFIC, 1990.

[Sac92] B.Buchberger, G.E.Collins, M.J.Encarnacion, H.Hong, J.R.Johnson, W.Krandick, R.Loos, A.M.Mandache, A.Neubacher, H.Vielhaber, *SACLIB User's Guide*, RISC-Linz Tech. Report, 1992.

[Tem92] M. Temperini, *Design and Implementation Methodologies for Symbolic Computation Systems*, Ext. Abs. of Ph.d. Thesis, University *"La Sapienza"*, Roma, Italy, 1992.

[Wan85] P. Wang, R. Pavelle, *MACSYMA from F to G*, J. of Symbolic Computation, 1,1, 1985.

[Wol91] S. Wolfram, *Mathematica, A System for Doing Mathematics by Computer*, second edition, Addison Wesley, 1991.

A Unified-Algebra-based Specification Language for Symbolic Computing

J. Calmet and I.A. Tjandra

Universität Karlsruhe
Institut für Algorithmen und Kognitive Systeme
Am Fasanengarten 5; 76128 Karlsruhe; Germany
e-mail: calmet@dkauni2.bitnet

Abstract. A precise and perspicuous specification of mathematical domains of computation and their inherently related type inference mechanisms is a prerequisite for the design and systematic development of a system for symbolic computing. This paper describes FORMAL, a language for giving modular and well-structured specifications of such domains and particularly of "mathematical objects". A novel framework for algebraic specification involving so-called "unified algebras" has been adopted, where sorts are treated as values. The adoption of this framework aims also at being capable of specifying polymorphism, unifying the notions of "parametric" and "inclusion" polymorphisms. Furthermore, the operational nature of the specification formalisms allows a straightforward transformation into an executable form.

1 Introduction

Nowadays, the demand of specifying *mathematical domains of computation* in symbolic computing grows rapidly due to the fact that correct nontrivial computations are strongly related to the proper domains. For these purposes a system called AXIOM [Sut91] has been designed based on the notions of domains and categories. Domains represent algebraic structures and categories designate collections of domains having common operations with stated mathematical properties. Using this concept, however, properties are handled more or less as comments, which have to be taken into consideration by the programmer who implements the function symbols. Moreover, many existing systems have been extended to be able to handle such domains of computation, e.g. Reduce's theory of domains [BHPS86], which are, in contrast to AXIOM, in some sense more limited.

The main goal of designing FORMAL consists in providing a tool for the specifications of mathematical domains, taking into account the properties of function symbols, as well as for the specifications of type inference engine involving parametric and inclusion polymorphism[1].

---

[1] We use the notions of parametric and inclusion polymorphism according to those introduced in [CW85].

Many specification languages have been developed, e.g. CLEAR [San84], ASL [Wir86], ACT-TWO [EM90], OBSCURE [LL87], ASF [BHK89] and OBJ [JKKM88]. The design of FORMAL takes into account the results achieved by these specification languages.

First of all, we want to depict the major prerequisites for achieving the goal. Accordingly, we have to provide a suitable formalism that can be adopted in order to design FORMAL.

Basically, the initial step of constructing a mathematical domain of computation comprises the determination of its *abstract specification* consisting of the declarations of the sorts, operations and properties under consideration. Such specifications of abstract mathematical domains can rely on the definition of abstract structures as, usually, found in any text-book on abstract algebra. Regarding these abstract structures, e.g. the abstract structure of **Module** that is based on additive Abelian group and possesses **Ring** as its formal parameter, it turns out that the specification language to be designed should be able to handle *structured* (or *modularly-built*) and *parameterized* specifications.

Moreover, it is very convenient to make use of properties imposed on an abstract structure, e.g, in order to draw a conclusion[2], whether two ground terms have the same value or two terms are equal in the initial algebra of the specification, or in order to find the value of a given term, etc. With regard to this requirement at a first glance one might have the impression that the *execution* of a specification could be very useful. A feasible way to do this job consists in *transforming* the specifications into a particular executable form, e.g. transformation of ASF-specifications into Prolog clauses [BHK89]. Hence, we have to take into consideration that the formalism of FORMAL should be tailored in such a way, or it should be *operational* enough, so that it allows a straightforward transformation.

An other important aspect to design the specification language concerns the *inclusion polymorphism* and also the *type inference engine*, as specifications of abstract mathematical domains and their interrelationships represent hierarchies involving multiple inheritance. Thus, the specification of a semantics of inclusion polymorphism involve operations that map sorts to sorts, e.g. sort constructors [SLC91]. Hence, we need a specification formalism allowing sorts to be treated as values. The main idea consists in representing the carrier of a structure including sorts, not only elements of data, e.g. homogeneous algebras. The problem of parametric polymorphism can, basically, be solved straightforward, i.e. there are no exceptional requirements to get this job done.

The prerequisites cited above motivated us to adopt the framework for algebraic specification involving so-called *"unified algebras"* [Mos89] that, actually, has been designed for the so-called action semantics [Mos92]. FORMAL is intended to be used as a tool, particularly, for specifying mathematical domains of computation which are embodied as parameterized modules which are similar to those basic concepts adopted in ACT-TWO [EM90] and OBSCURE [LL87]. The semantics of FORMAL is determined by a meaning function mapping models

---

[2] Assuming that the set of properties is complete.

into models. This kind of function is also be used in EXTENDED-ML [ST85]. In contrast to the subsort concept treated in OBJ we deal with subsorts using the partial ordering defined in unified algebras allowing the specification of type hierarchies with multiple inheritance and type constructions, particularly, needed in symbolic computing.

The paper is organized as follows: Section 2 is devoted to unified algebras based upon a category theoretical approach. In this section we describe basic notions and notations that we need in the remainder of the paper, relying on those introduced in [Mos89] where technical details may be found. Section 3 deals with the syntax of modules representing specifications. Section 4 deals with the semantics of modules as well as with the semantics of its module concepts to build up larger specifications from smaller parts. In this section notations are used, such as the category $Signu_{\mathcal{UNI}}$ and the functor $Modu_{\mathcal{UNI}}$ introduced in section 2. Finally, some concluding remarks are presented in section 5. This paper can be thought of as a report on the design of the specification language FORMAL pursuing the goals and the motivations described above. The application of this language, for instance, to specify an inclusion type polymorphism is beyond the scope of this paper. It can be found in [CT93]. The reader is assumed to be familiar with the basic notions of category theory [Mac71], lattice theory and order-sorted frameworks for algebraic specifications [Wir90, EM85].

2 Basic Notions and Notations

We describe merely those basic notions and notations of unified algebras in terms of the institution $\mathcal{UNI}$ that we need tor the remainder of this paper.

According to Goguen and Burstall [GB84, GB85] an *institution* $\mathcal{I}$ consists of (i) a category of "signatures", $Sign_{\mathcal{I}}$, (ii) a functor $Sen_{\mathcal{I}} : Sign_{\mathcal{I}} \longrightarrow Set$ giving the set of sentences[3] over a given signature, (iii) a functor $Mod_{\mathcal{I}} : Sign_{\mathcal{I}} \longrightarrow Cat^{op}$ giving the category of models of a given signature and (iv) a satisfaction relation $\models_{\mathcal{I}} \subseteq |Mod(\Sigma)| \times Sen(\Sigma)$ for each Σ in $Sign_{\mathcal{I}}$ such that for each signature morphism $\phi : \Sigma \longrightarrow \Sigma'$ in $Sign_{\mathcal{I}}$ the following satisfaction condition holds: for each m' in $|Mod_{\mathcal{I}}(\Sigma')|$ and each e in $Sen_{\mathcal{I}}(\Sigma)$:

$$m' \models_{\mathcal{I}} Sen_{\mathcal{I}}(\phi)(e) \quad :\Leftrightarrow \quad Mod_{\mathcal{I}}(\phi)(m') \models_{\mathcal{I}} e$$

A Σ-theory presentation is a pair (Σ, Γ) where Σ is in $Sign_{\mathcal{I}}$ and Γ is a set of sentences in $Sen_{\mathcal{I}}(\Sigma)$. The closure of (Σ, Γ) is (Σ, Γ') where Γ' is the set of those Σ-sentences that are satisfied by every Σ-model that happens to satisfy all of the sentences in Γ. A Σ-theory T is a Σ-theory presentation (Σ, Γ) that is its own closure.

Initially, we want to have a look at the institution $\mathcal{HFO}$ for homogeneous first-order specification. Let $Sym = \bigcup_{Sym_n, n \geq 0}$ with $Sym_i \cap Sym_j = \emptyset$, $i \neq j$ be the set of function and predicate symbols where n denotes the arity. Let Var be a set of variables, disjoint from Sym. A homogeneous first-order signature

---
[3] Also called properties.

Σ is a pair (F, Π) where $F, \Pi \subseteq Sym$. A homogeneous Σ-algebra A and a homogeneous Σ-homomorphism $h : A \longrightarrow B$ are defined as usual based on the notion of *homogeneous* first-order signatures. Henceforth, we assume that all signatures are homogeneous. For any signature morphism $\phi : \Sigma \longrightarrow \Sigma'$, $Mod_{\mathcal{HFO}}$ is defined as the forgetful functor induced by ϕ (using the notion of ϕ-reduct). It must be noted that it doesn't forget any values at all but only functions and predicates due to the homogeneity. $Sen_{\mathcal{HFO}}$ is the set of closed first-order formulae with function and predicate symbols from Σ and variables from Var. $\models_{\mathcal{HFO}}$ is defined as usual.

A unified signature Σ is a homogeneous first-order signature (F, Π^{o}) where $F^{o} = \{\bot, \_|\_, \_\&\_\} \subseteq F$ and $\Pi^{o} = \{\_=\_, \_\leq\_, \_:\_\}$ [4].

A unified Σ-algebra A is a homogeneous Σ-algebra such that:

- $|A|$ is a distributive lattice with $\_|\_^{A}$ as join, $\_\&\_^{A}$ as meet and $\bot^{A}$ as bottom.
- There is a distinguished set of values $E^{A} \subseteq |A|$ representing the elements of A.
- f^{A} is monotone w.r.t. the partial order of the lattice for every f in F.
- $x =^{A} y \quad :\Leftrightarrow \quad x$ ist identical to y.
- $x \leq^{A} y \quad :\Leftrightarrow \quad x|^{A} y = y$, i.e. $\leq^{A}$ is the partial order of the lattice.
- $x :^{A} y \quad :\Leftrightarrow \quad x \in E^{A}$ and $x \leq^{A} y$

$Mod_{\mathcal{UNI}}(\Sigma)$ is the full subcategory of $Mod_{\mathcal{HFO}}(\Sigma)$ whose objects are the unified Σ-algebras. $Sen_{\mathcal{UNI}}(\Sigma)$ is the restriction of $Sen_{\mathcal{HFO}}(\Sigma)$ to universal Horn clauses and, finally, $\models_{\mathcal{UNI}}$ is the restriction of $\models_{\mathcal{HFO}}$ to unified algebras and universal Horn clauses. Now let's look at some (nice) properties of the institution $\mathcal{UNI}$.

$\mathcal{UNI}$ is a *liberal* institution in the sense that for every theory morphism $\phi : T \longrightarrow T'$ the forgetful functor $\_|_{\phi} : Mod_{\mathcal{UNI}}(T') \longrightarrow Mod_{\mathcal{UNI}}(T)$ has a left adjoint.

For any unified signature Σ and set of unified Σ-sentences Γ, the class of unified Σ-algebras that satisfy Γ has an *initial algebra*.

A Σ-data constraint is a pair $(\phi : T'' \longrightarrow T', \theta : \Sigma' \longrightarrow \Sigma)$ where Σ' is the signature of T. A Σ-algebra A satisfies this constraint iff $A|_{\phi}$ is a model of T' and ϕ-free in the sense that it is naturally isomorphic to the free model over their ϕ-reduct. As ϕ-reduct never forget any values, models that are ϕ-free do not exist, except in trivial cases. To obtain a useful version of data constraint for unified algebras the functor $\_\dagger_{\phi}$, that also forget values, is defined. Henceforth, we replace $\_|_{\phi}$ by $\_\dagger_{\phi}$.

3 The Syntax of FORMAL

The goal of this section is to provide an overview of the specification language FORMAL. A specification is represented by a module. The module concepts to

---

[4] Here we use the mixfix notation where the number and positions of arguments are indicated by occurrences of a place holder "$\_$".

build up large specification from smaller ones are illustrated by some examples, cf. Figure 2. The Module **Boolean** (or **SemiGroup**) is specified as a *basic module* possessing the constants **Boolean**, **true** and **false** whose relationships are embodied by the relators =, | and :, cf. section 2. It also possesses the function symbols **and**, **or** and **not** that are represented in the *Operation* part together with their functionalities. The properties of these function symbols are expressed by means of Horn Clauses with equality. The Module **Rng** and **Leftmodule** are specified by using the module concepts **Union** and **Rename**, and **Union**, **Rename** and a formal parameter **Rng**, respectively.

Let *Module* be the set of modules. The syntax of modules is defined by the grammar given in figure 1.

⟨*Module*⟩	::=	(**Module** ⟨*ModuleId*⟩⟨*Basic*⟩)
⟨*Basic*⟩	::=	⟨*ModuleId*⟩ |
		(⟨*Import*⟩ ⟨*Basic*⟩) |
		(**Union** ⟨*Basic*⟩ ⟨*Basic*⟩) |
		(**Composition** ⟨*Basic*⟩ ⟨*Basic*⟩) |
		(⟨*Basic*⟩ ⟨*Renaming*⟩ ⟨*Ignoring*⟩ ⟨*Body*⟩)
⟨*Import*⟩	::=	(**Use** ⟨*BasicList*⟩) | λ
⟨*BasicList*⟩	::=	⟨*Basic*⟩ | ⟨*Basic*⟩ ⟨*BasicList*⟩
⟨*Renaming*⟩	::=	(**Rename** ⟨*FunctionList*⟩ ⟨*FunctionList*⟩) | λ
⟨*Ignoring*⟩	::=	(**Ignore** ⟨*FunctionList*⟩) | λ
⟨*Body*⟩	::=	(**Define** ⟨*ConstantSymbols*⟩ ⟨*OperationSymbols*⟩ ⟨*Properties*⟩) | λ
⟨*ConstantSymbols*⟩	::=	(**Constants** ⟨*Constants*⟩) | λ
⟨*Constants*⟩	::=	(⟨*Relator*⟩ ⟨*Identifier*⟩ ⟨*Term*⟩) | ⟨*Identifier*⟩ |
		⟨*Constants*⟩ ⟨*Constants*⟩
⟨*Relator*⟩	::=	: | =< | =
⟨*OperationSymbols*⟩	::=	(**Operations** ⟨*Operations*⟩) | λ
⟨*Operations*⟩	::=	(⟨*Identifier*⟩ ⟨*Functionalities*⟩) | ⟨*Operations*⟩ ⟨*Operations*⟩
⟨*Functionalities*⟩	::=	(⟨*Term*⟩ ⟨*FctArrow*⟩ ⟨*Identifier*⟩) |
		⟨*Functionalities*⟩ ⟨*Functionalities*⟩
⟨*FctArrow*⟩	::=	-> | -?>
⟨*Properties*⟩	::=	(**Clauses** ⟨*Clauses*⟩)
⟨*Clauses*⟩	::=	⟨*Formula*⟩ | (⟨*Formula*⟩ ⟨*FormulaList*⟩) |
		(**Imply** ⟨*Formula* ⟨*Clauses*⟩⟩ |
		(**Imply** ⟨*Formula*⟩(**And** ⟨*ClauseList*⟩))
⟨*Formulalist*⟩	::=	⟨*Formula*⟩ | ⟨*Formula*⟩ ⟨*FormulaList*⟩
⟨*ClauseList*⟩	::=	⟨*Formula*⟩ | ⟨*Formula*⟩ ⟨*Clauselist*⟩
⟨*Formula*⟩	::=	(⟨*Relator*⟩ ⟨*Term*⟩ ⟨*Term*⟩)

Fig. 1. The Syntax of Modules

The grammar does not further define the syntax of ⟨*FunctionList*⟩, ⟨*Formula*⟩ and ⟨*Term*⟩ as they depend on the user-defined specifications of modules.

```
(Module
 Boolean
 (Define
  (Constants
   (= Boolean (| true false))
   (: true Boolean)
   (: false Boolean))
  (Operations
   (or ((Boolean Boolean) -> Boolean))
   (and ((Boolean Boolean) -> Boolean))
   (not (Boolean -> Boolean)))
  (Clauses
   (= (or true true) true)
   (= (or true false) false)
   (= (or false true) false)
   (= (or false false) false)
   (= (and true true) true)
   (= (and false true) false)
   (= (and true false) false)
   (= (and false false) false)
   (= (not true) false)
   (= (not false) true))))

(Module
 SemiGroup
 (Define
  (Constants
   SemiGroup)
  (Operations
   (o (SemiGroup SemiGroup) -> SemiGroup))
  (Clauses (Imply
   (: (a b c) SemiGroup)
   (= (o (o a b) c)
      (o a (o b c)))))))

(Module
 Monoid
 (Union
  (SemiGroup)
  (Define
   (Constants
    (=< Monoid SemiGroup)
    (: Neutral Monoid))
   (Clauses (Imply
    (: a Monoid)
    (and (= (o Neutral a)
            a)
         (= (o a Neutral)
            a)))))))

(Module
 Group
 (Union
  (Monoid)
  (Define
   (Constants (=< Group Monoid))
   (Operations (inv Group -> Group))
   (Clauses (Imply
    (: (a b) S)
    (and (= (op (inv a) a)
            Neutral)
         (= (op a (inv a))
            Neutral)))))))
```

```
(Module
 AbelianGroup
 (Union
  (Group)
  (Define
   (Constants
    (=< AbelianGroup Group)
    (Clauses (Imply
     (: (a b) AbelianGroup)
     (= (o a b)
        (o b a)))))))

(Module
 Rng      (*a help definition for Ring*)
 (Union
  (AbelianGroup
   (Rename (AbelianGroup Neutral o inv)
           (AddAbelianGroup 0 + -)))
  (SemiGroup
   (Rename (SemiGrouup o)
           (MultSemiGroup *)))
  (Define
   (Constants
    Rng)
   (Clauses (=< Rng AddAbelianGroup)
            (=< Rng MultSemiGroup)
    (Imply
     (: (a b c) Rng)
     (and (= (* a (+ b c))
             (+ (* a b) (* a c)))
          (= (* (+ a b) c)
             (+ (* a c) (* b c)))))))))

(Module
 LeftModule
 (Use Rng)
 (Union
  (AbelianGroup
   (Rename (AbelianGroup Neutral o inv)
           (AddAbelianGroup 0 + -)))
  (Define
   (Constants
    (=< LeftModule AddAbelianGroup))
   (Operations
    (* (Rng LeftModule) -> LeftModule))
   (Clauses (Imply
    (: (x y) Rng)
    (Imply
     (: (a b) LeftModule)
     (and (= (* x (* a b))
             (* (* x a) b))
          (= (* x (+ a b))
             (+ (* x a) (* x b)))
          (= (* (+ x y) a)
             (+ (* x a) (* y a)))))))))))
```

Figure 2: Simple Examples of Modules

The symbol -?> is used to specify a partial function. The three relator symbols coincide with those as defined in section 2.

Use and **Define** indicate the imported and the exported signatures of a module, respectively. The union of two modules is denoted by the module concept **Union**. It builds the union of the imported and exported signatures. It should be noted that ambiguities have to be avoided, e.g. name duplication. The module concept **Composition** offers a possibility to compose two modules. It is comparable to the concept **enrich** of CLEAR [San84] or composition of ACT-Two [EM90]. The composition of modules **m1** and **m2** takes the imported signature of **m2** as its imported signature and the exported signature of **m1** as its exported signature. The module concept **Rename** is used to rename only particular exported operation and constant symbols of a module. The module concept **Ignore** allows to get rid of some exported operation and constant symbols. Using this module concept one can simplify a module by forgetting or hiding some operation and constant symbols that, for instance, no longer need to be exported. The module concept **Submodule** and **Quotient** are omitted in this paper.

4 The Semantics of FORMAL

The semantics of a module **m** is given by the signature function $S[m]$ consisting of $S_i[m]$ in $Sign_{u\mathcal{N}\mathcal{I}}$, the imported unified signature of **m**, and $S_e[m]$ in $Sign_{u\mathcal{N}\mathcal{I}}$, the exported unified signature of **m**, and the meaning function $B[m]$ presenting a partial function $B[m] : Mod_{u\mathcal{N}\mathcal{I}}(S_i[m]) \rightsquigarrow Mod_{u\mathcal{N}\mathcal{I}}(S_e[m])$. Furthermore, let $InS[m]$ be the set of inherited signature, i.e. $S_i[m] \cap S_e[m]$. The inductive construction of modules is determined by these two functions.

A syntactically correct module **m** in $Module$ is said to be a basic module iff it does not contain the module concepts **Union**, **Composition**, **Rename** and **Ignore**. The inductive construction of modules is as follows:

- Basic modules are in $Module$ with the signature and meaning functions as described above.
- Let **m1** and **m2** be in $Module$ and (**Module m** (**Union m1 m2**)) be a union of the two modules with the following functions:
$$S[m] = \underbrace{(S_i[m1], S_e[m1])}_{S[m1]} \cup \underbrace{(S_i[m2], S_e[m2])}_{S[m2]}$$

$$B[m](A) = \begin{cases} B[m1](A\dagger_{\iota_{i1}}) \cup B[m2](A\dagger_{\iota_{i2}}) \\ \quad \text{if } B[m1](A\dagger_{\iota_{i1}}) \text{ and } B[m2](A\dagger_{\iota_{i2}}) \text{ are defined} \\ \text{Undefined} \quad \text{otherwise} \end{cases}$$

where
$\iota_{i1} : S_i[m1] \hookrightarrow S_i[m]$ and
$\iota_{i2} : S_i[m2] \hookrightarrow S_i[m]$
are inclusion signature morphisms.
m is in $Module$, if the following conditions are satisfied:
- $B[m](A)\dagger_{\iota_{e1}} = B[m1](A\dagger_{\iota_{i1}})$
- $B[m](A)\dagger_{\iota_{e2}} = B[m1](A\dagger_{\iota_{i2}})$

where
$\iota_{e1} : S_e[m1] \hookrightarrow S_e[m]$ and
$\iota_{e2} : S_e[m2] \hookrightarrow S_e[m]$
are inclusion signature morphisms.
Accordingly, $B[m]$ is a unique unified $S_e[m]$ Algebra iff the following conditions are satisfied:

- $S_e[m1] \cap S_e[m2] \subseteq S_i[m1] \cap S_i[m2]$
- $S_e[m1] \cap S_i[m2] \subseteq S_i[m1]$
- $S_e[m2] \cap S_i[m1] \subseteq S_i[m2]$

− Let **m1** and **m2** be in *Module* and **(Composition m1 m2)** be a composition of both modules with
$S[m] = (S_i[m2], S_e[m1])$
$$B[m](A) = \begin{cases} B[m1](B[m2](A)) \\ \quad \text{if } A' := B[m2](A) \text{ and } B[m1](A') \text{ are defined} \\ \text{Undefined} \quad \text{otherwise} \end{cases}$$
m is in *Module*, iff the exported signature of **m2** coincides with the imported signature of **m1** or if the following conditions are satisfied:

- $S_e[m2] = S_i[m1]$
- $S_i[m2] \cap S_e[m1] \subseteq S_i[m1]$

− Let **m1** be in *Module* and
(Module m (m1 (Rename $\langle RenamingList1 \rangle$ $\langle RenamingList2 \rangle$)))) be a module built from **m1** by renaming particular operation and constant symbols. Furthermore, let $\phi_e : S_e[m1] \longrightarrow S_e[m]$ be a unified signature morphism that can be constructed by using $\langle RenamingList1 \rangle$ and $\langle RenamingList2 \rangle$ and ϕ be the pair of signature morphism $(id(S_i[m1]), \phi_e)$ with
$S[m] = \phi(S[m1]) = (S_i[m1], \phi_e(S_e[m1]))$
$$B[m](A) = \begin{cases} B[m1](A){\restriction}_{\phi_e^{-1}} & \text{if } B[m1](A) \text{ is defined} \\ \text{Undefined} & \text{otherwise} \end{cases}$$
where $\phi_e^{-1} : \phi_e(S_e[m1]) \hookrightarrow S_e[m1]$
m is in *Module* iff **m** satisfies the following conditions:

- $\phi_e{\restriction}_\iota = id(InS[m1])$
  where $\iota : InS[m1] \hookrightarrow S_e[m1]$ is an inclusion signature morphism.
- ϕ_e is injective.
- $S_i[m1] \cap \phi_e(S_e[m1]) \subseteq S_e[m]$

− Let **m1** be a module and **(Module m (m1 (Ignore $\langle FunctionList \rangle$))))** be a module that is constructed by hiding particular constant and operation symbols with
$S[m] = (S_i[m1], S_e[m1] \backslash \langle FunctionList \rangle)$
$$B[m](A) = \begin{cases} B[m1](A){\restriction}_\iota & \text{if } B[m1](A) \text{ is defined} \\ \text{Undefined} & \text{otherwise} \end{cases}$$
where $\iota : S_e[m1] \backslash \langle FunctionList \rangle \hookrightarrow S_e[m1]$ is an inclusion signature morphism.
m is in *Module* iff the following condition holds:

- $S_e[m1] \backslash \langle FunctionList \rangle \in Sign_{UNI}$

Now, we introduce some additional definitions in order to outline some basic properties of the module concepts described above.

A module morphism $h : m \longrightarrow m'$ where m and m' are in $\mathcal{M}odule$ is a pair (h_i, h_e) of signature morphisms $h_i : \mathcal{S}_i[m] \longrightarrow \mathcal{S}_i[m']$ and $h_e : \mathcal{S}_e[m] \longrightarrow \mathcal{S}_e[m']$ with $h_i\dagger_{InS[m]} = h_e\dagger_{InS[m]}$.

Let m, m' be in $\mathcal{M}odule$, $h : m \longrightarrow m'$ be a module morphism, $\mathcal{S}_i[m] = (F, \Pi)$, $A \in Mod_{\mathcal{UNI}}(\mathcal{S}_i[m])$ with the signature Σ_A, $B \in Mod_{\mathcal{UNI}}(\mathcal{S}_i[m'])$ with the signature Σ_B, $\phi_A : \mathcal{S}_i[m] \longrightarrow \Sigma_A$ and $\phi_B : \mathcal{S}_i[m'] \longrightarrow \Sigma_B$ be unified signature morphisms. Further, let $B_{h(x)} := \{y \mid y \in |B| \wedge y : \phi_B(h(x))\}$ and $A_x := \{y \mid y \in |A| \wedge y : \phi_A(x)\}$. A family $r = (r_x)_{x \in Constants(\mathcal{S}_i[m])}$ of surjective and partial functions $r_x : B_{h(x)} \rightsquigarrow A_x$ is said to be h-Representation of A by B (denoted by $B \xrightarrow{r} A$) iff for each operation $f \in F_n$ and every $\pi \in \Pi_n$, $n \geq 0$:

$$d_x(h(f)^B(a_1, \cdots, a_n)) = \begin{cases} f^A(d_x(a_1), \cdots, d_x(a_n)) & \text{if } d_e(a_i)\ 1 \leq i \leq n \text{ is def.} \\ \text{Undefined} & \text{otherwise} \end{cases}$$

$d_x(h(\pi)^B(a_1, \cdots, a_n))$ implies $\pi^A(d_x(a_1), \cdots, d_x(a_n))$.

Let $Rep_h(B, A)$ be all h-representations of A by B and $R_h(B, A)$ be a class of h-representations, i.e. $R_h(B, A) \subseteq Rep_h(B, A)$.
For a module m, a module m' in $\mathcal{M}odule$ is said to be in $Mod_{\mathcal{UNI}}(\mathcal{S}_e[m])$[5] or m' is an implementation of m by means of h and R_h (denoted by $m' \overset{h, R_h}{\Longrightarrow} m$) if for each r in R_h the following conditions are satisfied:

- $\mathcal{B}[m'](B)$ is defined, whenever $\mathcal{B}[m](A)$ is defined
- If $\mathcal{B}[m](A)$ is defined then there are partial functions:
  $r'_x : \mathcal{B}[m'](B)_{h(x)} \rightsquigarrow \mathcal{B}[m](A)_x$, $x \in Constants(\mathcal{S}_e[m])$, an extension of r_x

Using the definitions described above we can state the following theorems:

If (Module m (Union m1 m2)) *is in* $\mathcal{M}odule$, m1' *is in* $Mod_{\mathcal{UNI}}(\mathcal{S}_e[m1])$ *and* m2' *is in* $Mod_{\mathcal{UNI}}(\mathcal{S}_e[m2])$ *then* (Module m' (Union m1' m2')) *is in* $Mod_{\mathcal{UNI}}(\mathcal{S}_e[m])$.

If (Module m (Composition m1 m2)), (Module m (Composition m1 m2)) *are in* $\mathcal{M}odule$ *and* $h1 : m1 \longrightarrow m1'$, $h2 : m2 \longrightarrow m2'$ *are module morphisms with* $h1_i = h2_e$ *then* $h = (h2_i, h1_e)$ *is a module morphism form* m *to* m'.

If (Module m (m1 (Rename $\langle RenamingList1 \rangle$ $\langle RenamingList2 \rangle$))), (Module m' (m1' (Rename $\langle RenamingList1' \rangle$ $\langle RenamingList2' \rangle$)))) *are in* $\mathcal{M}odule$ *with the corresponding signature morphisms* $\phi_e : \mathcal{S}_e[m1] \longrightarrow \mathcal{S}_e[m]$ *and* $\phi'_e : \mathcal{S}_e[m1'] \longrightarrow \mathcal{S}_e[m']$ *then one can construct a module morphism* $h : m \longrightarrow m'$ *by means of the module morphism* $h_1 : m1 \longrightarrow m1'$ *in such away that the following diagram commutes:*

---

[5] An additional condition to $\models_{\mathcal{UNI}}$

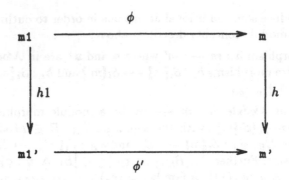

If (Module m (m1 (Ignore $\langle FunctionList \rangle)))$,
(Module m' (m1' (Ignore $\langle FunctionList' \rangle)))$ *are in* $\mathcal{M}odule$ *and* $h1 : m1 \longrightarrow m1'$ *is a module morphism, such that* $h_e(\mathcal{S}_e[\![m1]\!] \setminus \langle FunctionList \rangle) \cap \langle FunctionList' \rangle = \emptyset$ *then one can construct a module morphism* $h : m \longrightarrow m'$ *under certain conditions*

5 Conclusion

We have given an overview of the specification language FORMAL. Among the desired characteristical features of FORMAL are: (i) structured or modularly-built and parameterized specifications are fundamentals, (ii) the operational nature of the specification formalisms might allow a straightforward transformation and, finally, (iii) to specify sorts constructors it is essential to be capable of treating sorts also as values. The prerequisites cited above motivated us to adopt the framework for algebraic specification involving so-called "unified algebras" [Mos89] that, actually, has been designed for the so-called action semantics [Mos92]. In contrast to the specification concept of AXIOM, in this specification language the set of function properties of a specification is taken into account.

FORMAL is part of a programming environment for symbolic computing involving an interactive acquisition component, a transformation component, a hybrid knowledge representation system and a query processing component. The environment is implemented in Common Lisp. The syntax of FORMAL (strongly influenced by Lisp) is adapted to this environment. Even though an interactive graphical user interface for acquiring specifications is being developed. Using this graphical user interface, the lisp-like syntax of FORMAL that is not very convenient for the user will disappear in the front-end system.

The emphasis of this paper is the formal semantics of FORMAL. An overview of the transformation of modules into an executable form using the knowledge representation system MANTRA [CTB91] can be found in [CTH93].

We use FORMAL also to specify a semantics of inclusion polymorphism and a semantics of parametric polymorphism as well [CT93].

Acknowledgements: We are indebted to K. Homann, M. Kreidler, M. Funke and M. Lobmeyer for constructive suggestions regarding FORMAL. We owe also thanks to the referees for their hints to improve this paper.

References

[BHK89]   J.A. Bergstra, J. Heering, and P. Klint. *Algebraic Specification*. Addison-Wesley Publishing Company, 1989.

[BHPS86] R.J. Bradford, A.C. Hearn, J.A. Padget, and E. Schrufer. Enlarging the REDUCE domain of computation. In *SYMSAC 1986*, pages 100 – 106. ACM, New York, 1986.

[CT93]   J. Calmet and I.A. Tjandra. Specifying type polymorphisms in symbolic computing with FORMAL. *In preparation*, 1993.

[CTB91]   J. Calmet, I.A. Tjandra, and G. Bittencourt. Mantra: A shell for hybrid knowledge representation. In *IEEE-Conference on Tools for AI*, pages 164 –171. IEEE, IEEE Computer Society Press, 1991.

[CTH93]   J Calmet, I.A. Tjandra, and K. Homann. Unified domains and abstract computational structures (to be published). In J. Calmet and J. Campbell, editors, *Int. Conference on Ai and Symbolic Computations*. Springer-Verlag, LNCS, 1993.

[CW85]   L. Cardelli and P. Wagner. On understanding types, data abstraction and polymorphism. *Computing Survey*, 17(4):471 – 522, 1985.

[EM85]   H. Ehrig and B. Mahr. *Fundamental of Algebraic Specification 1 – Equations and Initial Semantics*. EATCS Monograph on Theoretical Computer Science, vol 6. Springer-Verlag, 1985.

[EM90]   H. Ehrig and B. Mahr. *Fundamental of Algebraic Specification 2*. Monograph on Theoretical Computer Science, vol 21. Springer-Verlag, 1990.

[GB84]   J.A. Goguen and R.M. Burstall. Introducing institution. In *Logics of Programming Workshop*, pages 221 – 256. Springer-Verlag, 1984.

[GB85]   J.A. Goguen and R.M. Burstall. Institutions: Abstract model theory for computer science. Technical report, Stanford University, CSLI-85-30, 1985.

[JKKM88] J.P. Jounnaud, C. Kirchner, H. Kirchner, and A. Megrelis. OBJ: Programming with equalities, subsorts, overloading and parameterization. In J. Grabowski, P. Lescanne, and P. Wechler, editors, *Algebraic and Logic Programming*, pages 41 – 52. Spriner-Verlag, LNCS 343, 1988.

[LL87]   T. Lehman and J. Loeckx. The specification language OBSCURE. In D. Sanella and A. Tarlecki, editors, *Recent Trends in Data Types Specification*, pages 131 – 153. Springer-Verlag, LNCS 332, 1987.

[Mac71]   S MacLane. *Categories for the Working Mathematician*. Graduate Text in Mathematics. Springer-Verlag, 1971.

[Mos89]   P.D. Mosses. Unified algebras and institutions. In *Logics in Computer Science*, pages 304 – 312. IEEE Press, 1989.

[Mos92]   P.D. Mosses. *Action Semantics*. Cambridge Tracks in Theoretical Computer Science. Cambridge University Press, 1992.

[San84]   D Sanella. A set-theoretic semantics of CLEAR. *Acta Informatica*, 21(5):443 – 472, 1984.

[SLC91]   P Schnoebelen, D Lugiez, and H Comon. A rewrite-based type discipline for a subset of computer algebra. *Journal of Symbolic Computation*, 11:349 – 368, 1991.

133

[ST85]   D. Sanella and A. Tarlecki.   Program specification and development in
         STANDARD-ML. In *12th ACM Symposium on Principle of Programming Languages, New Orleans, USA*, pages 67 – 77. ACM Press, 1985.

[Sut91] R.S. (Ed.) Sutor. AXIOM *User's Guide*. The Numerical Algorithm Group
         Limited, 1991.

[Wir86] M. Wirsing. Structured algebraic specification: A kernel language. *Theoretical Computer Science*, 42:124 – 249, 1986.

[Wir90] M. Wirsing. Algebraic specification. In J. VanLeeuwen, editor, *Handbook of Theoretical Computer Science*, volume B, pages 677 – 788. Elsevier Science Publishers B.V., 1990.

An Order-sorted Approach to Algebraic Computation[*]

Anthony C. Hearn[1] and Eberhard Schrüfer[2]

[1] RAND, Santa Monica CA 90407-2138, E-mail: hearn@rand.org
[2] GMD, D-53731 Sankt Augustin, Germany, E-mail: schruefer@gmd.de

Abstract. This paper presents the prototype design of an algebraic computation system that manipulates algebraic quantities as generic objects using order-sorted algebra as the underlying model. The resulting programs have a form that is closely related to the algorithmic description of a problem, but with the security of full type checking in a compact, natural style.

1 Introduction

The semantic model of REDUCE 3 was developed more than twenty years ago. In the intervening years, extensions were made to allow for problem domains that were not initially considered. As a result, the model has become quite complicated, making it hard to maintain, difficult to explain, and rather inefficient. It also does not scale well as more and more contributors add code of increasing complexity. We therefore decided to develop a new model that, building on past experience, presents a simple, unified model to the user. This allows for the development of robust, compact programs that have a natural mathematical syntax and semantics.

A new model can also take into account the considerable progress that has been made during the past decade in programming system design. In particular, the concepts of abstract datatyping and object-oriented programming are based on the recognition that programs should be organized around the objects they manipulate rather than the manipulations themselves. Other important characteristics include operator overloading and polymorphism, that make code easier to read, write and maintain; locality of storage, that makes it easier to support distributed and concurrent programming; and inheritance, that provides for more effective code reuse. Such methodology is particularly important in algebraic computation, where many different objects must be considered in a wide variety of representations. It is not a unambiguous task to use such techniques, however, since the term "object-oriented programming" means different things to different people, and there is little current standardization of the techniques involved. A recent classification [11], for example, compares twenty three different models that have been called object-oriented.

Given the alternative styles that were possible, we decided not to lock ourselves into an existing "off-the-shelf" object-oriented or datatyping language model, since we guessed, quite correctly in retrospect, that we would be making many changes in the form of our model as the work proceeded. On the other hand we have had long

---

[*] This work was supported in part by the U.S. National Science Foundation under Grant No. CCR-8910826.

experience with Lisp as a prototyping language, and know the intimate workings of REDUCE that is based on this language. This made Lisp a natural choice for this work.

One important early decision, which influences much of the design, was the choice of dynamic typing for the evaluation strategy. We believe that algebraic computation semantics should be based on strong typing. However, static typing (which many people incorrectly think of as a characteristic of strong typing) is too restrictive a requirement for a system with a rich spectrum of types. This means that type inferencing should be done only at run time when there is insufficient information for static resolution.

Our design choices were also influenced by the belief that an algebra system should be based as much as possible on a rigorous semantic model, rather than relying too heavily on the heuristic approaches typical of many algebra systems. We also believe that formal algebraic specification methods offer the best approach in this regard. Our experiments demonstrate that it is possible to specify algebraic algorithms in a manner consistent with the approaches used for algebraic specifications. Since we hope to apply formal verification methods to our programs, the more consistent we can be with algebraic specification practice, the easier this task should be.

Unlike other systems such as Axiom [10], Views [1] and Weyl [13], our current prototype does not support dynamic type generation. The inclusion of such a capability can have a marked effect on both the efficiency and complexity of the resulting system. We investigate, however, on-the-fly instantiation in defining datatypes (see Section 5). So far, we have not felt constrained by this choice.

Most existing algebra systems which support datatyping are based on the notion of *many-sorted algebras*, i.e. they view datatypes as (heterogeneous) algebras on multiple carriers [2]. Recent studies in the algebraic specification community, especially that of Goguen and his collaborators [8], propose a generalization of many-sorted algebras called *order-sorted algebras*[3]. We decided, after a careful study of their properties, that such algebras represented the best model for algebraic computation currently available. We make the case for this choice in a later section.

2 Syntax

We have, where possible, adapted the existing REDUCE language to support the current prototype. This means we have not had to invest time in defining new language constructs except in those cases where it is essential or desirable. From the user's point of view, the syntactical style for a straightforward "algebraic calculator" manipulation then is little changed from REDUCE 3. The normal mode of operation is to carry out a calculation by a functional transformation. For example,

```
solve({x^2+y=2,y^2-x^3=3},{x,y});
```

computes the solutions of the coupled nonlinear equations $x^2 + y = 2$ and $y^2 - x^3 = 3$.

---

[3] Goguen emphasizes in several papers (e.g., [7]), that the word *type* is overworked in computer science. He prefers the word *sort* when talking about the datatypes themselves. However, we have decided that the use of *type* for this purpose is so ingrained in the computing community that it would be confusing to use *sort* instead.

The main syntactical difference is in the handling of types. In REDUCE 3, what little typing information existed appeared as declarations. In the current prototype, we allow for the specification of the type of an object by means of a following colon-separated expression. Thus, whereas in REDUCE 3 one could say in a program block:

```
begin scalar x,y; integer n; ... end
```

we now say[4]

```
begin local x,y:generic, n:int; ... end
```

Similarly, a procedure definition can include typing information as follows:

```
procedure factorial n:posint;
    for i:=1:n product i;
```

Notice that we have not provided a target datatype for the factorial definition. Since typing inference is done dynamically, it is not necessary to specify the type of the output. However, such information can also be included; it can then be used for static code optimizations.

We are also experimenting with a syntax in which an end-of-line can serve as a statement delimiter. With this syntax, the above factorial definition takes the form:

```
procedure factorial n:posint
    for i:=1:n product i
```

We shall use this style for all subsequent programming examples in this paper[5].

3 Order-sorted Algebras

Order-sorted algebras generalize many-sorted algebras by adding the requirement of a *partial ordering* on the types, which is interpreted as subset inclusion of the carrier sets. Order-sorted algebra resolves overloading of operator symbols by the requirement that the result of an operation has to coincide whenever the arguments are related by the partial order. A considerable literature about them exists [8] so we shall not describe the methodology in detail here. However, since this theory may be unfamiliar to many readers, we shall explain the salient features through an example, namely the manipulation of multivariate polynomials. REDUCE, like most algebra systems, has many different representations for polynomials. However, the default representation is one that manipulates these objects in a *canonical form*. Much of the REDUCE 3

---

[4] The type *generic* in this example represents a universal or "top" type to which all types can be lifted. Most typing systems include this notion, either implicitly or explicitly. In practice we never meet this type during execution, since, as we explain later, all objects are reduced to their floor types before use.

[5] In our current notation, we use the meaning of an expression to decide if it continues on the next line rather than a continuation marker. Thus $a + b$ is complete, whereas $a + b+$ continues on the next line.

polynomial package uses this representation, and it has proved to be quite flexible and extensible over the years [3].

A polynomial in this representation takes the form:

$$u(x_1, ..., x_n) = \sum_{\nu=0}^{n} u_\nu(x_2, ..., x_n) x_1^\nu$$

where x_1 is ordered ahead of the other x_i. If there is only one variable, u_ν is a constant.

The current REDUCE canonical polynomial representation, or *standard form*, has the structure:

```
<standard_form>
     ::= <zero> | <domain_element> | <standard_term> . <reductum>

<zero> ::= NIL

<domain_element> ::= <nonzero_integer>

<reductum> ::= <standard_form>

<standard_term> ::= <standard_power> . <standard_form>

<standard_power> ::= <kernel> . <positive_integer>

<kernel> ::= <identifier>
```

Thus a *standard_term* represents one term in the above sum, and a *domain_element* a constant term. It is also necessary to provide constraints on the left and right fields of the various tuples in the above to make the representation unique. In particular, the *reductum* in the standard form definition must contain terms with a lower power than that in the *standard_term*.

For simplicity in this discussion we have limited the data structures *domain_element* and *kernel* to *nonzero_integer* and *identifier* respectively, even though more complicated structures are possible in general [3]. Relatively few changes are required to make this representation consistent with an order-sorted approach. A suitable form for this purpose is:

```
<standard_form>
     ::= <zero> | <domain_element> | <nonconstant_standard_form>

<zero> ::= 0

<domain_element> ::= <integer>

<integer> ::= <nonzero_integer> | <zero>

<nonzero_integer> ::= <positive_integer> | <negative_integer>

<nonconstant_standard_form>
     ::= <standard_term> | <extended_standard_form>
```

```
<extended_standard_form>
    ::= <standard_term> . <nonconstant_standard_form>
                       | <standard_term> . <domain_element>

<standard_term> ::= <standard_power> | <extended_standard_term>

<extended_standard_term>
    ::= <standard_power> . <nonconstant_standard_form>
                         | <standard_power> . <nonzero_integer>

<standard_power> ::= <kernel> | <extended_standard_power>

<extended_standard_power> ::= <kernel> . <positive_integer>

<kernel> ::= <identifier>
```

The relations between these datatypes can be expressed as follows:

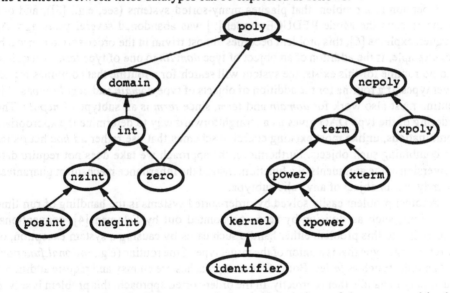

The type names in this diagram are obvious abbreviations of those we used in the BNF description. This diagram is in fact a DAG (directed acyclic graph) showing the *order-sorted relationships* between the various types of expressions that comprise a polynomial. The arrows in this DAG model an inclusion relationship among the various datatypes. For example, nonzero integers are subtypes of domain elements, which in turn are subtypes of polynomials. These relationships may be expressed by writing $nzint < domain < poly$. Note that in this diagram, *multiple inheritance* also occurs because some types have more than one supertype. For example, $zero < int$ and $zero < poly$.

As pointed out by Goguen and Wolfram [7], order-sorted algebra provides a semantic foundation for abstract datatypes with both inheritance and coercion. The main new feature is that it formally models an inclusion relationship among the various datatypes.

Under type inheritance, a function may accept arguments of lower types. Thus a routine for adding *polys* must also be able to handle the *zero*, *domain* and *ncpoly* subtypes. The code for these subtype operations must be also able to handle their subtype operations and so on. The results of such operations could be of the most general type possible for the operation (e.g., *poly* in the case of polynomial addition), or it could in principal be of a lower type. To make this explicit, our evaluation model reduces the datatype of the result of any operation to the lowest, or floor, datatype. This reduction is unique, and is straightforwardly done by following the downward paths in the datatype DAG[6]. Such reductions may include a change in representation of the object (e.g., the variable *a* may have a different representation from the polynomial *a*). Almost always, such a change in representation simplifies the data structure used to represent the object, unlike traditional coercion, which often complicates the form of the object. Notice also that such reductions are only possible in a system with dynamic inferencing; it is impossible in a statically typed system to infer for example that $a + 1 - a$ is of (ground) type *posint* unless one actually does the calculation.

Coercion is a problem that plagues many-sorted systems (see, e.g., [5]), and one of the reasons the Mode REDUCE model [9] was abandoned several years ago. As Goguen explains [6], this problem becomes almost trivial in the order-sorted approach. For example, if the addition of an object of type *domain* to one of type *term* is required and no routine for this exists, the system will search for a routine that combines higher level types. If a routine for the addition of objects of type *domain* and *ncpoly* exists, this routine must also work for *domain* and *term*, since *term* is an subtype of *ncpoly*. The ordering of the type DAG gives us a straightforward way to determine the appropriate combinations, unlike most existing coercion schemes that use rather *ad hoc* heuristics for combining such objects. Furthermore, the approach we take does not require data conversion of the arguments of a function. Instead the inheritance mechanism guarantees we may use an object of anv valid subtype.

Another problem easily solved by order-sorted systems is the handling of run time exceptions, such as division by zero. As pointed out by Fateman [4], conventional approaches to this problem either handle such cases by causing a system exception, or by returning a type that is a union of the target type of the routine (e.g., *rational_function*) and an object such as *failed*. Both of these approaches are clumsy and require additional machinery to handle them correctly. In the order-sorted approach, this problem is solved by defining a routine for division by a *nzint* but not by an *int*. If a division by zero occurs at run time, the system will not find a routine for such division, or for any type to which zero can be lifted. The division fails with a straightforward typing error.

4 Parametrized Datatypes

The ability to parametrize datatypes is of great importance in an algebra system with its many different families of types. However, unlike many abstract datatyping algebraic systems, we believe it is not necessary to allow arbitrary objects, such as individual integers, as parameters of a type. Instead we limit such parameterized types to types

---

[6] Each arc in the DAG is directed, since the datatypes are partially ordered, so the notion of "downward" and "upward" paths in the DAG is unambiguous.

that take types as arguments, such as *matrix(poly)* or *list(list(int))*. Instead of allowing *matrix(3,4)* as a parameterized type, we consider the dimensions of the matrix, in this case {3, 4}, as an *attribute* of the particular instance of that type, that can then be tested with appropriate functions. Such attributes are stored in an *environment field* of that object. *Type constraints* can also be used as tests on these attributes. As a result, it is not necessary to handle the plethora of types that results from allowing sizes and similar attributes as parameters of a type.

5 New Datatype Definition

Our discussion so far uses a programming style that is quite similar to the *algebraic mode* of REDUCE 3, which is the REDUCE style closest to conventional algebraic manipulation. In particular, most operators are used generically. REDUCE 3 includes in addition another programming level known as *symbolic mode*, which is essentially Lisp code in a different syntax, and is the mode in which most of the core REDUCE system is written. Advanced users also use this mode to make their programs run more efficiently. One goal we have set for the new prototype is to produce a system for algebraic calculations where the code written in the more natural (i.e., generically-typed) algebraic mode, is as efficient as that written in the REDUCE 3 symbolic mode. In other words, we want there to be no performance penalty in using the abstract datatyping constructs for code development. Although we are not at that stage yet, the code we have produced is comparable in efficiency with REDUCE 3 algebraic mode code for polynomial operations. One advantage of the new approach is that the regularity of the evaluation process eliminates much of the redundancy that occurred in the older model because of the *ad hoc* way in which that model developed.

A further goal is to provide facilities for the introduction of new datatypes and their operations in a style similar to an algebraic specification. Such a definition uses a more abstract level of definition than the algebraic programming style of REDUCE 3. In addition to being closer to the mathematics being described, such a definition brings all the advantages of a rule-driven specification. In particular, optimizations for particular machine architectures, such as distributed or parallel machines are easier, since many of the specific operational details are eliminated.

We shall once again use our polynomial example to illustrate this style. The following experimental code is a fairly complete definition of a polynomial datatype equivalent to that in Section 3 and the operations of addition and multiplication on such objects.

```
subtypes zero domain ncpoly < poly, nzint zero < int < domain,
         term xpoly < ncpoly, power xterm < term,
         kernel xpower < power, identifier < kernel,
         posint negint < nzint

ranks    u + v : {poly,poly} -> poly symmetric,
         u + v : {poly,zero} -> poly,
         u + v : {xpoly,poly} -> poly,
         u + v : {xpoly,xpoly} -> poly,
         u + v : {xterm,xpoly} -> xpoly when lpow u > lpow v,
         u + v : {xterm,domain} -> xpoly,
```

```
             u*v : {poly,poly} -> poly symmetric,
             u*v : {poly,zero} -> zero,
             u*v : {poly,xpoly} -> poly,
             u*v : {xterm,domain} -> xterm,
             u*v : {xterm,xterm} -> xterm,
             u*v : {xpower,xpoly} -> xterm when mvar u > mvar v,
             u*v : {xpower,domain} -> xterm,
             u^n : {kernel,posint} -> xpower,
             u = v : {kernel,kernel} -> bool,
             u = v : {xpower,xpower} -> bool,
             u > v : {kernel,kernel} -> bool,
             u > v : {xpower,xpower} -> bool,
             lt u  : {xpoly} -> xterm,
             lpow u : {xpoly} -> xpower,
             lpow u : {xterm} -> xpower,
             ldeg u : {xpoly} -> posint,
             ldeg u : {xterm} -> posint,
             mvar u : {xpoly} -> kernel,
             mvar u : {xterm} -> kernel,
             mvar u : {xpower} -> kernel,
             lc u  : {xpoly} -> poly,
             lc u  : {xterm} -> poly,
             red u : {xpoly} -> poly,
             red u : {xterm} -> zero,
             domainp u : {poly} -> bool,
             u:poly -> domain when domainp u,
             u:poly -> zero when u = 0,
             u:ncpoly -> term when red u = 0,
             u:term -> power when lc u = 1,
             u:power -> kernel when ldeg u = 1,
             u:int -> nzint when u neq 0,
             u:nzint -> posint when u > 0

    (u:zero + v:poly):poly => v

    (u:xpoly + v:poly):poly => lt u + (red u + v)

    (u:xpoly + v:xpoly):poly =>
       if lpow u = lpow v then
          lpow u * (lc u + lc v) + (red u + red v) else
       if lpow u > lpow v then lt u + (v + red u) else
          lt v + (u + red v)

    (u:zero * v:poly):zero => u

    (u:xterm * v:domain):term => lpow u*(lc u*v)

    (u:poly * v:xpoly):poly =>  u*lt v + u*red v
```

```
(u:xterm * v:xterm):xterm =>
  if mvar u = mvar v then
    mvar u ^ (ldeg u + ldeg v)*(lc u*lc v) else
  if mvar u > mvar v then lpow u*(v*lc u) else
    lpow v*(u*lc v)
```

The *subtypes* definition is a convenient way of introducing the topology of the relevant DAG. The *when* clause in the rank definitions is an example of a constraint declaration. Such rules only fire when the constraint in the *when* clause is satisfied. The keyword *symmetric* after the target type declares the operator to be symmetric on its arguments.

At the present time, a datatype introduced in this manner has an unacceptable performance level — approximately two orders of magnitude slower than an equivalent symbolic mode definition that eliminates most type checking. A specification like the above is therefore usually accompanied by another which optimizes such calculations by mapping the above expressions into more compact and efficient items of the underlying implementation language. For example, if we represent *type* and *value* as a Lisp list, the term $2 * x^3$ would have the default representation (xterm (times (xpower (expt (variable x) (posint 3))) (posint 2))) which could be mapped into the more compact standard REDUCE format (xterm ((x . 3) posint . 2)). This possibility underlines the flexibility and power of this approach.

There are many other opportunities to optimize code generated from datatype definitions since all information is available about the structures being manipulated. For example if an operation is not defined for a particular datatype combination but is defined for all subtypes of the given type, it is possible to construct automatically the required function. Referring back to our datatype DAG, if we know how to add an object of type *domain* to one of type *term*, and also one of type *domain* to one of type *xpoly*, the system can construct a rule for adding terms of type *domain* and *ncpoly* by producing code looking like:

```
(u:domain)+(v:ncpoly) =>
    if liftable(v,term) then (u:domain)+(v:term) else
        (u:domain)+(v:xpoly)
```

A more compact definition, but one that is perhaps not as transparent, is:

```
(u:domain)+(v:ncpoly) =>
    if liftable(v,term) then u+v else u+v
```

This works because the type inferencing scheme knows that the type of v in the right hand side of the conditional expression is either *term* in the first consequent or *xpoly* in the second.

6 Future Work

The model we have described in this paper is now working effectively, and could at this stage be used as an alternative interface to the current REDUCE 3 system, with entries to the current code supported by additional interface code to each package. The semantics of the new model is well-structured and easy to describe, and therefore a user can write his or her own code with some confidence that it will work as expected. However, we have not yet achieved our goal of providing the same efficiency in the fully typed code that one can get by direct calls to specific procedures at the Lisp level in REDUCE 3. We expect that to come as we optimize some of the features of the prototype that were initially implemented in a very general way initially to give us the maximum flexibility (such as the organization of the DAG of datatypes). Other optimizations will come from recognizing those cases where static analysis improves code efficiency. For example, in many cases it is possible to determine automatically an upper bound on the resulting type and therefore reduce the amount of work done dynamically. So if it can be determined statically that two arguments of an operator are of ceiling type *poly* it would be possible to replace the construct with a call to an instance of that operator that has polynomial arguments. Given that such operations must also work for all relevant subtypes, the code will still work when called dynamically on arguments whose type has been reduced to something lower than *poly*.

Where appropriate, one could also enforce complete static type checking, adding *retracts* (i.e. functions which statically assign a lower related type to an expression and dynamically disappear when the type reduces properly) when a particular typing decision must be deferred until run time. For example, an integer division where the two arguments are *ints* could call the divisor function with a non-zero second argument after wrapping around that argument a retract which tests for this type and gives a typing error otherwise.

Another issue to consider is the evaluation strategy for type raising and lowering. Typically types need to be raised on entry to a procedure and lowered on exit. The question is whether that should be done as part of a function definition, which can create more complexity in the definition, or as part of the general function evaluation protocol, which can lead to inefficiencies because such general checking is often redundant. At the moment we are following the latter protocol since we believe that the extra programming effort involved in writing code to satisfy the former protocol would be counterproductive.

There is also a possibility of using *lazy type lowering*; in other words, of deferring the lowering of types until it is absolutely necessary. This is similar in some ways to the deferment of gcd calculations in polynomial arithmetic. In that case, it is often more efficient to compute with rational expressions that are not in lowest terms, and only do a gcd calculation when that expression is displayed or used as argument to another operator. In the typing case, one could for example defer the type lowering until one could not find a direct typing match for a particular operation. At that point, types could be lowered and the type match then completed.

The current prototype has a rather primitive notion of a module. We intend to provide over time a more general facility for module composition. This will include various forms of module importation and module parametrization. Also requirement interfaces ("views", "theories" in OBJ) will be supported. Finally, we plan to study rigorous semantic models for the object-oriented paradigm such as hidden-sorts [7] and sheaf semantics [12].

References

1. S.K. Abdali, G.W. Cherry, N. Soiffer, *An Object Oriented Approach to Algebra System Design*, Proc. 1986 Sym. on Symbolic and Algebraic Computation, pp. 24-30, ACM (1986)
2. G. Birkhoff and J.D. Lipson, *Heterogeneous Algebras*, Journal of Combinatorial Theory, 8:115–133, 1970
3. R.J. Bradford, A.C. Hearn, J.A. Padget, E. Schrüfer, *Enlarging the REDUCE Domain of Computation*, Proc. 1986 Sym. on Symbolic and Algebraic Computation, pp. 100-106, ACM (1986)
4. R. J. Fateman, *Advances and Trends in the Design and Construction of Algebraic Manipulation Systems*, Proc. International Sym. on Symbolic and Algebraic Computation, Tokyo, pp. 60-67, ACM (1990)
5. A. Fortenbacher, *Efficient Type Inference and Coercion in Computer Algebra*, Proc. International Sym. DISCO '90, pp. 56-60, Springer, Lecture Notes in Computer Science, vol. 429
6. J. Meseguer and J. Goguen, *Order-sorted Algebra Solves the Constructor-selector, Multiple Representation and Coercion Problems*, Technical Report SRI-CSL-90-06, SRI International, Computer Science Laboratory, 1990
7. J.A. Goguen, D.A. Wolfram, *On Types and FOOPS*, in W.K.R. Meersman, S. Khosla (editors), Object Oriented Databases: Analysis, Design and Construction, pp. 1-22, North Holland (1991), Proc., IFIP TC2 Conference, Windermere, UK
8. J.A. Goguen, J. Meseguer, *Order-sorted Algebra I: Equational Deduction for Multiple Inheritance, Overloading, Exceptions and Partial Operations*, Theoretical Computer Science (1992), vol. 105, no.2, pp. 217-274
9. A.C. Hearn, *A Mode Analyzing Algebraic Manipulation Program*, Proc. ACM '74, pp. 722-724, ACM (1974)
10. R. Jenks, R. Sutor, *AXIOM The Scientific Computation System*, Springer (1992)
11. D. E. Monarchi and G. I. Puhr, *A Research Topology for Object-oriented Analysis and Design*, Comm. ACM, 35(9):35–47, September 1992
12. D.A. Wolfram, J.A. Goguen, *A Sheaf Semantics for FOOPS Expressions*, Proc. ECOOP'91 Workshop on Object-based Concurrent Computation Springer (1992), Lecture Notes in Computer Science, vol. 612
13. R. E. Zippel. *The Weyl Computer Algebra Substrate*, Technical Report 90–1077, Dept. Computer Science, Cornell University, Ithaca, NY, 1990

Variant Handling, Inheritance and Composition in the ObjectMath Computer Algebra Environment

Peter Fritzson, Vadim Engelson, Lars Viklund
Programming Environments Laboratory
Department of Computer and Information Science
Linköping University, S-581 83 Linköping,
Sweden

Email: {petfr,vaden,larvi}@ida.liu.se
Phone: +46 13 281000, Fax: +46 13 282666

Abstract. ObjectMath is a high-level programming environment and modeling language for scientific computing which supports variants and graphical browsing in the environment and integrates object-oriented constructs such as classes and single and multiple inheritance within a computer algebra language. In addition, composition of objects using the part-of relation and support for solution of systems of equations is provided. This environment is currently being used for industrial applications in scientific computing. The ObjectMath environment is designed to handle realistic problems. This is achieved by allowing the user to specify transformations and simplifications of formulae in the model, in order to arrive at a representation which is efficiently solvable. When necessary, equations can be transformed to C++ code for efficient numerical solution. The re-use of equations through inheritance in general reduces models by a factor of two to three, compared to a direct representation in the Mathematica computer algebra language. Also, we found that multiple inheritance from orthogonal classes facilitates re-use and maintenance of application models.

1 Background

The goal of the ObjectMath project is to develop a high-level programming environment that enhances the program development process in scientific computing, initially applying the environment to advanced machine element analysis (a machine element can loosely be defined as "some important sub-structure of a machine"). There is a clear need for such tools as the current state of the art in the area is still very low-level. Most scientific software is still developed in FORTRAN the traditional way, manually translating mathematical models into procedural code and spending much time on debugging and fixing convergence problems. See [7] for a detailed discussion. The ObjectMath programming environment is centered around the ObjectMath modeling language which combines object-oriented constructs with computer algebra. In this paper we describe the ObjectMath language and report experiences from using it for modeling and analyzing realistic machine element analysis problems.

The current practice in mechanical analysis software modeling and implementation can be described as follows: Theory development is usually done manually, using only pen and paper. Equations are simplified and rewritten by hand to prepare for solution of relevant properties. This includes a large number of coordinate transformations, which are laborious and error prone. In order to perform numerical computations the mathematical model is implemented in some programming language, most often FORTRAN. Existing numerical subroutines might be used, but data-flow between routines must still be written by hand. Tools such as finite element analysis (FEM) or multibody systems analysis programs can at best be used for limited subproblems as the total computational problem usually is too complex. The program development process is highly iterative. If correlation with experiments is not achieved the theoretical model has to be refined, which subsequently requires changes in the numerical program. Numerical convergence problems often arise, since the problems usually are non-linear. Often as much as 50-75% of the total time of a project is spent on writing and debugging FORTRAN programs.

The ideal tool for modeling and analysis in scientific computing should eliminate these low-level problems and allow the designer to concentrate on the modeling aspects. Some of the properties of a good programming environment for modeling and analysis in scientific computing are:

- The user works at a high level of abstraction.
- Modeling is done using formulae and equations, with good structuring support (such as object-oriented techniques).
- Support for symbolic computation is provided. One example is automatic symbolic transformation between different coordinate systems.
- The environment should provide support for numerical analysis.
- The environment should support changes in the model. A new iteration in the development cycle should be as painless as possible.

Symbolic computation capabilities provided by computer algebra systems [4] are essential in a high-level programming environment for scientific computing. Some existing computer algebra systems are Macsyma [15], Reduce [9], Maple [2], or Mathematica [18]. However, their support for structuring complex models is too weak.

In the following sections we describe the ObjectMath programming environment and language, especially with respect to uses of inheritance, composition and variants. Finally, we compare ObjectMath with related work and present conclusions.

2 The ObjectMath Programming Environment

The ObjectMath programming environment is designed to be easy to use for application engineers, e.g. in mechanical analysis who are not computer scientists. It is interactive and includes a graphical browser for viewing and editing inheritance hierarchies, an application oriented editor for editing ObjectMath equations and formulae, the Mathematica computer algebra system for symbolic computation, support for generation of numerical code from equations, an interface for calling external functions, and a class library. The graphical browser is used for viewing and editing ObjectMath inheritance hierarchies. ObjectMath code is automatically translated into Mathematica code and symbolic computations can be done interactively in Mathematica. Figure 1 shows the environment during a typical session.

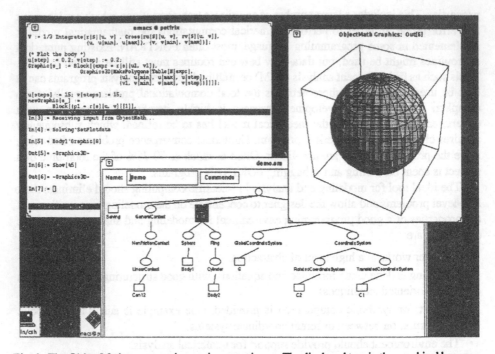

Fig. 1. The ObjectMath programming environment in use. The displayed tree in the graphical browser window shows the inheritance hierarchy of classes in the model, the text windows to the left show the currently edited class definition and the Mathematica window for symbolic computations, whereas the visualized object (Body1 in the window at upper right) is instantiated from a specialized Sphere class.

The programming environment currently runs on Sun workstations under the X window system. Recently additional capabilities has been added to the environment, such as multiple inheritance, composing objects of parts, variant handling of models and classes and extended code generation support. This is described later in the paper. A description of the implementation of an earlier version of ObjectMath can be found in [17]. Initial experience from using an early version of ObjectMath is reported in [8].

3 The ObjectMath Language

ObjectMath is both a language and a programming environment. The current ObjectMath language has recently been enhanced with features for *multiple inheritance* and modeling *part-of* relations between objects. Both of these features has turned out to be important in realistic application models. An early version of the ObjectMath language only supported single inheritance [16].

The ObjectMath language is an hybrid modeling language, combining object-oriented constructs with a language for symbolic computation. This makes ObjectMath a suitable language for implementing complex mathematical models, such as those used in machine element analysis. Formulae and equations can be written with a notation that closely resembles conventional mathematics, while the use of object-oriented modeling makes it possible to structure the model in a natural way.

We have chosen to use an existing computer algebra language, Mathematica, as a basis for ObjectMath. Building an object-oriented layer on top of an existing language is not ideal, but this decision was made to make it possible to quickly implement a stable prototype that could be used to test the feasibility of our approach. The current ObjectMath language is a prototype and we plan to redesign the whole language in the future, making it independent of Mathematica.

Mathematica was chosen over other similar systems partly because it was already in use by our industrial partner, and partly because of its excellent support for three-dimensional graphics. The relationship between Mathematica and ObjectMath can be compared to that between C and C++. The C++ programming language is basically the C language augmented with classes and other object-oriented language constructs. In a similar way, the ObjectMath language can be viewed as an object-oriented version of the Mathematica language.

3.1 Object-Oriented Modeling

When working with a mathematical description that consists of hundreds of equations and formulae, for instance one describing a complex machine element, it is highly advantageous to *structure* the model. A natural way to do this is to model machine elements as *objects*. Physical bodies, e.g. rolling elements in a bearing, are modeled as separate objects. Properties of objects like these might include a surface description, a normal to the surface, forces and moments on the body, and a volume. These objects might define operations such as finding all contacts on the body, computing the forces on or the displacement of the body, and plotting a three-dimensional picture of the body.

Abstract concepts can also be modeled as objects. Examples of such concepts are coordinate systems and contacts between bodies. The coordinate system objects included in the ObjectMath class library define methods for transforming points and vectors to other coordinate systems. Equations and formulae describing the interaction between different bodies are often the most complicated part of problems in machine element analysis. This makes it practical to encapsulate these equations in separate *contact objects*. One advantage of using contact objects is that we can substitute one mathematical contact model for another simply by plugging in a different kind of contact object. The rest of the model remains completely unchanged. When using such a model in practice, one often needs to experiment with different contact models to find one which is exact enough for the intended purpose, yet still as computationally efficient as possible. The ObjectMath class library contains several different contact classes.

The use of *inheritance* facilitates *reuse* of equations and formulae. For example, a cylindrical roller element can inherit basic properties and operations from an existing general cylinder class, refining them or adding other properties and operations as necessary. Inheritance may be viewed not only as a sharing mechanism, but also as a concept specialization mechanism. This provides another powerful mechanism for structuring complex models in a comprehensive way. Iteration cycles in the design process can be simplified by the use of inheritance, as changes in one class affects all objects that inherits from that class. *Multiple inheritance* facilitates the maintenance and construction of classes which need to combine different orthogonal kinds of functionality.

The *part-of* relation is important for modeling objects which are *composed* of other objects. This is very common in practice. For example, a car is composed of parts such as wheels, motor, seats, brakes, etc. This modeling facility was missing from the first version of ObjectMath, which caused substantial problems when applying the system to more complicated applications. Note that the notions of *composition* of parts, and *inheritance* are quite different and orthogonal concepts. Inheritance is used to model specialization hierarchies, whereas composition is used to group parts within container objects while still preserving the identity of the parts. Thus, composition has nothing to do with specialization. Sometimes these concepts are confused; there have been attempts to use inheritance to implement composition, usually with disastrous results for the model structure.

Object-oriented techniques make it practical to organize repositories of reusable software components. The ObjectMath class library is one example of such a software component repository. It contains general classes, for instance material property classes, different contact classes and classes for modeling simple bodies such as cylinders and spheres.

3.2 ObjectMath Classes and Instances

A *CLASS* declaration declares a class which can be used as a template when creating objects. ObjectMath classes can be parameterized. The ObjectMath *INSTANCE* declaration is, in a traditional sense both a declaration of class and a declaration of one object (instance) of this class. This makes the declaration of classes with singleton instances compact.

An array containing a symbolic number of objects can be created from one *INSTANCE* declaration by adding an index variable in brackets to the instance name. This allows for the creation of large numbers of nearly identical objects, for example the rolling elements in a rolling bearing. To represent differences between such objects, functions (methods) that are dependent upon the array index of the instance can be used. The implementation makes it possible to do computations with a symbolic number of elements in the array.

The bodies of ObjectMath *CLASS* and *INSTANCE* declarations contain formulae and equations. Mathematica syntax is used for these. Note that the Mathematica context mark, `, denotes remote access, i.e. X`y is the entity y of the object X.

There are some limitations in the current prototype implementation of the ObjectMath language for ease of implementation. Most notable is that a mechanism for enforcing encapsulation is missing. Encapsulation is desirable because it makes it easier to write reusable software since the internal state of an object may only be accessed through a well defined interface.

On the other hand, access to equations defined in different classes is necessary when specifying computer-algebra related transformations, substitutions and simplifications. Such symbolic operations on equations need to access the source representation of those equations. Since transformations are specified manually and executed interactively, equations should not be hidden from the user. Enforcing strict encapsulations of equations is only possible in systems which do not allow manual manipulation of equations.

3.3 Single Inheritance Examples: Cylinders and Spheres

In this section we use some of the classes of physical objects from an ObjectMath model to exemplify the ObjectMath language. In addition to classes describing bodies with different geometry depicted in the inheritance hierarchy of Figure 2, there are additional classes which describe interactions between bodies and coordinate systems. The full inheritance hierarchy is displayed in the lower part of Figure 1 and the full class definitions are available in Appendix A. Note that the inheritance hierarchy usually is edited graphically so that the user does not have to write the class headers by hand.

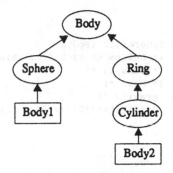

Fig. 2. An inheritance hierarchy of classes for modeling bodies with different geometries such as cylinders and spheres.

At the top of this inheritance sub-hierarchy is the class Body, which contains definitions and functions common to all bodies. This includes the function r [S] [u_, v_] which describes the geometry of a body through a parametric surface; partial differentials of this surface, a general formula for the volume of a body, a function for plotting 3D graphic images of bodies, etc. This class has two parameters: S which is the name of the body-centered coordinate system, and B which is the set of bodies this body is interacting with. For example, when inherited down to Body1, the parametric surface function is instantiated to r [C1] [u_, v_] since C1 is the body-centered coordinated system of Body1.

```
CLASS Body(S, B)
(* Geometry defined through a parametric surface *)
 r[S][u_, v_];
 r[s_][u_, v_] := S`TransformPoint[r[S][u, v], s];
 ...
(* Partial differentials of surface *)
 ru[S][u_, v_] := D[r[S][u1, v1], u1] /. { u1 -> u, v1 -> v };
 rv[S][u_, v_] := D[r[S][u1, v1], v1] /. { u1 -> u, v1 -> v };
(* Volume of body *)
 V := 1/3 Integrate[r[S][u, v] . Cross[ru[S][u, v], rv[S][u, v]],
                {u, u[min], u[max]}, {v, v[min], v[max]}];
(* Graphic method for plotting bodies *)
 Graphic[s_] := ...
(* Forces and moments, equations for equilibrium, etc ...)
 ...
END Body;
```

The class Sphere contains a specialization of the parametric surface function to give the special geometry of a sphere. Finally the class and instance Body1 instantiates a specific sphere, which is in contact with the cylinder. It actually rests on top of the cylinder, exerting a contact force on the cylinder, as is shown in Figure 3. The class Body is also specialized as class Ring, which is further specialized as class Cylinder and instance Body2. The definitions of these classes can be found in Appendix A.

```
CLASS Sphere(S, B) INHERITS Body(S, B)
  R;  (* Radius *)
  ...
  u[min] := 0; u[max] := Pi;    v[min] := 0; v[max] := 2 Pi;
  r[S][u_, v_] := R * { Sin[u]*Cos[v],  Sin[u]*Sin[v],  Cos[u] };
  ...
END Sphere;

INSTANCE Body1 INHERITS Sphere(C1, {Body2})
  Con[Body2] := Con12;  (* Define contact from this body to Body2 *)
  rho;                  (* Density *)
  m := rho V;           (* Mass *)
  (* External force from gravity *)
  F$[S1][Ext][1] := 0;   F$[S1][Ext][2] := 0;   F$[S1][Ext][3] := - g m;
END Body1;
```

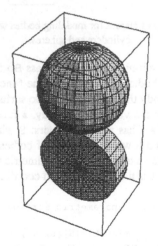

Fig. 3. The sphere rests on top of the cylinder.

3.4 Examples of Multiple Inheritance

Multiple inheritance is useful when combining orthogonal concepts. This is exemplified in Figure 2.

The filled lines denote single inheritance, whereas the dotted lines denote additional inheritance, i.e. we have multiple inheritance. Since material properties and geometry are orthogonal concepts there are no collisions between inherited definitions in this example.

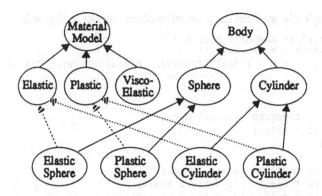

Fig. 4. Multiple inheritance hierarchy of bodies of different materials and geometries.

If instead we are forced to use a single-inheritance hierarchy as in Figure 5, we are have to repeat the equations describing material properties twice. This is bad model engineering since it would force us to repeat any changes to the material model twice. Also, this precludes the creation of pure material library classes which can be combined with other classes.

Here, the material equations describing elasticity or plasticity have to be repeated twice. This model structure is harder to maintain when changes are introduced into the model.

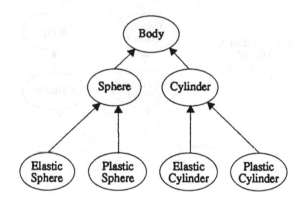

Fig. 5. Single inheritance version of the material-geometry model of Figure 2.

The general form of multiply inheriting class declarations follows below:

```
CLASS Child INHERITS Parent1,Parent2,... ParentN
...
END Child;
```

If there are conflicts between inherited definitions, e.g. if they have the same name, definitions from Parent1 will override definitions from Parent2, which will override definitions from Parent3, etc. The special case when only Parent1 is present corresponds to single inheritance.

Some example classes from the material-geometry model in Figure 2:

```
CLASS Sphere(S,B) INHERITS Body(S,B)
  R;  (* Radius - a variable *)
  r[S][u_, v_] := R * { Sin[u]*Cos[v],  Sin[u]*Sin[v],  Cos[u] };
  ....
END Sphere;

CLASS Elastic INHERITS Material_Model
  Force := k1 * delta;
END Elastic;

CLASS Plastic INHERITS Material_Model
  Force := k2 * Limit   /; delta > Limit;
  Force := k2 * delta   /; delta <= Limit;
END Plastic;

CLASS Elastic_Sphere INHERITS Sphere, Elastic
  ...
END Elastic_Sphere;
```

Another useful case of *multiple-inheritance* is shown below, where an integration method is inherited into classes from two separate inheritance hierarchies: a hierarchy of contact classes containing integrated forces and moments between bodies, and classes describing bodies themselves including integrated moments of inertia, mass and volumes.

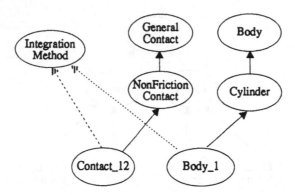

Fig. 6. Example of multiple inheritance of a numerical integration method into two different classes. Here to be used for integrating forces or volumes. One class contains contact equations; another contains volumes, moments and equilibrium equations.

The entities inherited from class Integration_Method will typically be a combination of entities such as procedural code, transformation rules (e.g. for symbolic differentiation), etc.

3.5 Modeling Part-Of Relations

As mentioned previously, the *part-of* relation is important for modeling objects which are composed of other objects, also noting that this concept is orthogonal to the concept of *inheritance* which is used to represent specialization. For example, a bicycle contain parts such as wheels, frame, pedals, etc. A rolling bearing contain inner ring, outer ring, rolling elements, lubrication fluid, etc.

The ObjectMath syntax for expressing composition using the *part-of* relation is exemplified below for a `Bicycle` class:

```
CLASS Bicycle(C,P)
  ...
  PART frontwheel INHERITS Wheel(P);
  PART rearwheel INHERITS Wheel(P);
  PART frame INHERITS Body;
  ...
END Bicycle;
```

Another example is a small section of a class from a rather advanced model of four-body interaction developed by our industrial partner:

```
CLASS FourCylCurvSegBody(cBody, cmBody, cRef, cInert) INHERITS
                        DynRigidBody(cBody, cmBody,cRef, cInert)
  ...
  PART sRaceF INHERITS CylinderSeg(cBody);
  PART sRaceB INHERITS CylinderSeg(cBody);
  PART sSegL  INHERITS CylinderSeg(cBody);
  PART sSegR  INHERITS CylinderSeg(cBody);
  ...
END FourCylCurvSegBody;
```

4   Variants of Classes

During the development of complex mathematical models there is often a need to explore different variants of solution strategies and formulations of equations. One would like to experiment with alternative ways of expressing equations and transformations within a certain class and still keep the previous version of the class definition in the model. Each new variant of a class can of course be tried out by creating an entirely new model where all classes except one are identical compared to the previous model.

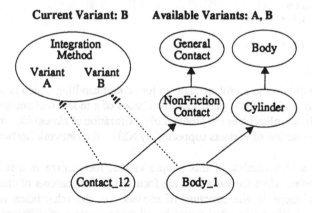

Fig. 7. The same example as in Figure 6 but with two available variants A and B. The currently active variant of the model is B, which will select Variant B of Integration Method and the default of other classes.

However, this solution is very undesirable from a software engineering point of view since it leads to a proliferation of almost identical models. For example, if a class within one model is modified, one will have to manually modify a large number of other, almost similar, models. This is both error-prone and cumbersome. The updating problem can of course be remedied by creating library modules which can be imported into models. The library solution is often not desirable however, since it implies additional restructuring work in creating libraries, especially if many classes are used only within a single model.

In order to provide a convenient solution to the variant problem we have introduced a mechanism in the ObjectMath environment which allows several variants of classes within a model.

The user specifies the names of all allowed variants within the model. For each class, it is possible to have one or more variants of the class body labeled using some of the specified variant names. There is also the notion of *current variant* of the model, which tells which variant of a class body should be selected when several are available. The most common case is that only one default unlabeled variant is available for a class, as is the case for all classes except Integration_Method in Figure 7. Then the single variant should of course be selected for inheritance or symbolic computations. If several variants are available, but no one matches, the unlabeled default should be selected. If no unlabeled variant is available the first one should be selected.

This mechanism is currently implemented as conditional inclusion of class-definitions using C-preprocessor style directives as shown below. If VariantA is the currently active model variant then the first definition of Force is selected; if VariantB is active then the second definition is selected, otherwise the third. An alternative implementation instead of the conditional inclusion mechanism would be the keep each variant separate. This has the advantage of looking somewhat cleaner, but the disadvantage of creating problems to keep the common parts of the variants consistent when the model is modified, which is why we did not select this alternative.

```
CLASS Plastic INHERITS Material_Model
#ifdef VariantA
  Force := k2 * LimitA - 35
#elif VariantB
  Force := k3 * Limit +100
#elif 1
  Force := k2 * Limit
#endif
END Plastic;
```

In this way we obtain a graceful mechanism for variant handling which is not noticed when it is not needed, and can be applied to a few classes of a model without disturbing the rest of the model. It also eliminates the problem of proliferation of almost identical models. This is similar to the notion of variants supported by NSE – the Network Software Environment [3].

However, a shortcoming of this simple variant mechanism is that it only handles variants of models when the variations are focussed to the contents of classes. It does not cope with variations in the structure of models, i.e. the inheritance and composition structure. This might be handled better by allowing models of different structure to be defined as different configurations of class definitions imported from a class library, rather extending the current variant mechanism.

5 Code Generation from ObjectMath

There is a clear need to be able to generate efficient numerical implementations from ObjectMath models. The ObjectMath system provides two alternative mechanisms for code generation. The first mechanism will translate a set of function definitions in ObjectMath syntax to corresponding C++ code, either by translating the code in a function as it is, or applying all available symbolic transformations producing a very large expression, which then is optimized using common subexpression elimination and used.

The second mechanism makes it possible to include hand-written C++ functions in ObjectMath classes as shown below. This is described in somewhat more detail in [16].

```
CLASS FindingContacts(B, Body1, Body2)
  INHERITS GeneralContacts(B, Body1, Body2)
  ...
  (* Declaration of a method written in C++ *)
  EXTERNAL "C++" ppa[su1_, v1_];
  ...
END FindingContacts;
```

The first translation mechanism is now in regular use by our industrial partner SKF Engineering & Research Centre to translate the computational parts of whole models (i.e. not the parts which specify symbolic transformations) into C++ code. For example, a two-body interaction model of approximately 150 kbytes of ObjectMath code was translated into 450 kbytes of Mathematica code, of which the computational part was translated into 440 kbyte of C++ code. Still, this C++ code is rather compact due to the use of overloading matrix and vector operations on arithmetic operators such as "*" and "+". The re-use due to inheritance of ObjectMath classes and composition of parts makes the model approximately a factor of three more compact than if it would have been represented directly in Mathematica code.

The translator to C++ code is implemented in Mathematica. The whole translation process for the twobody model takes approximately 2 hours on a Sun Sparcstation ELC (a 20 Mips workstation). Most of the time is spent on symbolic transformations and simplifications, and on common subexpression elimination on the very large expressions generated during the transformations. Separate procedures consisting of many small statements are usually generated from such large expressions. To be able to generate efficient type-correct C++ code and to resolve some ambiguities, it was necessary to add type declarations of variables and functions to ObjectMath models.

Our aim is to eventually provide fully automatic translation of whole models. Currently, the user has to write calls to explicitly invoke the translator on each ObjectMath function, expression, or set of equations that should be translated into C++. In order to achieve this goal, additional type and structure information in the form of declarations will have to be added to models, as well as extending the capabilities of the translator itself.

6 Applications of ObjectMath

So far the ObjectMath environment has been tested and evaluated by modeling and analyzing several different problems, two of which are briefly described here:

- A three-dimensional model describing a rolling bearing, see Figure 8.

• An advanced surface description model, used for rolling bearing analysis.

The rolling bearing example was designed to be used as a realistic but manageable test case for the ObjectMath language and environment. It consists of over 200 equations and can easily be extended, for instance with more realistic contact models. Fritzson *et. al.* describes both the mathematical model [6] and the ObjectMath implementation [8]. Figure 8 shows a three-dimensional view of the bearing, automatically generated from equations in the ObjectMath model.

The bearing consists of an inner ring, an outer ring, and a number of rolling elements in between. Each of these correspond to an ObjectMath instance declaration, except for the rolling elements, where the ObjectMath feature of declaring an array of instances is used.

The second implemented ObjectMath model mentioned here is being used in the development of an advanced surface description model which is used to model the interaction between bodies. If ObjectMath had not been available, this implementation would have been developed by hand and coded in FORTRAN.

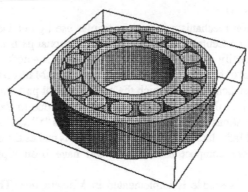

Fig. 8. A view of the example rolling bearing, automatically generated from an equational description.

7 Related Work

Object-oriented modeling has been used for a number of different application areas, and mathematical modeling in general [13]. ObjectMath also has a lot in common with conventional computer algebra systems (in particularly Mathematica). But, as mentioned in the introduction, these systems usually lack capabilities that are important for advanced scientific computing, most notably good structuring support, e.g. as provided by object oriented notions.

There are other computer algebra systems which are to some extent based on object-oriented concepts. One of the better known examples is AXIOM [11] and its predecessor SCRATCHPAD [10] which have type systems which in some sense are object-oriented, even though the language constructs provided are different from the ones usually found in object-oriented languages. Another example is the TASSO system [12]. However, the object-oriented notions supported by the ObjectMath language is primarily intended for modeling in scientific computation, for instance in engineering applications. Most other computer algebra systems employing object-oriented techniques focuses on using these techniques for modeling mathematical objects, implementing algebraic algorithms etc.

Omola [1] is an object-oriented modeling language with single-inheritance, also supporting composition. It is primarily designed as a simulation modeling language for continuous time systems, especially in the application area of control engineering. It contains no computer algebra support. A similar system, called ASCEND, is described by Piela *et. al.* [14].

8 Conclusions

There is a strong need for efficient high-level structuring tools and languages in scientific computing. We feel that the ObjectMath system is highly successful in satisfying part of this need. Complex mathematical equations and functions can be expressed at a high level of abstraction rather than as procedural code. ObjectMath integrates computer-algebra language features with object-oriented notions such as multiple inheritance, classes, as well as composition for modeling structured objects. Object-oriented notions allow better structure of models and permit reuse of equations. Variant support permits convenient incorporation of multiple solution strategies and different equational representations within a single model. Our conclusion that these facilities are important is supported by the successful modeling and analysis of several realistic applications.

9 Acknowledgments

Henrik Nilsson improved the english in this paper. Johan Herber has implemented the Emacs support in ObjectMath. Joakim Malmén has done most of the implementation work on the most recent version of the ObjectMath to C++ translator. Dag Fritzson has provided mathematical expertise and developed the majority of the currently available ObjectMath models. The ObjectMath parser is partly based on a Mathematica parser in Common Lisp written by Richard Fateman [5].

This work is supported by The Swedish Board for Technical and Industrial Development (NUTEK), in part under the Esprit project PREPARE. It is also supported in part by SKF Engineering & Research Centre, Nieuwegein, The Netherlands.

10 References

[1] Mats Andersson. Omola – an object-oriented language for model representation. Licentiate thesis, Department of Automatic Control, Lund Institute of Technology, P.O. Box 118, S-221 00 Lund, Sweden, May 1990.

[2] Bruce W. Char, Keith O. Geddes, Gaston H. Gonnet, Benton L. Leong, Michael B. Monagan, and Stephen M. Watt. *Maple V Language Reference Manual*. Springer-Verlag, 1991.

[3] Willaim Courington, Jonathan Feiber, and Masahiro Honda. NSE highlights. In Mark Hall and John Barry, editors, *The Sun Technology Papers*. Springer-Verlag, 1990.

[4] J.H. Davenport, Y. Siret, and E. Tournier. *Computer Algebra – Systems and Algorithms for Algebraic Computation*. Academic Press, 1988.

[5] Richard J. Fateman. A Mathematica parser in Common Lisp. Personal Communications. Computer Science Division, Dept. of Electrical Engineering and Computer Science, University of California, Berkeley, California, 1991.

[6] Dag Fritzson and Peter Fritzson. Equational modeling of machine elements – applied to rolling bearings. Technical Report LiTH-IDA-R-91-05, Department of Computer and Information Science, Linköping University, S-581 83, Linköping, Sweden, March 1991.

[7] Peter Fritzson and Dag Fritzson. The need for high-level programming support in scientific computing applied to mechanical analysis. *Computers & Structures*, 45(2):387–395, 1992. Also as technical report LiTH-IDA-R-91-04,Department of Computer and Information Science, Linköping University, S-581 83, Linköping, Sweden.

[8] Peter Fritzson, Lars Viklund, Johan Herber, and Dag Fritzson. Industrial application of object-oriented mathematical modeling and computer algebra in mechanical analysis. In Georg Heeg, Boris Magnusson, and Bertrand Meyer, editors, *Technology of Object-Oriented Languages and Systems – TOOLS 7*, pages 167–181. Prentice Hall, 1992.

[9] A. C. Hearn. *REDUCE-3 User's Manual, version 3.3*. The Rand Corporation, Santa Monica, Califoria, USA, 1987. Publication CP78 (7/78).

[10] Richard D. Jenks. A primer: 11 keys to new SCRATCHPAD. In John Fitch, editor, *Proceedings of EUROSAM 84/International Symposium on Symbolic and Algebraic Computation*, July 1984.

[11] Richard D. Jenks and Robert S. Sutor. *AXIOM – The Scientific Computation System*. Springer-Verlag, 1992.

[12] C. Limongelli, A. Minola, and M. Temperini. Design and implementation of symbolic computation systems. In P. W. Gaffney and E. N. Houstis, editors, *Programming Environments for High-Level Scientific Problem Solving*, pages 217–226. North-Holland, 1992. Proceedings of the IFIP TC2/WG 2.5 Working Conference on Programming Environments for High-Level Scientific Problem Solving.

[13] Thomas W. Page, Jr., Steven E. Berson, William C. Cheng, and Richard R. Muntz. An object-oriented modeling environment. In *OOPSLA'89 Conference Proceedings*, pages 287–296, 1989.

[14] P. C. Piela, T. G. Epperly, K. M. Westerberg, and A. W. Westerberg. ASCEND: An object-oriented computer environment for modeling and analysis: The modeling language. *Computers & Chemical Engineering*, 12(7):53–72, 1991.

[15] Symbolics Inc. *MACSYMA Reference Guide*, 1985.

[16] Lars Viklund and Peter Fritzson. An object-oriented language for symbolic computation – applied to machine element analysis. In Paul S. Wang, editor, *Proceedings of the International Symposium on Symbolic and Algebraic Computation*, pages 397–405. ACM Press, 1992.

[17] Lars Viklund, Johan Herber, and Peter Fritzson. The implementation of ObjectMath – a high-level programming environment for scientific computing. In Uwe Kastens and Peter Pfahler, editors, *Compiler Construction – 4th International Conference, CC '92*, volume 641 of *Lecture Notes in Computer Science*, pages 312–318. Springer-Verlag, 1992.

[18] Stephen Wolfram. *Mathematica – A System for Doing Mathematics by Computer*. Addison-Wesley Publishing Company, second edition, 1991.

Appendix A. The bearing model expressed in the ObjectMath language

This appendix contains the full model SphereCylinder referred to earlier in this paper.

```
MODEL SphereCylinder;
PACKAGES "Vectors`", "Subst`", "Misc`", "Plot`";

g; (* Gravity constant *)

CLASS AbstractCoordinateSystem
  ...
END AbstractCoordinateSystem;

CLASS CoordinateSystem(Reference, A, R) INHERITS AbstractCoordinateSystem
  ...
END CoordinateSystem;

CLASS TranslatedCoordinateSystem(Reference, R) INHERITS
  ...
END TranslatedCoordinateSystem;

INSTANCE C1 INHERITS TranslatedCoordinateSystem(G, {0,0,d})
END C1;

CLASS RotatedCoordinateSystem(Reference, Phi) INHERITS ...
  ...
END RotatedCoordinateSystem;

INSTANCE C2 INHERITS RotatedCoordinateSystem(G, {Pi/2,0,0})
END C2;

CLASS GlobalCoordinateSystem INHERITS AbstractCoordinateSystem
 FromGlobal := {this};
END GlobalCoordinateSystem;

INSTANCE G INHERITS GlobalCoordinateSystem
END G;

CLASS Body(S, B)
 (* Surface *)
u[min]; u[max]; v[min]; v[max]; (* virtual *)
r[S][u_, v_]; (* virtual *)
r[s_][u_, v_] := S`TransformPoint[r[S][u, v], s];

 (* Partial differentials of surface *)
ru[S][u_, v_] := D[r[S][u1, v1], u1] /. { u1 -> u, v1 -> v };
rv[S][u_, v_] := D[r[S][u1, v1], v1] /. { u1 -> u, v1 -> v };

 (* Normal of surface *)
n[S][u_, v_] := Cross[ru[S][u, v], rv[S][u, v]] /
        AbsVector[Cross[ru[S][u, v], rv[S][u, v]]];
n[s_][u_, v_] := S`TransformVector[n[S][u, v], s];

 (* Volume of body *)
V := 1/3 Integrate[r[S][u, v] . Cross[ru[S][u, v], rv[S][u, v]],
        {u, u[min], u[max]}, {v, v[min], v[max]}];

 (* Plot the body *)
u[step] := 0.2; v[step] := 0.2;
Graphic[s_] := Block[{expr = r[s][u1, v1]},
        Graphics3D[MakePolygons[Table[N[expr],
                {u1, u[min], u[max], u[step]},
                {v1, v[min], v[max], v[step]}]]]];

 (* Coordinate system to solve equilibrium in *)
S1 := S; (* May be overridden in subclasses *)

 (* Contact objects *)
```

```
Con[b_]; (* virtual *)

(* Forces and moments *)
F[S1][b_] := Array[F$[S1][b], 3];
F[s_][b_] := AccessMember[S1, "TransformVector"][F[S1][b], s];
M[S1][b_] := Array[M$[S1][b], 3];
M[s_][b_] := AccessMember[S1, "TransformVector"][M[S1][b], s];

(* External loading *)
F[S1][Ext] := Array[F$[S1][Ext], 3];
M[S1][Ext] := Array[M$[S1][Ext], 3];
p[S1][Ext] := {0, 0, 0};

(* Equilibrium *)
Eq[1] := Plus @@ F[S1] /@ B + F[S1][Ext] == {0, 0, 0};
Eq[2] := Plus @@ ( M[S1][#] +   Cross[AccessMember[AccessMember[#, "Con"]
      [this], "p"][S1], F[S1][#]] &) /@ B + M[S1][Ext] + Cross[p[S1][Ext],
         F[S1][Ext]]== {0, 0, 0};
END Body;

CLASS Ring(S, B) INHERITS Body(S, B)
 (* Variables *)
 Rb; (* Inner radius *)
 Ra; (* Outer radius *)
 L; (* Width *)

 (* Definition of parametric surface *)
 u[min] := 0; u[max] := 4;   v[min] := 0; v[max] := 2 Pi;

 x1[u_] := Rb + (Ra - Rb) u;
 x2[u_] := Ra;
 x3[u_] := Rb + (Ra - Rb) (3 - u);
 x4[u_] := Rb;

 x[u_] := x1[u] /; 0 <= u < 1;
 x[u_] := x2[u] /; 1 <= u < 2;
 x[u_] := x3[u] /; 2 <= u < 3;
 x[u_] := x4[u] /; 3 <= u <= 4;

 z1[u_] := 1 / 2 L;
 z2[u_] := -L (u - 3 / 2);
 z3[u_] := - 1 / 2 L;
 z4[u_] := L (u - 7 / 2);

 z[u_] := z1[u] /; 0 <= u < 1;
 z[u_] := z2[u] /; 1 <= u < 2;
 z[u_] := z3[u] /; 2 <= u < 3;
 z[u_] := z4[u] /; 3 <= u <= 4;

 (* The is the complete rotation surface *)
 r[S][u_, v_] := { x[u] Cos[v],   x[u] Sin[v],   z[u] };

 (* Partial differentials of surface *)
 Dx[u_] := x1'[u] /; 0 <= u < 1;
 Dx[u_] := x2'[u] /; 1 <= u < 2;
 Dx[u_] := x3'[u] /; 2 <= u < 3;
 Dx[u_] := x4'[u] /; 3 <= u <= 4;

 Dz[u_] := z1'[u] /; 0 <= u < 1;
 Dz[u_] := z2'[u] /; 1 <= u < 2;
 Dz[u_] := z3'[u] /; 2 <= u < 3;
 Dz[u_] := z4'[u] /; 3 <= u <= 4;

 x' := Dx;   z' := Dz;
 u[step] := 1; v[step] := Pi/15;   u[steps] := 4; v[steps] := 31;
END Ring;

CLASS Cylinder(S, B) INHERITS Ring(S, B)
 Rb := 0;   u[max] := 3;
END Cylinder;
```

```
INSTANCE Body2 INHERITS Cylinder(C2, {Body1})
  Con[Body1] := Con12;
END Body2;

CLASS Sphere(S, B) INHERITS Body(S, B)
  (* Variables *)
  R; (* Radius *)
  (* Definition of parametric surface *)
  u[min] := 0; u[max] := Pi;
  v[min] := 0; v[max] := 2 Pi;
  r[S][u_, v_] := R { Sin[u] Cos[v], Sin[u] Sin[v], Cos[u] };
  u[step] := Pi/15; v[step] := Pi/15;  u[steps] := 16; v[steps] := 31;
END Sphere;

INSTANCE Body1 INHERITS Sphere(C1, {Body2})
  Con[Body2] := Con12;
  rho;          (* Density *)
  m := rho V;   (* Mass *)
  (* External force from gravity *)
  F$[S1][Ext][1] := 0;   F$[S1][Ext][2] := 0;   F$[S1][Ext][3] := - g m;
  (* No external moment *)
  M[S1][Ext] := {0, 0, 0};
END Body1;

CLASS GeneralContact(S, Body1, Body2)
  (* Variables: *)
  delta;    u[Body1]; v[Body1];    u[Body2]; v[Body2];
  (* The contact point: *)
  p[S] := Body1`r[S][u[Body1], v[Body1]] + 1/2 delta Body1`n[S][u[Body1],
  v[Body1]];
  p[s_] := S`TransformPoint[p[S], s];
  (* Equations *)
  Eq[1] := Body1`n[S][u[Body1], v[Body1]] + Body2`n[S][u[Body2], v[Body2]]
             == { 0, 0, 0 };
  Eq[2] := delta Body1`n[S][u[Body1], v[Body1]] ==
       Body2`r[S][u[Body2], v[Body2]] - Body1`r[S][u[Body1], v[Body1]]];
  Eq[3] := Body1`F[S][Body2] + Body2`F[S][Body1] == { 0, 0, 0 };
  Eq[4] := Body1`M[S][Body2] + Body2`M[S][Body1] == { 0, 0, 0 };
END GeneralContact;

CLASS NonFrictionContact(B, Body1, Body2) INHERITS GeneralContact(B, Body1,
Body2)
  k[delta_];    (* The contact stiffness function *)
  (* Equations: *)
  Eq[10] := Body1`F[S][Body2] == k[delta] Body1`n[S][u[Body1], v[Body1]];
  Eq[11] := Body1`M[S][Body2] == { 0, 0, 0 };
END NonFrictionContact;

CLASS LinearContact(B, Body1, Body2) INHERITS NonFrictionContact(B, Body1,
Body2)
  k[delta_] := Cstiff[delta] delta - 1;
  Cstiff[delta_] := Csoft /; delta > 0;
  Cstiff[delta_] := Chard /; delta <= 0;
END LinearContact;

INSTANCE Con12 INHERITS LinearContact(C1, Body1, Body2)
END Con12;

INSTANCE Solving
  SetPlotdata := ( Body1`R := 1; Body2`Ra := 1; Body2`L := 0.6;
          C1`d := Body1`R + Body2`Ra;
          );
  ClearPlotdata := ( Clear[Body2`Ra, Body2`L, Body1`R, C2`d]; );
  Force := Flatten[Solve[{ Body1`Eq[1], Body2`Eq[1], Con12`Eq[3],
          g!=0, Body1`rho!=0, Body1`R!=0 },
          Body2`F[C2][Ext],
          { Body2`F$[C2][Body1][2], Body1`F$[C1][Body2][3] } ]];
END Solving;
```

Matching and Unification for the Object-Oriented Symbolic Computation System
AlgBench

Georgios Grivas* and Roman E. Maeder

Theoretical Computer Science
ETH Zurich
{grivas, maeder}@inf.ethz.ch

Abstract. *Term matching* has become one of the most important primitive operations for symbolic computation. This paper describes the extension of the object-oriented symbolic computation system *AlgBench* with pattern matching and unification facilities. The various pattern objects are organized in subclasses of the class of the composite expressions. This leads to a clear design and to a distributed implementation of the pattern matcher in the subclasses. New pattern object classes can consequently be added easily to the system. Huet's and our simple mark and retract algorithm for standard unification as well as Stickel's algorithm for associative commutative unification have been implemented in an object-oriented style. Unifiers are selected at runtime. We extend *Mathematica*'s type-constrained pattern matching by taking into account inheritance information from a user-defined hierarchy of object types. The argument unification is basically instance variable unification. The improvement of the pattern matching operation of a rule- and object-based symbolic computation system with unification in an object-oriented way seems to be very appropriate.

1 Introduction

While pattern matching is not a necessary requirement in symbolic or functional programming, it is however nowadays a part of modern symbolic and functional languages such as *Mathematica*[1] [19], Axiom[2] [11], Standard ML [15], Miranda[3] [17] and Haskell [10]. Actually, *Mathematica* provides a more sophisticated pattern matcher than any of the symbolic and functional languages. It offers type-constrained matching, conditional rules, associative, commutative and identity matching, pattern alternatives, repeated patterns etc.

On the other hand there are also relatively efficient systems that do not support this declarative kind of programming (e.g. Maple[4] [5]). This gives the

* Research supported by the Swiss National Science Foundation
[1] *Mathematica* is a registered trademark of Wolfram Research, Inc.
[2] Axiom is a registered trademark of The Numerical Algorithms Group Ltd.
[3] Miranda is a registered trademark of Research Software Ltd.
[4] Maple is a registered trademark of Waterloo Maple Software.

indication that sometimes the procedural way is more efficient. The main reason for this is the relatively bad efficiency of pattern matching algorithms used in existing symbolic computation systems.

AlgBench [13] is a workbench for the design, implementation and performance measurement of algorithms for symbolic computation, including computer algebra. It provides an interpreter for a symbolic language and a skeleton for implementing data types and algorithms in all areas of symbolic computation. An object-oriented design makes incorporation of new code easy. A compiled implementation language (AT&T C++ V2.0) was chosen to allow measurements of efficiency of algorithms and data structures down to machine level. The provided data types are those of a symbolic system: symbols, numbers, and composite expressions (isomorphic to LISP lists). We adopted from *Mathematica* its syntax and evaluation model (*infinite evaluation*).

In *Mathematica*, the Prolog-like pattern matching and the database of rules arise from its designers' goal: a system must fit naturally the way a scientist works. Looking up tables of mathematical functions in handbooks is an important (and tedious) aspect of this work. A rule-based system can implement the implicit pattern matching that goes on during this activity. A symbolic computation system should provide a database of the necessary handbooks of mathematical functions and a fast search procedure for a formula in the database. If this system uses infinite evaluation (see Sect. 3), it additionally imitates the method a scientist uses and at the same time offers a natural and easy way for the specification of mathematical formulas.

AlgBench provides, like *Mathematica*, a functional language with pattern matching and term rewriting facilities [14]. Type-based dispatch of operations is in *Mathematica* nicely integrated with the pattern matcher, as mentioned also in [8]. On the other hand there is no way to define hierarchies and inheritance. We show in this paper an extension of this type-constrained pattern matching which takes into account inheritance information from user-defined hierarchies.

We aim to provide a powerful inheritance-based pattern matcher and unifier for *AlgBench* and to enrich its language through unification and subtyping. Our main contribution in relation to the existing symbolic computation systems is the support of the various special unifiers, their runtime selection, as well as their clear design due to object-oriented techniques.

The organization of this paper is as follows: In Sect. 2 we briefly discuss efficient data representation, which has a great impact on the performance of the pattern matching process. In Sect. 3 we replace the pattern matching operation in *AlgBench*'s term rewriting environment with unification. In Sects. 3.1, 3.2 and 3.3 an overview of the design and implementation of pattern matching and standard and associative unification in an object-oriented way is given. In Sect. 3.4 we show a mechanism for selecting the special unifiers. Finally, in Sect. 4 the construction of hierarchies at the user level as well as using this information for the unification of types is illustrated with examples.

2 Efficient Data Representation

A term (concrete syntax) has one obvious representation, namely it can be represented as a linear array (abstract syntax) whose elements are taken from function symbols, pattern variables, and constants (symbols, strings or numbers). Commas and parentheses may be eliminated from a term without introducing any ambiguity. This *array representation* is equivalent to *tree representation*, in which a function symbol stands for the root of a subtree whose children represent the function's arguments. Pattern variables and constants end up as the leaves of such a tree. A more concise representation, well known from graph theory is the *direct acyclic graph (DAG) representation*, where the work to unify shared subterms is performed only once.

AlgBench has like *Mathematica* the known problem listed in [8], concerning the tree representation of expressions. For example, Infinity − Infinity evaluates to Indeterminate, but f[Infinity] − f[Infinity] gives 0. A DAG representation would speed up the pattern matching operation. This constitutes the greatest part of the efficiency achieved in the linear unification algorithms (e.g. the Paterson-Wegman algorithm [6] and the Huet algorithm [18]) compared with Robinson's original algorithm [16]. However, the previous problems become greater. For instance we could not get the right result if we simplified the pattern expression x_ − x_ which may be instantiated with Random[] − Random[]. The result should not be 0 in this case, forbidding evaluation of the uninstantiated pattern expression. With the unique storage of pattern variables we condemn our system to simplify the x_ − x_ pattern expression always to zero. It concerns exactly the same variable. The same problems remain in languages with feature structures (e.g. Login [2]), since all occurrences of the same variable in a given term are represented with *coreferences* [12], in other words with a pointer to that object. For a logic programming language this might be tolerable, but not for a symbolic one.

3 From Matching to Unification

AlgBench and *Mathematica* unify diverse programming styles (procedural, functional, rule-based and APL-like). All procedure and function definitions are rewrite rules, each of which has a pattern on the left-hand side and a replacement on the right-hand side. The functions themselves can be defined as lambda expressions.

Reduction or term rewriting and pattern matching are central elements in most modern functional languages. During term rewriting, the left-hand side of a rule is matched against a part of an expression to produce bindings for the variables in the right-hand side of the rule. This part is then rewritten by the bound right-hand side of the matched rewrite rule. In *AlgBench* the evaluation proceeds in this way as long as more user-defined or built-in rules are found (*infinite evaluation*). All other programming styles are implemented on top of this paradigm. However, most functional languages have a one-step evaluator instead.

Resolution and unification are central operations in logic languages. During resolution, the left-hand side of a clause is unified with a goal, producing bindings for the variables in both the clause and the goal. Unification is on this count from the programmer's point of view more flexible than matching. In unification the pattern variables can be unified and bound to other pattern variables. That means they can hold values of any type, including patterns themselves. When a pattern variable is instantiated, all the variables bound to it see its instantiation value. Variables can be passed as function arguments or as arguments of compound expressions inside the function definition.

Term rewriting is not as general as resolution. *Narrowing* [9] is a generalization of term rewriting. The idea to replace pattern matching with unification in a term rewriting environment, i.e. to implement narrowing, is legitimate in order to get the output substitution produced by unification. For this purpose we implemented various syntactical unifiers (see Sect. 3) in the kernel and a type unifier at the top-level (see Sect. 4).

3.1 Pattern Object Classes

In traditional symbolic computation systems, the evaluator or interpreter is a routine that dispatches on the type of the expression. In an object-oriented system this can be implemented much cleaner as a *virtual* (or *dynamically bound*) method. The pattern matchers and unifiers suggested here are such virtual methods as well.

Patterns are classes of expressions. They contain pattern object classes which represent sets of possible expressions. Since our system is object-oriented it is obvious that the pattern object classes are implemented in this way too. Every pattern object class is a subclass of the class of the composite expressions. This makes the system extensible concerning new pattern object classes. The complete hierarchy of expressions is similar to the one found in LISP systems. There is an abstract base class expr, from which string, symbol, number, and normal are derived. The "normal" (composite) expressions are written in *Mathematica*'s syntax as $head[e_1, \ldots, e_n]$, $n \geq 0$ and correspond to LISP lists ($head\ e_1 \ldots e_n$). The class normal is again an abstract class. This organization minimizes duplication of code and simplifies the evaluator considerably, as described in [13].

The pattern objects listed in Table 1 are supported. Two patterns with the same name must be unified with the same expressions.

The unifier is declared as virtual method

```
virtual int expr :: unify(expr * e, environment &env);
```

in the abstract base class of the expressions. The specialized matchers for the subclasses string, symbol and number are trivial: only literal matching is carried out. All unifiers are applied to a pattern in the form pat->unify(e, env) and give true when the pattern is unifiable with the expression e. In this case the bindings in the environment are updated. The default does in the base class literal matching and is inherited by all expression classes that do not override it.

Table 1. Valid pattern objects

input/output form	internal form	description
_	Blank[]	anonymous pattern variable
_type	Blank[type]	variable constrained to type
var:pat	Pattern[var, pat]	a pattern named var
var_	Pattern[var, Blank[]]	short form of var:_
Literal[pat]	Literal[pat]	unifies like pat (but is not evaluated)
pat /; cond	Condition[pat, cond]	unifies only if cond evaluates to true
anything else		literal matching

The only nontrivial implementations are for the classes of pattern objects and for normal expressions. The specialized unifiers of the pattern classes blank, condition and literal are given below. The bindings of pattern variables to matched terms are kept in the *environment*, organized as a stack of bindings. This environment defines methods for looking up the binding of a variable, adding a new binding, and for marking and retracting it. [5]

The pattern Blank[] always matches. Blank[hh] matches if the head of e is equal to hh. hh is an instance variable of objects of the class Blank. The pattern Literal[pat] matches if pat matches. The pattern Pattern[name, pat] matches if pat matches and if the matched value can be bound to name. If name was already bound, its value must be the same as e (or unify with e in the case of two-way matching).

The pattern Condition[pat, cond] matches if pat matches and if cond evaluates to True after substituting all pattern variables by their values. If the condition is not satisfied, the matching fails and any new bindings of pattern variables must be undone. The following C++ code makes these ides clear.

```
int condition::unify(expr * e, environment & env)
{
    void *marker = env.mark();  /* mark stack of bindings */
    if (pat->unify(e, env)) {   /* matching itself succeeds */
        /* test condition */
        expr condi = makesubsts(cond, env);
        condi = (*rec_eval) (condi);       /* evaluate it */
        if (condi == steTrue) return True; /* it worked */
    } /* else: condition not true or no match */
    env.retract(marker);        /* undo new bindings */
    return False;
}
```

3.2 Standard Unification

We propose a mark and retract algorithm for standard matching of simple expressions. Since in a rule-based evaluator most patterns are of the simple form

---

[5] The operations on the environments are explained below (Sect. 3.2).

$f[var_1, var_2, \ldots, var_n]$ a straightforward algorithm works very well in practice. For more complicated expressions one of the asymptotically better methods can be used. The mark and retract algorithm is based on a stack-oriented environment mechanism. The idea is similar to the procedure activation mechanism of procedural languages. The environment contains the bindings of pattern variables to the matched terms. Bindings are created when a pattern object of type Pattern[var, pat] is encountered, as seen above. The unifier for composite expressions marks the current environment and recursively tries to unify the head and the elements of the expression. If this fails, the environment is restored to its old value, otherwise the unification succeeds with the new bindings left on the environment. We do not implement this code directly as a method for normal expressions because it will become only one of several possible unifiers.

```
int stackunif(normal *p, normal *en, environment &env)
{
    if (p->Length() != en->Length())
        return False;              /* lengths differ */
    void *marker = env.mark();     /* mark stack of bindings */
    for (int i = 0; i <= p->Length(); i++) {
        /* loop over head and elements */
        if (!(*p)[i]->unify((*en)[i], env)) {/* no match */
            env.retract(marker);          /* undo bindings */
            return False;
        }
    }
    return True;   /* all elements unified */
}
```

To enable unification instead of simple one-way matching we exchange the pattern with the subject expression in the unifier's argument list:

```
int expr::unify(expr *e, environment &env)
{
    if (e->isA(pnormal))
        return e->unify(this, env); /* it is a pattern */
    else return sameQ(this, e);     /* a literal */
}
```

Consequently we do not loose efficiency in those cases where only matching is necessary.

Two almost linear algorithms for standard unification (or unification under the empty theory) have been examined. First, we examined an improved version of the *Paterson-Wegman algorithm* [6] . This algorithm is linear in the sum of vertices V and edges E in the DAG, hence $O(V + E)$. The basic idea of the Paterson and Wegman algorithm is to unify first the more simple subexpressions and because of that to avoid the exponential growth of the expressions being unified. This is accomplished with parent pointers and DAG representation of terms.

This requirement is however not available in the *AlgBench* system, because only the symbols are stored in a symbol table, not the whole expressions.

Second, J. Vitter and R. Simons [18] presented an improved version of the *Huet algorithm* with complexity $O(\alpha(2E, V) E + V)$, where α is the inverse of the Ackermann function i.e. it has the nice property to grow very slowly (see [18] for a proof). Its basic concept is to build equivalence classes for subexpressions through union-find trees (see [1] for details). Its advantage regarding *AlgBench*, that works with both DAG and non-DAG representation of terms, was decisive for our choice. On the other hand it fits well with a future introduction of a DAG representation too. Another reason was the simple parallelization of the algorithm contrasted with Paterson-Wegman's.

We implemented the abstract data structure of the union-find tree as a class with the following interface:

```
class unionfind
{
public:
    unionfind(expr*);     // create a tree from an expression
    ~unionfind();         // destructor
    void        unite(unionfind *);    // union of two trees
    unionfind *find();   // find root of a tree
    int         rootQ(); // is this the root of its tree?
    unionfind *class;    // class representative of this tree
    expr        *element; // element in the tree
private:
    int         count;   // count for weighted union
    unionfind  *father;  // pointer to father node (NULL => root)
};
```

The functions unite and find are realized as methods of the class unionfind. This is an important conceptual difference compared to the procedural solution of [1]. The Huet algorithm in [18] is formulated for functions and variables. Therefore we extended it suitably. We view all symbols, strings, numbers and pattern objects as variables and when they have to be unified with an expression, the specialized unifier for the corresponding class is called. As a result the algorithm works also for new coming classes of pattern objects. All other composite expressions are viewed as functions.

3.3 Associative Commutative Unification

The associative commutative unification (or unification under the A, C and A+C theory) is the unification problem in the presence of associative and—or commutative functions. The improved version of Stickel's algorithm [7] was chosen for implementation in *AlgBench*. As it is common in such algorithms [4] the problem is reduced to the solution of (in this case linear homogeneous) diophantine equations. For the declaration of associative and commutative functions we provide – like in *Mathematica* – the command SetAttributes[*symbol*, *attribute*], where the attribute is called Flat for the associative case and Orderless for the commutative case.

```
In[1]:= SetAttributes[f, Flat]
In[2]:= SetAttributes[f, Orderless]
In[3]:= Unify[f[g[x_, y_], a_, 3], f[g[1, 2], 3, 2]]
Out[3]= {y -> 2, x -> 1, a -> 2}
```

3.4   Runtime Selection of the Special Unifiers

In addition to the preceding, a mechanism in the system's kernel selects the right unification algorithm to be called. We distinguish at the moment three cases: complicated expressions, simple expressions and expressions with *Flat* and—or *Orderless* attributes. Later we want to support expressions that contain inheritance information. The criteria for simple vs. complicated terms include the length and the number of nested expressions within the term. In the simple case we choose the mark and retract algorithm, in the case of complicated expressions Huet's algorithm and in the case of associative commutative functions Stickel's algorithm. The main implementation of the unify method of normal expressions is the selection mechanism of the special unifiers:

```
int normal::unify(expr *e, environment &env)
{
    if (!e->isA(normal)) return False; /* different types */
        normal &en = (normal) e;          /* downcast */
    if (Head()->isA(symbol)) {            /* head is symbol */
        symbol &sh = (symbol) Head();     /* downcast */
        if (sh->hasAttributes(Flat | Orderless))
            return stickelacunifier(this, en, env);
        else if (complicated(this))
            return huetunifier(this, en, env);
    } /* fall through */
    return stackunif(this, en, env);
}
```

The last call treats also recursively the case of a composite head of a pattern, for example f[a]g[b][x_].

We put at the user level the command Unify[$expr_1$, $expr_2$]. The system selects then the right unifier for composite expressions and gives the substitution list: {var_1 -> val_1, ..., var_n -> val_n}. With unification a non-ground expression can rewrite either to a ground term, such as:

```
In[1]:= Unify[f[x_, 2, g[3, 5]], f[a, 2, g[y_, z_]]]
Out[1]= {z -> 5, y -> 3, x -> a}
```

or to a non-ground term:

```
In[2]:= Unify[f[x_, g[a, y_]], f[1, g[a, h[z_^2]]]]
Out[2]= {x -> 1, y -> h[z_^2]}
```

The unification algorithm alone (i.e. without the resolution mechanism) gives the possibility to program in the following way: Find a possible value for y of f[1,y_] matching the pattern f[x_,2].

```
In[4]:= f[1, y_] /. f[x_, 2] :> x + 1
Out[4]= 2, {x -> 1, y -> 2}
```

4 Inheritance-based Unification

Subtyping is a substitutability relationship. Intuitively one type is a subtype of another if an object of a subtype can stand in for an object of a supertype. Objects of a subtype consequently support at least the same number of operations as the objects of the supertype. In object-oriented programming a subtype has all fields and operations of its supertype as well as additional fields and operations. In other words, *subtyping* is the distribution of types into a generalization/specialization hierarchy. However, in practice it is often necessary not only to add new fields and operations in a data type but also to limit the inheritance of some of them. If a language supports dynamic binding, the operations (or methods) that are overridden are not a part of the subtype anymore. P. America [3] demonstrates that *inheritance* (also called *subclassing*), which deals with code sharing among classes, is not always subtyping, which has to do with specialization in behavior of objects. In this sense subclasses inherit the implementation and subtypes inherit the interface.

H. Ait-Kaci and R. Nasr [2] presented a logic programming language *Login*, which incorporates inheritance-based information (in form of an IS-A taxonomy) immediately into the Huet unification algorithm. Because of our object-oriented design, we do not need to integrate inheritance information into the matching process for the data types provided by the system. The type checking is carried out automatically by the virtual function mechanism. The idea of inheritance-based unification is however well suited for the object-oriented *AlgBench*.

S. Wolfram describes in his *Mathematica* book [19] the various interpretations one can give to a part of an expression. The meaning of the head of a subexpression can be a function, a command, an operator or an object type. The elements are then the arguments of the function or of the command, the operands of the operator, or the fields of the object type structure. Type-checking is done at the user level as a special type of pattern matching, e.g. f[x_Integer] := 0. However this is not sufficient.

We rely on the last interpretation and we see the heads of rewrite rules as object types. In our first attempt we implemented in *AlgBench* a package, that creates with the subtype command the inheritance hierarchy and then takes into account this information for the object type unification. The non-typed objects are unified with the kernel unifiers. We give the user the opportunity to define inheritance hierarchies of types via a subtype command: Subtype[*list1*, *list2*], where *list1*, *list2* are lists of object types. Hence, every object type t in this command may be head of a composite expression, where t_ denotes that it can be bound within the unification process. Consider the following example:

Example 1.

```
In[1]:= Subtype[f,{g, h, i}]
In[2]:= Subtype[h, j]
In[3]:= Subtype[{g, i}, k]
In[4]:= f[x_] := x
In[5]:= g[y_] := x+y
```

We express the type extension in the definition of g, where the arguments of f are inherited implicitly.

Instead of allowing the pattern variables to have any value, we restrict them to values that are elements of a small finite set; a *lattice*. In our case this lattice is *AlgBench*'s class hierarchy, the system lattice. With the subtype command we allow user-defined type hierarchies within composite expressions, the user-defined lattice.

Definition 1. *Two object types F, G unify iff there exists a non empty H such that H is the greatest lower bound of F and G with the following binding list:* $\{F \rightarrow H, G \rightarrow H\}$

This means in the previous example:

```
In[1]:= Unify[i[j_], g[h_]]
Out[1]= {i -> k, g -> k, h -> j}
```

The heads of the composite subexpressions (object types) are unified as a *greatest lower bound (GLB)* lattice operation. This notion is an extension of the type-constrained matching used in *Mathematica*. Consider the rules and the type hierarchy of Example 1. We can do the following:

```
In[1]:= k[ x_f ] := x + 1
In[2]:= i[g[a, b]]
Out[2]= a + b + 1
```

Since there is no rule for i, the types of k, i are unified with the GLB operation to k. The types f, g are unified for their part to g. As there is a rule for g, we get after a couple of rewrite steps a+b+1.

All rules of the supertype are inherited after the rules of the subtype as in the following example:

Example 2. Consider the rules for the abstract logarithm function:

```
In[1]:= log[x_, y_] := log[x] + log[y]
In[2]:= log[x_^n_] := n log[x]
In[3]:= log[1] = 0
```

We specialize the logarithm function with the logarithms of base 2 (ld) and e (ln):

```
In[4]:= Subtype[log, {ld, ln}]
In[5]:= ld[2] = 1; ln[e] = 1
```

Now we get:

```
In[6]:= ld[2^k]
Out[6]= k
```

Since there is no rule for matching the pattern ld[2^k] the inherited rules are examined, and following rewrite steps are carried out: ld[2^k] $\longrightarrow$ k ld[2] $\longrightarrow$ k 1 $\longrightarrow$ k.

Reexamining Example 1 at the beginning of this section, we see that we need an extension of the subtype command, where it is specified which arguments of the subtype are projected to the supertype: Subtype[h1[$expr_1$], h2[$expr_2$]]. The user gives the arguments that have to be inherited, otherwise we select the arguments from left to right and cut the rest. Consider the following example:

Example 3.

```
In[1]:= f[x_] := x^2
In[2]:= Subtype[f, g]        /* there is no rule for g */
In[3]:= g[a,b]               /* select from left-to-right a and cut b */
Out[3]= a^2
In[4]:= Subtype[f[y_], g[x_, y_]]
In[5]:= g[a, b]              /* select b and cut a */
Out[5]= b^2
```

We realized the above extension by producing and storing a new global rule. That means for the subtype command Subtype[f[y_], g[x_, y_]] from Example 3, we store the rule g[x_, y_] := f[y] and we update the inheritance hierarchy appropriately.

5 Conclusions and Future Work

The expressive power of *AlgBench* is increased with the support of one- and two-way pattern matching (unification). The distribution of the pattern objects classes as subclasses of the composite expressions stems from the clean object-oriented design of *AlgBench*. This makes the addition of new pattern object classes very easy. The construction of hierarchies is introduced and the user is able to write programs with both typed and untyped objects. A type unifier is implemented at the top-level, which takes into account the inheritance information. Several special unifiers are implemented in the kernel in an object-oriented style: the mark and retract algorithm for simple terms, the improved version of

Huet's algorithm for complicated terms as well as the improved version of Stickel's algorithm for associative commutative unification. The availability of them and of their selection mechanism make the system more powerful and faster.

In a second attempt to realize an inheritance-based unifier, this time in the kernel, we will revise appropriately the current top-level unifier. A big problem of the existing symbolic computation systems is –as we already mentioned in the Introduction– the bad efficiency of pattern matching algorithms used in. The elimination of large case statements is what we pay for more elegant programming. We plan to compile the patterns and rules to such case statements in order to regain the lost efficiency.

Acknowledgements. We would like to thank O. Gloor and S. Missura for reviewing an early draft of this paper. Many thanks go also to the anonymous referees for their useful comments and to Claude Kirchner for further suggestions. Parts of the mentioned unifier implementations were realized by M. Ochsner and M. Werner.

References

1. A. Aho, J. Hopcroft, and J. Ullman. *The Design and Analysis of Computer Algorithms.* Addison-Wesley, 1974.
2. H. Ait-Kaci and R. Nasr. Login: A logic programming language with built-in inheritance. *J. of Logic Programming*, 3:185–215, 1986.
3. P. America. Inheritance and subtyping in a parallel object-oriented language. In *ECOOP'87*, 1987.
4. H. Buerckert. Opening the AC-unification race. *J. of Automated Reasoning*, 4, 1988.
5. B.W. Char, K.O. Geddes, G.H. Gonnet, M.B. Monagan, and S.M. Watt. *MAPLE: Reference Manual.* University of Waterloo, 1988. 5th edition.
6. D. de Champeaux. About the Paterson-Wegman linear unification algorithm. *J. of Computer and System Sciences*, 32, 1986.
7. F. Fages. Associative commutative unification. *J. of Symbolic Computation*, 3:257–275, 1987.
8. R. Fateman. A review of Mathematica. *J. of Symbolic Computation*, May 1992.
9. M. Fay. First order unification in equational theories. In *4th CADE'79*, 1979.
10. P. Hudak, S. Peyton-Jones, and P. Wadler. Report on the programming language Haskell. Technical report, Department of Computer Science, Yale University, August 1991.
11. R. Jenks and R. Sutor. *AXIOM, The Scientific Computation System*, Springer-Verlag, 1992.
12. M. Kay. Functional grammar. 5th Annual Meeting of the Berkeley Linguistics Society, Berkeley, California, 1979.
13. Roman E. Maeder. AlgBench: An object-oriented symbolic core system. In J. H. Davenport, editor, *Proc. of Design and Implementation of Symbolic Computation Systems DISCO '92*. To appear.
14. Roman E. Maeder. *Programming in Mathematica.* Addison-Wesley, second edition, 1991.

15. R. Milner, M. Tofte, and R. Harper. *The definition of Standard ML*. MIT Press, Cambridge, MA, 1990.
16. J.A. Robinson. A machine-oriented logic based on the resolution principle. *J. of the ACM*, 1965.
17. D. Turner. Miranda: A non-strict functional language with polymorphic types. In *Functional Programming Languages and Computer Architecture FPCA '85*, September 1985.
18. J. Vitter and R. Simons. New classes for parallel complexity: A study of unification and other complete problems for P. *IEEE Transactions on Computers*, C-35(5), May 1986.
19. Stephen Wolfram. *Mathematica: A System for Doing Mathematics by Computer*. Addison-Wesley, second edition, 1991.

A Type System for Computer Algebra

Philip S. Santas

Institute of Scientific Computation

ETH Zurich, Switzerland.

email: santas@inf.ethz.ch

Abstract. We examine type systems for support of subtypes and categories in computer algebra systems. By modelling representation of instances in terms of existential types instead of recursive types, we obtain not only a simplified model, but we build a basis for defining subtyping among algebraic domains. The introduction of metaclasses, facilitates the task, by allowing the inference of type classes. By means of type classes and existential types we construct subtype relations without involving coercions.

1 Introduction

Type theories for strongly typed computer algebra systems like Axiom [JeSu92, JTWS85], or object oriented languages [Gro91, Gold89] usually take one object - a rational number, say, with methods and functions for reading and updating its numerator and denominator or performing addition among rationals - to be an element of a recursively defined type [BrMi92] like:

```
Rational :: Class = Class [
    numerator: Self -> Integer,
    denominator: Self -> Integer,
    SetNumDenom: Self -> Integer -> Integer -> Self,
    + : Self -> Self -> Self,
    * : Self -> Self -> Self,   etc. ]
```

where Self stands for the type under scope (here Rational).

This encoding hides the fact that Rationals may have an internal state that is shared by all its methods: the responsibility for building a new instance of Rational in response to a *SetNumDenom* call is placed within *SetNumDenom* itself. The notation used in this paper makes the representation component of instances implementation of the representation, as it has been proposed by Pierce and Turner [PiTu92]:

```
Rational = Class (Rep::Any)
    numerator: Rep->Integer,
    denominator: Rep->Integer,
    SetNumDenom: Rep->Integer->Integer->Rep
    * : Rep->Rep->Rep, etc. ]
```

This method gives the caller of *SetNumDenom* the responsibility for transforming the value returned by *SetNumDenom* into a new Rational instance.

Although this notation may seem strange, it offers much in terms of *simplicity*: the entire development, including the notions of *Self* ($ in Axiom), *Dynamic* (in Eiffel [Mey91]), etc., can be carried out without the need of recursive types, which are difficult to formulate and implement, or extensible records [Car88, Mit88] which are implementation dependent; the reason we want to avoid these difficulties is that recursive definitions involve reference semantics, which destroy many of the appealing properties of type systems based on subtyping.

Furthermore, existential types hide the implementation of the objects, allowing us to deal with abstract types independently of implementation details without any cost in type safety. In section 6 we present some examples of the advantages of this approach.

This paper briefly examines ideas developed by Cardelli, Cook, Mitchell, Jenks, Pierce, Trager and the designers of the Axiom programming language. We observe that these concepts are not adequate for the properly typed modelling of the relations among simple algebraic constructs like the domains of integers and rationals. We extend the calculus of categories and metaclasses defined in [JeSu87] and [San93] respectively, and we give theoretical documentation of the subtyping among categories and metaclasses. By using metaclasses, we allow a type inference system to construct type classes and insure subtyping among domains in terms consistent with the classical algebraic notation.

The examples given are coded in $k - bench$ language, which is influenced by *Axiom* and *Haskell*, but with a simpler type system. Although demonstration of simplicity is out of the scope of this paper, we claim that our type inference mechanism reduces the need for retractions (and to some extent even that of coercions, although we will not remove them) in Axiom, and the coding of Haskell-like type classes, since they are inferred automatically. More elaborate details and examples can be found in [San93].

2 Objects in Computer Algebra

Most computer algebra operations involve purely functional objects: functions or methods operate on objects and return a new object with a new internal state, instead of updating the state of an older object. This fact has to be seen in contrast to object orieneted programming where there is an in-place updating of the internal state of an object: an object must be a reference. Additionally, many functional languages like Scheme implement closures in a similar way, allowing side effects[1]. Due to the poor properties of references and side effects however, most type theoretic accounts of object oriented programming [CaWe85, Car88, Car92, Ghe91, MiMM91, Pier91] deal with purely functional objects and employ techniques used for purely functional closures.

It is a common practice in systems like *Axiom* to have a special type which represents the internal state of the objects of a certain domain: if *Rep* is the

---

[1] Objects are only one of the many possible implementations of closures.

type of the internal state, then the type of the functions applied to the rational numbers of our initial example are[2]:

```
RationalFun := Rep::Any +> [
    new: Integer -> Integer -> Rep,
    numerator: Rep -> Integer,  etc.
    + : Rep -> Rep -> Rep,
    * : Rep -> Rep -> Rep,
    / : Rep -> Rep -> Rep,
    0 : Rep,
    1 : Rep    ]
```

RationalFun is actually a functor, ie. a *function from types to types* that specifies the visible behaviour of the functions on rationals, in terms of the **abstract representation type**. An object satisfying this specification consists of a list of functions of type *RationalFun(Rep)* for some concrete type *Rep*, paired with a state of type *Rep*: both are surrounded with an abstraction barrier that protects the internal structure[3] from access except through the above specified functions[4]. This encapsulation is directly expressed by an **existential type** [PiTu92]:

```
Rational := (Rep::Any +> Class( Rep, RationalFun Rep )) SomeRep
```

Abstracting *RationalFun* from *Rational* yields a higher order type operator that, given a specification, forms the type of objects that satisfy it.

```
DeclareDomain := (Fun: Any->Any) +>
            (((Rep: Any) +> Class(Rep,Fun Rep)) SomeRep)
```

Here *Any* $\rightarrow$ *Any* is the category of functions from types to types, ie. the **category of functors**. The type of *Rational* objects can now be expressed by applying the *DeclareDomain* constructor to the specification *RationalFun*:

```
Rational := DeclareDomain(RationalFun)
```

or the shortcut:

```
Rational :: RationalFun
```

In order to give proper treatment to the interaction between representations and subtyping in the following sections, it is necessary to separate *Rational* into the specifications of its functions and the operators which capture the common structure of all object types. This separation is also important for the semantical construction of categories and the definition of the internal structures of the types.

---

[2] The operator $+>$ denotes λ application: $x +> e$ stands for $\lambda x.e$
[3] By internal structure we mean either the internal state, or the hidden functions that operate on this state, or any other operation that we want to hide from the external environment.
[4] Axiom allows the explicit exporting of the representation, but in the general case it follows a similar encapsulation policy [JeSu92]. Here we extend this technique to apply to any form of data structure.

Rationals are created using the function *box*, which captures the semantics of the pointer implementations of abstract structures in *SML* and dynamic objects in object oriented programming. A rational number with representation (x:Integer, y:Integer), internal state (5,2) and method implementations:

```
SetNumDenom := s:(x:Integer, y:Integer) +> (m:Integer) +> (n:Integer) +>
                    (s.x := m; s.y := n)
numerator := s:(x:Integer, _:Integer) +> s.x
denominator := s:(_:Integer, y:Integer) +> s.y
etc.
```

can be created as:

```
r1 := box (coerceTo (
  [Rep := (x:=5, y:=2),
    [SetNumDenom:=s:(x:Integer,y:Integer) +> (m:Integer,n:Integer) +>
                    (x:=m; y:=n),
      numerator:= s:(x:Integer,_:Integer) +> s.x,
      etc. ]],
  Rational ))
```

The *coerce* function here is only a syntactic construct, which shows to the compiler, how to view the introduced list; in effect it is the identity function.

The *box* function is helpful for the implementation of the *new* function and the + and * operators which return new instances of Rational:

```
new:=(initX:Integer) +> (initY:Integer) +>
  box ( coerceTo ([Rep:=(x:=initX,y:=initY),...],Rational))
```

Unlike Axiom and some statically typed object oriented languages, the elements of an object type may have *different internal representations*, and different internal representations of their functions. For example, a rational with representation type ($x : Integer$) might be implemented as follows[5]:

```
r2 := box (coerceTo (
  [Rep := (x:=5),
    [SetNumDenom:= s:(x:Integer) +> (m:Integer) +> (_:Integer) +> (x:=m),
    numerator := s:(x:Integer) +> s.x,
    denominator := s:(x:Integer) +> 1,
    etc. ]],
  Rational ))
```

and the definition of *new* changes accordingly. Notice that the functions *SetNumDenom, numerator, denominator*, have the *same* signature as in the previous implementation. Like this we can obtain static type checking even without having all the information about the internal structure of the objects in question [San93].

This variability will be helpful for defining type classes in section 6. The constants 0 and 1 can be implemented in any of the two ways, having x assigned to 0 and 1 respectively, while the y field can be 1.

---

[5] this example implements integers which are also rational numbers: we will use it again later on

However it is useful to have all these definitions in the class *Rational*. In such case we assume a default representation for all objects of class *Rational*:

```
Rep:=(x:Integer, y:Integer)
RationalClass := () +> coerce (
  [SetNumDenom := s:Rep +> (m:Integer) +> (n:Integer) +> (x:=m; y:=n),
   numerator := s:Rep +> s.x,
   + := s1:Rep +> s2:Rep +> box(...),
   etc. ],
 RationalFun Rep )
```

The *box* operator becomes:

```
p1:= box [(x:=5,y:=2), rec [RationalFun Rep] RationalClass]::RationalFun
```

Suppose that we invoke the function *SetNumDenom* passing to it the object *p1*. We have to open *p1*, and make sure that the only functions applicable to its *Rep* are the ones declared in *RationalFun*. Consequently we produce a new value of type *Rep*, which is *reboxed* as a rational number which has access to the methods of *RationalClass*, and hides *Rep*.

The same process happens for functions like + and *. On the other hand, the functions *numerator* and *denominator* do not return any new object, so no reboxing is necessary. This optimization comes in accordance with static type-checking: even if the internal structure is unknown, its type is given and the operations *numerator* and *denominator* return an unambigous result.

3 Metaclassing

We have seen that the instances of a class may have various representations, while it is desirable to have an abstraction on the level of functions defined for the instances of a class. Mitchell and Plotkin [MiPl88] proposed that the *interface part of a data type be viewed as a type, while the implementation part as a value of that type*[6]. According to this analysis a class comprises:

- the internal representation of its instances
- operations for analyzing and manipulating this representation
- an abstraction barrier that prevents any access to the representation except by means of the given operations

However some of the functions defined in our example with rationals, like *numerator, denominator, SetNumDenom* and *new* deal only with the representation of its instances: these functions form the type of the representation; other

---

[6] We name this value a Class. Note that although a Class is a value, it is not implied that programs should have classes as run-time entities. In our system we use classes for type inference, creating the typeclasses, and we are not concerned with their run-time behaviour. However it is possible in future versions of *Axiom* to have either classes or types as run-time entities or first class values.

functions like $+$, $*$, $/$, and the constants 0 and 1 are expected to be declared in a transparent fashion to any form of implementation.

It is also the case that the latter set of operations can be included in the specification of many other types, independently of the operations which manipulate the representation. For instance, $(+)$ can be defined for Matrices or Booleans, without any need for them to include operations like *numerator* or *denominator*.

We can transliterate this, introducing metaclasses:

```
RationalMeta := Self::Any +> [
    + : Self -> Self -> Self,
    - : Self -> Self,
    * : Self -> Self -> Self,
    / : Self -> Self -> Self,
    0 : Self,
    1 : Self ]
```

This specification can be viewed as a coding for the *algebraic structure Field* (we can write $Field := RationalMeta$). $Self^7$ is an existential type and is bound to the representation of the classes which are declared as instances of Field:

```
Rational :: Field
```

One might argue that the definition of Metaclasses is expressed as Categories in *Axiom*: this would had been true if *Self* were bound to the class which is instance of, say, *Field*. However in our case, *Self* is bound to the representation of the class; the positive side effect of this is that it does not introduce recursive definitions in the system, permitting us to define subtyping in the next sections. Additionally it captures a wider set of types as we saw in the previous examples, since they are dealing with existential types. Unfortunately this means that although type checking can be static, binding of functions has to be dynamic, except if we remove subtyping from the type system. This is not bad, because we avoid spending computational resources on performing coercions, since we actually deal with uncoerced objects by simply binding them to a function[8]. Optimizations such as unboxed objects can be performed in many cases. In general the scheme we present here is expected to increase both the speed and the type safety of an Axiom-like system.

Metaclasses have otherwise the same properties as Categories in Axiom: their degree of abstraction permits the modelling of algebraic structures in a definitely abstract level, without involving representation issues.

4 Subtyping

Descriptions of sets of entities can be arranged into useful classification hierarchies: for example the set if integers can be seen as a subset of the rational numbers. Moreover, this fact supports a useful kind of reasoning: if X is an integer, then

---

[7] In Axiom it is coded as $

[8] The call of coercions in the *Axiom* interpreter is dynamic too!

X must also be a rational number, and every interpretation of a rational number should be true for X.

In typed λ-calculi, an analogous sort of reasoning is provided by introducing a new judgement form $S \sqsubseteq T$, or more generally, $\Gamma \vdash S \sqsubseteq T$ ie. S is a subtype of T under the assumptions Γ.

The assumptions Γ in the above statement must include the preconditions $S :: M$ and $T :: M$, ie. the two classes must be instances of a common metaclass, in the scope of which, the subtyping is assumed[9]. For example it is valid to make the judgement $Integer \sqsubseteq Rational$:

$$Integer :: Ring, Rational :: Ring \vdash$$
$$Integer \sqsubseteq Rational$$

The semantics of $S \sqsubseteq T$ are included in the following statement:

If $S \sqsubseteq T$, then an instance of S may safely be used in any context where an instance of T is expected.

For example the function

```
foo(x:Rational) := 3/5 + x*8 + x**2
```

can be safely applied to the integer argument 4 because it is possible to view any integer as a rational.

More important, since integers form a *subring* of rationals, and the definition of *foo* includes operations defined for rings, we can do this substitution under the assumption that both *Integer* and *Rational* form a ring.

This leads to the conclusion:

$$\Gamma \vdash x : S, \Gamma \vdash S \sqsubseteq T \Rightarrow \Gamma \vdash x : T$$

However, it is not clear how an integer is a rational concerning its implementation in computers: in the abstract world of mathematical objects, the subtype relationship is usually considered to hold; but in computer programs, the internal representations of integers and rationals is in almost all the cases completely different: the set of machine integers is not a subset of the set of the records which represent rationals (Axiom implementation). However on most machines, every representable integer can be converted to a record representation so it matches with a rational, without loss of information. In this sense, two formal accounts can be given for the semantics of subtyping. In the simpler view, the syntactic subtype relation $S \sqsubseteq T$ is interpreted as asserting that the semantic domain denoted by S is included in that denoted by T. In the more general view[10], $S \sqsubseteq T$ is interpreted as a canonical coercion function from the domain denoted by S to the one denoted by T.

The second view can be formally given as:

---

[9] Here we assume subtyping without coercions. In case of coercions, no such pecondi-
tions are necessary.

[10] B. Pierce calls this more refined view, but we dissagree: the previous case can be
simulated with a coercion which is the identity function

$$x : S, \text{coerce} : S \hookrightarrow T \Rightarrow \text{coerce}(x) : T$$

and the coercions under assumptions are denoted as

$$\Gamma \vdash x : S, \Gamma \vdash \text{coerce} : S \hookrightarrow T$$
$$\Rightarrow \Gamma \vdash \text{coerce}(x) : T$$

ie.

$$x : S, {}_\Gamma coerce : S \hookrightarrow T \Rightarrow {}_\Gamma coerce(x) : T$$

For the moment we assume that the representation of the objects is not important for the subtyping (we elaborate further on this in section 6); this permits us to use the $\sqsubseteq$ symbol instead of *coerce*. We will refrain from using *coercions* for the definition of subtyping, restricting ourselves to natural subtyping, which is our point of interest.

The subtype relation is reflexive and transitive:

$$\Gamma \vdash T \sqsubseteq T$$
$$\Gamma \vdash S \sqsubseteq T, \Gamma \vdash T \sqsubseteq U \Rightarrow \Gamma \vdash S \sqsubseteq U$$

The rule for record types is:

$$\Gamma \vdash S_i \sqsubseteq T_i, n :< m$$
$$\Rightarrow [k_1 : S_1 ... k_m : S_m] \sqsubseteq [l_1 : T_1 ... l_n : T_n]$$

A record type with fields labelled $[k_1...k_m]$ is a subtype of any record type with a smaller collection of fields, where the type of each shared field in the *less informative* type is a supertype of the one in the *more informative* type.

The rule for function types comes in accordance with the rule for records:

$$\Gamma \vdash T_1 \sqsubseteq S_1, \Gamma \vdash S_2 \sqsubseteq T_2$$
$$\Rightarrow \Gamma \vdash S_1 \to S_2 \sqsubseteq T_1 \to T_2$$

The elements of an arrow type are functions. In our schema we represent functions as values, therefore they must have a type. Although algebraically the above subtype relation does not form a subset relation, it poses no problem in our type theory: The type $S_1 \to S_2$ places a stronger constraint on the behaviour of its instances than $T_1 \to T_2$, if it either:

- demands that they behave properly on a larger set of inputs ($T_1 \sqsubseteq S_1$) or
- demands that their results fall within a smaller set of outputs ($S_2 \sqsubseteq T_2$)

or possibly both.

It is incorrect, however, to say that $S_1 \to S_2$ describes a smaller set of functions than $T_1 \to T_2$: From the above rules we can derive that

$$\Gamma \vdash S_1 \to S_2 \sqsubseteq T_1 \to S_2$$

Algebraically this might seem strange conclusion, since the cardinality of $S_1 \to S_2$ is bigger than that of $T_1 \to S_2$; this implies that we cannot have a traditional injective coercion from the former to the latter.

On the other hand, we observe thet there is a natural coercion from $S_1 \to S_2$ to $T_1 \to S_2$:

$$\frac{\Gamma \vdash f : S_1 \rightarrow S_2}{\Gamma \vdash (f \circ (id : T_1 \rightarrow S_1)) : T_1 \rightarrow S_2}$$
$$\Gamma \vdash f : T_1 \rightarrow S_2$$

The same results can be obtained while using coercion from T_1 to S_1.

The rule for function types interprets the rule for records in the following way: If we think of a record as a function from labels to values, a record S represents a stronger constraint than T on the behaviour of such function, if S describes the function's behaviour on a larger set of labels or gives a stronger description of its behaviour on some of the labels also mentioned by T [Pier92].

5 Subtyping and Metaclasses

We define the metaclass Ring as:

```
Ring   := Rep::Any +>  [
              + : Rep -> Rep -> Rep,
              - : Rep -> Rep,
              * : Rep -> Rep -> Rep,
              0 : Rep,
              1 : Rep    ]
```

We observe that the definition of $Field$ we gave earlier is:

```
Field := Rep::Any +>  [
              operations for Rings,
              / : Rep -> Rep -> Rep ]
```

By the rules of record and function subtyping (the former has more fields and the type of the common fields agree), we have

$$Field \sqsubseteq Ring$$

This means that *every instance of the metaclass Field, can be viewed as an instance of Ring*, which comes in accordance with the algebraic concepts.

A better interpretation is: *Every instance of an instance of Field, is also an instance of an instance of Ring.* Here we have avoided to use classes in the subtyping relation, providing a scheme where we can define hierarchies of metaclasses, without using classes as computational values ie. without need for coercions among classes, which would increase the complexity of our semantics, and of the inference mechanism. In this way we have established subtyping among types of metaclasses. Unfortunately we cannot say the same for the instances of metaclasses.

If we recall the definition of the type $RationalFun$ from section 2, we can introduce in a similar way the type of integers which form a Ring. For convenience in the initial steps we assume the same representation for integers as for rationals:

```
IntegerFun := (Rep::Any) +> [
  new: Integer -> Integer -> Rep
```

```
numerator: Rep -> Integer,
denominator: Rep -> Integer,
SetNumDenom: Rep -> Integer -> Integer -> Rep,
+ : Rep -> Rep -> Rep,
- : Rep -> Rep,
* : Rep -> Rep -> Rep,
0 : Rep,
1 : Rep     ]
```

We observe that it is not the case that *IntegerFun* ⊑ *RationalFun*. Even worse, it is more likely to built an order of the form: *RationalFun* ⊑ *IntegerFun*, which is against the algebraic intuitions. The reason is that the specification of integers does not include the record field /, leaving *IntegerFun* with one field less that *RationalFun*.

However, we could define blindly the operation / on integers:

$$/ : Rep \rightarrow Rep \rightarrow Rep$$

which is unacceptable in algebraic terms; therefore we reject it, since it does not bring anything useful even in type-theoretic terms (we could say in such case that *IntegerFun* ⊑ *RationalFun*, but the other way around would have also been possible).

The fact that the type of the *Rep* in both *Rational* and *Integer* is the same, implies that rationals and integers must share the same representations whenever we assume subtyping without coercions. Although the actual representation used by an implementation of integers is usually different from the representation used by an implementation of rationals, the specifications of integers and rationals require that the *types* of their representations be *identical* to their common hidden representation types, whatever these may be.

Part of what it means for integers to be a subtype of rationals in the traditional sense (here we assume object oriented approach), is that we should be able to apply functions to integers as if they were rationals. However, this is not the case: here we were forced to introduce a new function, in order to be allowed to do this.

The only obvious solution left is the use of coercions, as we saw in the previous section. We could define a coercion from integers to rationals, and this would make any other discussion on the topic obsolete. The problem here is that coercions are expensive in computational resources, and in some cases they can introduce inconsistencies. If we can manage to define subtyping for common representations, without using coercions, we make a step towards defining type-consitency for algebraic relationships as we shall see in the next sections. In the rest of the cases we can still use coercions.

6 Type Classes

In this section, we are concerned with modelling subtyping without use of coercions. We introduce the concept of type classes, which has some similarities with the homonymous concept in Haskell. The semantics and the formal definitions for type classes are provided in Appendix A. The reader can conclude that type classes are distinct from metaclasses. The reader who is familiar with the type classes in *Haskell* will note that the semantics of type classes in our system are considerably different, although they can have the same implementation; additionally, our type classes are not defined in the language, but on its type system, reducing the complexity of the former.

Consider the function

```
dblSqrd(x: Rational) := (x+x)*(x+x)
```

If we discard the previous forced subtype relation between integers and rationals, we have seen that there is no obvious way, that *dblSqrd* can accept integer arguments, without coercing integers to rationals.

For our further elaboration of this issue we will assume the results from [San93] concerning specializations of subtyping among algebraic domains. We introduce additionally type classes in order to provide the facility of viewing integers as subring of rational:

```
RationalRing := (Rep::Any) +>  [
               + : Rep -> Rep -> Rep,
               - : Rep -> Rep,
               * : Rep -> Rep -> Rep,
               0 : Rep,
               1 : Rep    ]
```

This declares that any eventual subtype of rational which happens to be an instance of Ring implements the operations declared in the type class RationalRing. Since Integer forms a subring of Rational, from the definition

```
IntegerRing := (Rep::Any) +>
               [same as RationalRing]
```

we may conclude:

$$Integer\,Ring \sqsubseteq Rational\,Ring$$

The declaration of *Integer* as instance of the type class *IntegerRing* is straightforward:

```
Integer :: IntegerRing
```

while for Rationals we need one additional type class:

```
RationalField := (Rep::Any) +>  [
               same as RationalRing,
               / : Rep -> Rep -> Rep ]
```

```
Rational :: RationalField
```

Using type classes the system infers the most general signature for *dblSqrd*:
$$\Gamma, \forall a :: Rational\,Ring \vdash dblSqrd : a \to a \to a$$

This means that *dblSqrd* can receive as argument an instance of any the types of *RationalRing*, including *Integer*, without any need to coerce it to Rational, in other words, we have managed to define a natural subtype relationship between integers and rationals, which comes in accordance with the algebraic semantics of the terms.

Additionaly, subtyping has been induced without the use of coercions. This accomplishment permits us to override many ambiguities included in a graph of coercions: Assume the types A, B, C and D, and the coercions

$$A \hookrightarrow C \hookrightarrow B,$$
$$A \hookrightarrow D \hookrightarrow B$$

A type system cannot in general prove whether this graph commutes, since the former path may have different semantics than the latter. However if we can perform the operations defined in B directly in A (assuming that instances of type A are passed as arguments), the ambiguity is resolved.

Finally it is worth noting that none of the forgoing would have been possible had we not introduced the type variable *Rep* in the definition of the types and classes. Suppose that Rational and Integer had the definitions

```
Rational := (Rep::Any) +> [ ...
    + : Rational -> Rational -> Rational
    etc. ]

Integer := (Rep::Any) +> [ ...
    + : Integer -> Integer -> Integer
    etc. ]
```

This would influence the definitions of RationalFun and IntegerFun respectively, and finally the definitions of RationalRing and IntegerRing: there would had been no way to derive a subtype relationship among types which would include the terms *Rational* $\rightarrow$ *Rational* $\rightarrow$ *Rational* and *Integer* $\rightarrow$ *Integer* $\rightarrow$ *Integer*, since these two cannot form a subtype relationship, due to the introduction of the contravariance rule for function types. It is important to mention again that the types of the representations have to be the same in order to ensure subtyping, although their implementations do not need to be the same.

7   Conclusions

We have elaborated on type systems for symbolic computation, in respect to subtyping. For the purposes of our analysis we have introduced metaclasses, which not only solve the consistency problems between algebra and type systems [GrSa93], but participate in many cases in the formulation of subtyping among domains without need of coercions.

A type inference mechanism constructs type classes as a combination of metaclasses and domains, in order to resolve the conflicts introduced by the different definitions of subclass and subtype in algebra and type theory accordingly. This mechanism implements natural subtyping among domains, without invoking any coercion from one type to another. However in cases in which the subtype does not define operations of its supertype[11], coercions have to be called.

It is the subject of our current research to examine if coercions can form subtype relations among algebraic domains, without violating basic concepts of type theory. On the other hand, the approach we propose can be used in compilers for statically typed languages, like Axiom, which in its current version does not support subtyping by means of coercions. Although we often use existential types, the speed of the programs generated by a compiler which supports our type system should not be lower than the speed of object oriented programs, since both systems use dynamic binding, and avoid coercions; this means, that the speed should be higher than programs generated by the current version of Axiom.

8 Acknowledgements

I would like to thank R. Jenks, B. Trager, S. Watt and all the members of the Axiom group at the Thomas Watson Research Center for supporting this research, M. Bronstein and S. Spackman for providing useful guidelines and P. Grogono and K. Haughian for comments on early drafts of this paper.

References

[BrMi92]   K. Bruce, J. Mitchell. PER models of subtyping, recursive types and higher order poly morphism. 19th ACM Symposium on Principles of Programming Languages. New Mexico, Jan.1992.

[Car88]   L. Cardelli. Structural Subtyping and the notion of Power Type. 15th ACM Sumposium POPL. CA, Jan. 1988.

[Car92]   L. Cardelli. Typed Foundations for object-oriented programming. Tutorial at POPL, Jan. 1992.

[CaWe85]   L. Cardelli, P. Wegner. On understanding types, data abstraction, and polymorphism. Computing Surveys. 17(4). 1985.

[CoHiC90]   W. Cook, W. Hill, P. Canning. Inheritance is not Subtyping. 17thPOPL, ACM Press, Jan 1990.

[Ghe91]   G. Ghelli. Modelling features of object-oriented languages in second order functional languages with subtypes. in Foundations of OOP. Lecture Notes in Computer Science (489). Springer Verlag, 1991.

[Gold89]   A. Goldberg. Smalltalk-80. second edition, Addison-Wesley, 1989.

[Gro91]   P. Grogono. The Dee Report. Technical Report. CS, Concordia U. Montreal, Canada. Jan. 1991.

---

[11] This case never happens in reality: it is only the expectations of the user which cause this illusion. For instance, we have seen that integers without division do not form a subtype of rationals with division; furthermore, definition of division for integers is out of the context of algebra. Here we tried to find a consistent way to bind both worlds.

[GrSa93] P. Grogono, P.S. Santas. On Equality and Abstractions. To be published at JOOP.

[JeSu92] R.D. Jenks, R.S. Sutor. Axiom: The Scientific Computation System. NAG. Springer Ve rlag. 1992.

[JTWS85] R.D. Jenks, B. Trager, S. M. Watt, R S. Sutor. Scratchpad II Programming Language Manual, IBM, 1985.

[Mey91] B. Meyer. Eiffel: The language. Prentice Hall. 1991.

[Mit88] J. Mitchell. Polymorphic type Inference and containement. Information and Computation (76). 1988.

[MiMM91] J. Mitchell, S. Meldal, N. Madhav. An extension of SML modules with subtyping and inheritance. 18th Sumposium POPL, FL, Jan. 1991.

[MiPl88] J. Mitchell, G. Plotkin. Abstract Types have Existential Type. ACM Transactions on Programming Languages and Systems. 10(3), 1988.

[Pier91] B. Pierce. Programming with Intersection Types and Bounded Polymorphism. PhD thesis, CMU. Dec 1991.

[PiTu92] B. Pierce, D. Turner. Type Theoretic Foundations for OOP. Report. Dept of CS, U. of Edinbourgh. 1992.

[San93] P.S. Santas. Classes and Metaclasses. tech. report, Dept. of CS, ETH Zurich. 1993.

[San92] P.S. Santas. Multiple Subclassing and Subtyping for Symbolic Computation. Proc. of Workshop on Multiple Inheritance and Multiple Subtyping. ECOOP 92. Utrecht, Netherlands. June 1992.

[SuJe87] R.S. Sutor, R.D.Jenks. The Type Inference and Coercion facilities of the Scratchpad II Interpreter. IBM Tech. Report. March 1987.

A   Definition of Type Classes

The definition we have given for classes, assume that operations do not belong to a type, but to an algebra (that is a particular collection of types). A class combines one or more types for the implementation of its instances. A class's instances do not need to have a common internal structure, but they are elements of the types which a class assumes. Herewith we can define type classes, which form a refinement of classes.

Formally a type class has the following structure: $Class[T, B]$ in which T is the set of types and B is the behaviour of the class's instances. An instance of a type is by definition an instance to any of the classes in which this type belongs. The instanceOf relation (denoted by ::) represents in our definition membership in a set of instances and as such it is *irreflexive* and *non-transitive*.

$$(x :: C) \wedge (C :: M) \not\Rightarrow x :: M$$

The subclass relation (denoted by $\sqsubseteq$) is a *reflexive, antisymmetric* and *transitive* binary ordering relation in the partially ordered set of classes.

Subtyping can be seen in two ways (which are consistent with the definitions given in the previous sections):

- subtyping by means of subclassing:
  $(x :: C_1) \wedge (C_1 \sqsubseteq C_2) \Rightarrow x :: C_2$

- subtyping by means of coercions [JTWS85, For90]:
  $(x :: C_1) \wedge (\text{coerce} : C_1 \hookrightarrow C_2) \Rightarrow \text{coerce}(x) :: C_2$

Type classes C_1, C_2 are said to belong to the same **inheritance path** when one can derive through $\sqsubseteq$ or $\hookrightarrow$ relationships that $C_1 \sqsubseteq C_2$ or $C_1 \hookrightarrow C_2$ respectively.

Multiple subclassing can be used instead of multiple instantiation:

- $(x :: C) \wedge (C \sqsubseteq C_1) \wedge \ldots \wedge (C \sqsubseteq C_n) \Rightarrow x :: C_1 \wedge \ldots \wedge x :: Cn$

We avoided to use coercions here due to the possible inconsistencies in their implementations.

The above inductive definitions can be seen as the definition of *class intersection*. One may observe that they differ from the classical definition of set intersection. Since one cannot in general establish equality for instances of different classes, these definitions employ object identity. Under the usual definitions, however, such identity is only possible between members of the same type, thus class intersection corresponds to intersection of the sets of types implementing the classes. Since classes define behaviour by means of a set of axioms and operations among each class' instances, class intersection must produce behavioural *union*. Given the separation of classes from types, this definition has even constructive power, since an instance must be element of a particular type.

- Class Intersection:
  $x :: (C_1 \sqcap \ldots \sqcap C_n) \iff \forall_i x :: C_i$
  and
  $[T_1, B_1] \sqcap \ldots \sqcap [T_n, B_n] = [\bigcap_i T_i, \bigcup_i B_i]$
  where we write $[T, B]$ for the class implemented by each of the types $t \in T$ and supporting behaviours $b \in B$.

In the case that $T_1 \cap \ldots \cap T_n = \{\}$, the class intersection is $\perp$.
Similarly, we can define the union of classes as their superclass:

- Class Union:
  $x :: (C_1 \sqcup \ldots \sqcup C_n) \iff \exists_i x :: C_i$
  $[T_1, B_1] \sqcup \ldots \sqcup [T_n, B_n] = [\bigcup_i T_i, \bigcap_i B_i]$

Decision procedures for set/hyperset contexts.

Eugenio G. Omodeo[1] and Alberto Policriti[2]

[1] omodeo@assi.ing.uniroma1.it Università di Roma "La Sapienza",
Dip. di Informatica e Sistemistica. Via Salaria, 113. 00198-Roma.
[2] policrit@udmi5400.cineca.it Università di Udine,
Dip. di Matematica e Informatica. Via Zanon, 6–8. 33100-Udine (Italy).

Abstract. Pure, hereditarily finite, sets and hypersets are characterized both as an algorithmic data structure and by means of a first-order axiomatization which, although rather weak, suffices to make the following two problems decidable:
(1) Establishing whether a conjunction r of formulae of the form $\forall y_1 \cdots \forall y_m((y_1 \in w_1 \& \cdots \& y_m \in w_m) \to q)$, with q unquantified and involving only the relators $=, \in$ and propositional connectives, is satisfiable.
(2) Establishing whether a formula of the form $\forall y\, q$, q as above, is satisfiable.
Concerning (1), an explicit decision algorithm is provided; moreover, significantly broad sub-problems of (1) are singled out in which a classification —named the 'syllogistic decomposition' of r— of all possible ways of satisfying the input conjunction r can be obtained automatically. For one of these sub-problems, carrying out the decomposition results in producing a finite family of syntactic substitutions that generate the space of all solutions to r. In this sense, one has a unification algorithm.
Concerning (2), a technique is provided for reducing it to a sub-problem of (1) for which a decomposition method is available.

"There might exist axioms so abundant in their verifiable consequences, shedding so much light upon a whole discipline, and furnishing such powerful methods for solving given problems (and even solving them, as far as that is possible, in a constructivistic way) that quite irrespective of their intrinsic necessity they would have to be assumed at least in the same sense of any well established physical theory."
K. Gödel, 1947

1 Introduction

In undertaking the first axiomatization of set theory, around 1908, Zermelo could plainly refer to "set theory as it is historically given" (cf. [18]). Today one must distinguish between Set Theory, which is a wide and varied panorama, and several set languages embellishing the landscape, that have arisen in part from epistemological, in part from practical motivations. The meaning of such languages is usually committed to a system of axioms, subject to the deductive machinery of first-order logic. Sometimes an operational semantics is available.

Not only new postulates but, more interestingly, new first-class entities of discourse akin to sets have made their appearance in the theory: proper classes, 'extraordinary' sets or hypersets (cf. [2]), functions-as-rules (cf. [3]); while similar but usually less abstract entities, like multi-sets, single- and multi-valued maps, graphs, etc., have become common ingredients of any investigation in computer science. As the recent history of hypersets shows, there is no definite borderline between theory and practice: Aczel's interest in the semantics of concurrency soon led him to the discovery of a theory of 'non-well-founded sets' (cf. [1]) closely related to the one of [15], originated from a more speculative attitude; now applications of Aczel's theory are being found in the study of intensional semantics for description logics (cf. [11, 12]).

The main explanation of the success of set theory is often indicated in its high expressive power, which is demonstrated by the relative ease with which one can find set-theoretic equivalents of old-established mathematical notions and, accordingly, specify in purely set-theoretic terms a conjecture to be proved or a problem to be solved. To make a straightforward example, rendering as a set-theoretic problem the propositional satisfiability problem SAT (cf. [17]) requires but little ingenuity: one can translate SAT ($\&_{i=1}^n \bigvee_{j=1}^{m_i} \ell_{ij}$), where each ℓ_{ij} is a literal drawn from $\{A_1, \ldots, A_k, \neg A_1, \ldots, \neg A_k\}$, into the set equation $\{\{\emptyset, \{\emptyset\}\}\} = \{\{A_1, B_1\}, \ldots, \{A_k, B_k\}, \{\emptyset, L_{11}, \ldots, L_{1m_1}\}, \ldots, \{\emptyset, L_{n1}, \ldots, L_{nm_n}\}\}$, where $B_1, \ldots, B_k$ are new unknowns and, for $i = 1, \ldots, n$, $j = 1, \ldots, m_i$, and $h = 1, \ldots, k$, $L_{ij} =_{\text{Def}}$ if $\ell_{ij} = A_h$ then A_h else_if $\ell_{ij} = \neg A_h$ then B_h .[3]

One easily perceives that once a branch of Set Theory has been rendered constructive enough (a minimum effort expedient, for that sake, would be to just drop the infinity axiom, retaining all remaining axioms of the Zermelo-Fraenkel theory, which then can be interpreted in a universe of hereditarily finite sets), and once a suitable hybridization of its formalism with a computer programming language is achieved, a language will ensue with strong data structuring capabilities, and with a high power non only for the specification of problems but also for the support of solving methods. This has been convincingly argued in [3], and is demonstrated by the long life of the imperative set-based programming language described in [25], SETL (for an example of use of the latter, see Fig.1). Recently, the theme of embedding nested sets in a logic programming language has been tackled by many articles (cf., e.g., [14]).

To the automated deduction field, however, set theory poses a hard challenge. Neither a purely equational presentation of the axioms seems viable, nor the axioms can be reduced to Horn clauses, even in the weakest conceivable axiomatizations of sets (cf. [24]). Even if one resorts to a finite axiomatization (which is a debatable choice), the axioms must undergo a rather heavy 'preparation' (cf., e.g., [6]) before they are amenable to the common theorem-proving

---

[3] The two problems are clearly equivalent in the sense that one admits a solution if and only if so does the other. Moreover, the set assignments for $A_1, \ldots, A_k, B_1, \ldots, B_k$ that make the set equation true are in one-to-one correspondence with the truth-value assignments for $A_1, \ldots, A_k$ that make the given instance of SAT true.

techniques. However, the troubles seem to depend more on the limitations of the techniques in widespread use today, than on intrinsic difficulties with set theory.

We have treated in [24] the satisfiability problem for $\exists^*\forall$-sentences in varied theories of sets. At first, one may find non-surprising the conclusion that this problem is algorithmically solvable, in view of the low syntactic complexity of the sentences under study; however, when one comes to consider the quantificational complexity (cf. [10]) hidden in the axioms of the theory, which is at best $\exists z \forall u \forall y \exists w \forall v$, one realizes that this decidability result ensues from certain felicitous peculiarities shared by all sensible set theories. This result, mainly when referred to very weak theories of sets, sheds some light upon the area of automatic programming; on the other hand, when the theory is strong enough to ensure that every $\exists^*\forall$-sentence is either provable or refutable, the result has important consequences for the already mentioned emerging field of logic programming with sets. For instance, once the SAT problem is formulated as above in set-theoretic terms, it becomes a special subproblem of the set unification problem, which is in turn a special subproblem of the syllogistic decomposition (=set constraint handling) problem referred to the $\exists^*\forall$-sentences. Therefore, SAT can be automatically solved through the basic inference machinery of a logic programming language with sets.

We will see below that in spite of its innovative character, hyperset theory behaves with respect to the $\exists^*\forall$-sentences in much the same way as a rather conventional theory of sets. In either case, essentially by the same technique, an $\exists^*\forall$-sentence can be decomposed into satisfiable mutually incompatible disjuncts. This striking analogy between sets and hypersets also extends to broader families of formulae; while, for narrower families, it leads to simple yet paradigmatic solvable cases of the unification problem.

The paper is organized as follows. *Section 2* describes the algorithmic data structures 'hyperset' and 'set': the collection of sets forms a sub-universe of the collection of hypersets endowed with membership, and all entities in the latter universe are (in a sense to be clarified) hereditarily finite. *Section 3* abstracts from the properties of the said hypersets and sets an axiomatic view of entities of both kinds, very tightly tailored to make satisfiability problems below a certain complexity decidable. *Section 4* introduces another data structure, named 'flex', akin to sets/hypersets, which will be extensively used in the analysis of formulae regarding the latter entities. *Section 5* gives a technique for translating each formula $\forall x \, p$, with p unquantified, into an equivalent one involving restricted quantifiers only: such technique is legitimatized by a diagonal argument that proceeds quite similarly in the case of sets and of hypersets. *Section 6* gives an algorithm for deciding —both with respect to the theory of sets and to the one of hypersets— whether a given conjunction of restrictedly quantified formulae $\forall y_1 \cdots \forall y_m ((y_1 \in w_1 \& \cdots \& y_m \in w_m) \rightarrow q)$ is satisfiable or refutable; moreover, when the number m of quantifiers does not exceed 1 in any conjunct, a disjunctive normal decomposition method is

```
lbdMin2 := -- Of two pairs, this chooses the one with smaller second component.
    lambda(a,b); return if b(2)<a(2) then b else a end if; end lambda;
mApp := -- repeated application of 'binFunct'
    lambda(default, binFunct, compound);
    r := default; for c in compound loop r := binFunct(r,c); end loop;
    return r; end lambda;
procedure dijkstra(s,g);
-- 'g' is an oriented graph with distances on the arcs, represented as a set
-- of pairs [[x,y],d], where [x,y] is an arc and d is a positive number.
-- One wants to find the lengths of shortest paths leading from
-- a given node 's' to each one of the nodes accessible from 's' in 'g'.
    arcs := domain g;
    infty := 1 + #arcs * -- this 'infty' will serve in practice as infinity
        mApp(0,lambda(x,y); return x max y; end lambda, [d:d=g(a)]);
    s ?:= mApp([om, arb domain arcs], lbdMin2, {[x,y]:[y,x] in arcs})(2);
    -- If the start node s has been left undefined (i.e. s=om), then
    -- the smallest tail of an arc is taken as start node.
    unsettled := range arcs with s;
    -- Nodes whose distance from the start node must be found.
    dist := {[x,infty]: x in unsettled};
    -- This will become the distance of each node from the start node 's'.
    dist(s) := 0; -- the distance of the start node from itself is zero
    [latest,latestDist] := [s,0];
    -- Latest node whose distance from the start node was found,
    --     and distance of 'latest' from the start node.
    while latestDist /= infty and #unsettled /= 1 loop
        unsettled less:= latest;
        for x in arcs{latest} | x in unsettled loop -- revise distance
            dist(x) min:= dist(latest) + g(latest,x);
        end loop;
        [latest,latestDist] :=
            mApp([om, infty], lbdMin2, [ [x,dist(x)]: x in unsettled ]);
    end loop;
    return [ s, {[x,d]: d =dist(x) | d /= infty} ];
end dijkstra;
```

Fig. 1. SETL2 specification of the Dijkstra algorithm.

achieved. *Section 7* specializes the disjunctive decompositon method to cases of formulae meeting certain syntactic peculiarities; when, in particular, a conjunction of atoms of the four kinds $u = v$, $u \in v$, $u = \emptyset$, $u = v$ with z is given (where **with** denotes the element insertion operation and u, v, z stand for variables), the decomposition leads to a finite family of syntactic substitutions of the form $x_1 = y_{11}$ with$\cdots$with y_{1m_1} & $\cdots$ & $x_n = y_{n1}$ with$\cdots$with y_{nm_n} ($x_1, \ldots, x_n$ distinct variables). The latter result can be viewed as the positive answer to a unification problem naturally arising in the context of non-equational theories; furthermore, as will be apparent from the approach followed below, similar unification algorithms can be devised under richer axiomatizations than the ones to be examined here, to treat additional set-theoretic constructs (e.g. binary union).

It should be easy to adapt the methods in this paper to a hybrid framework where set constructs co-exist with free Herbrand functors. A hybridization of this kind has recently led to a proposal (cf. [13, 14]) of logic programming enriched with ordinary sets. An enhancement of that proposal with infinite rational terms (cf. [20]) and with non-well-founded sets is under way.

2 Hereditarily finite sets and hypersets

In the characterization we will give of it, a *hyperset* is just a peculiar *pointed graph*: this is to say, a directed graph G one of whose nodes, ν_*, has been singled out, and which has certain special features that we are about to describe. Without such features, a pointed graph will still *depict* a hyperset, but the latter will differ from the graph itself. We will regard as *sets* those hypersets in which no paths form cycles. This means among others that *self-loops*, i.e. arcs of the form $[\mu, \mu]$, admitted in hypersets, will be forbidden in sets.

To characterize *membership*, we will establish a one-to-one correspondence $\nu \mapsto G\lceil \nu$ from the nodes of any given hyperset $y = (G, \nu_*)$ into hypersets, and will convene that the images $G\lceil \nu$ of this mapping are all hypersets x such that a membership chain $x = x_0 \in \cdots \in x_n = y$ ($n \geq 0$) exists leading from x to y; in particular, $G\lceil \nu_*$ will coincide with (G, ν_*). Also, we will take the presence of an arc $[\mu, \nu]$ between two nodes μ, ν of (G, ν_*) as an indication that $G\lceil \mu \in G\lceil \nu$, and the absence of such an arc as an indication of the contrary. In the light of this interpretation, the above-made distinction between hypersets and sets is clearly meant to reflect the basic difference between the membership relation over hypersets and the more conventional membership relation over sets: the former is to contain cycles of all kinds, the latter can form no cycles at all.

We intend to take into account *hereditarily finite* hypersets and sets only. This reflects —still in the light of the said interpretation of nodes and arcs of a hyperset— into an assumption we will tacitly exploit throughout: namely that *each graph has finitely many nodes*.

All this said, let us now proceed to distinguish hypersets from among all graphs. Preliminarily, we settle a couple of minor technical points. For one thing, two pointed graphs that are isomorphic to each other will depict the same hyperset. In sight of this, it is convenient to adopt a criterion for choosing a *canonical representative* from the isomorphism class of any given pointed graph.

Definition 1. *A <u>numbering</u> of (G, ν_*) is a one-to-one function $\nu \mapsto \nu'$ from the nodes of G into consecutive natural numbers that sends ν_* to 0.*

Each numbering ' underline{induces} a pointed graph $(G', 0)$ isomorphic to $(G, \nu_\star)$: the nodes of G' are the '-images $0, 1, \ldots, n$ of the nodes of G, and its arcs are the pairs $[\mu', \nu']$ with $[\mu, \nu]$ arc of G.

An ordering of all finite binary relations over natural numbers must be fixed (it is immaterial ~~how~~), so it can be exploited to characterize the canonical representative of $(G, \nu_\star)$: actually, this will be the one graph induced by a numbering of $(G, \nu_\star)$ whose family of arcs is minimum with respect to the selected ordering. □

For another thing, any node μ of a pointed graph $(G, \nu_\star)$ such that no path of G leads from μ to $\nu_\star$ is redundant, in this sense: the subgraph of G that results from the withdrawal of all such nodes and relative arcs will depict the same hyperset as G.

Definition 2. *Let G be a graph and ν be one of its nodes. Consider the graph G_ν that results from G when all nodes whence ν cannot be reached in G, and all arcs entering such nodes, are dropped. By $G \lceil \nu$ we will denote the canonical representative of (G_ν, ν).* □

We now arrive at the only deep point in the characterization of hypersets:

Definition 3. *Let G be a graph. A bisimulation on G is a relation B between nodes of G such that for every pair ν_0, ν_1 of nodes with $\nu_0 B \nu_1$, corresponding to each arc $[\mu_b, \nu_b]$ that enters ν_0 or ν_1 in G there is at least one arc $[\mu_{1-b}, \nu_{1-b}]$ of G with $\mu_0 B \mu_1$.*

Those bisimulations on G that are contained in $\{[\mu, \mu] : \mu$ a node of $G\}$ are said to be trivial; if all of its bisimulations are trivial, then G is said to be uncontractible.

A pointed graph $(G, \nu_\star)$ with $(G, \nu_\star) = G \lceil \nu_\star$ is said to be a hyperset (and a set when G is acyclic) if and only if G is uncontractible. □

It easily turns out (cf. [1]) that for any graph G the following relation $\sim_G$ over the nodes of G:

$$\mu \sim_G \nu \text{ if and only if there is a bisimulation } B \text{ on } G \text{ such that } \mu B \nu,$$

is both an equivalence relation and a bisimulation. This enables one to associate with each node μ of G the hyperset $(G/\sim_G) \lceil \mu^{\sim_G}$, where $\mu^{\sim_G}$ is the $\sim_G$-class of μ, and the nodes of $G/\sim_G$ are all $\sim_G$-classes, while its arcs are the pairs $[M, N]$ with $\mu \in M$ and $\nu \in N$ for some arc $[\mu, \nu]$ of G. When $(G, \nu_\star)$ is a hyperset, $(G/\sim_G) \lceil \mu^{\sim_G}$ plainly coincides with $G \lceil \mu$ for every node μ; when $(G, \nu_\star)$ is a set, every pointed graph $G \lceil \mu$ is a set in its turn.

Definition 4. *Every pointed graph $(G, \nu_\star)$ is said to depict the hyperset $(G/\sim_G) \lceil \nu_\star^{\sim_G}$.* □

Membership between hypersets is defined as follows:

Definition 5. *Let x and y be hypersets, with $y = (G, \nu_\star) = (G, 0)$. Then $x \in y$ holds if and only if x equals $G \lceil \mu$ for some arc $[\mu, 0]$ entering 0 in G.* □

Notice that given an y from the denumerable collection of all pointed graphs with integer nodes, one can test whether y fulfills $y = G \lceil 0$, and whether y is a set or a hyperset. Also, the membership relation we have just defined is computable, and the collection of all (hyper)sets is infinite, because it comprises all pointed graphs $\vartheta_i = G_i\lceil 0$ $(i = 0, 1, 2, \ldots)$, where G_i has arcs $[i, i-1], \ldots, [2, 1], [1, 0]$.

Without proof, we state a lemma which is the key to show that sets/hypersets are closed under the *transitive closure*, *element removal* and *unionset* operations, and that they fulfill the so-called (cf., e.g., [21]) *separation* scheme.

Lemma 1. *Let G be an uncontractible graph. Whenever defined, $(G\lceil\nu)\lceil\mu$ coincides with $G\lceil\varrho$ for a suitable ϱ. No two distinct nodes μ', μ'' can fulfill $G\lceil\mu' = G\lceil\mu''$. If $\mu_1, \ldots, \mu_m$ ($m \geq 0$) are nodes of G, then a hyperset (actually a set when each $G\lceil\mu_i$ or the whole G is acyclic) exists whose only members are $G\lceil\mu_1, \ldots, G\lceil\mu_m$.* □

In order to prove that sets/hypersets are closed under the *element insertion* operation (see Lemma 6 below) and under the *powerset* operation, one more lemma is needed:

Lemma 2. *Given two hypersets $y' = (G', 0)$ and $y'' = (G'', 0)$, one can 'superpose' them to form an uncontractible graph G that contains G' and an isomorphic copy of G'' as subgraphs.* □

Before starting a new section, let us reflect for a moment on what has been achieved so far. A technique was specified for 'folding' any given pointed graph $(G, \nu_\star)$ into a structure, $(G/\sim_G)\lceil\nu_\star^{\sim_G}$, which we have convened to regard as a hyperset proper. This structure results from first discarding the nodes whence $\nu_\star$ is unreachable in G, and then fusing two nodes ν, μ into a single node whenever $G\lceil\nu, G\lceil\mu$ potentially depict the same thing.

In principle we could have followed the opposite approach (closer to [1]) of 'unfolding' $(G, \nu_\star)$ into a tree T with nodes roughly corresponding to the paths that lead to $\nu_\star$ in the original graph G. Then, after numbering the nodes of T in some very specific fashion, we could have elected T to be the entity depicted by $(G, \nu_\star)$. With this approach hypersets would have turned out to be trees, but would have been less easily amenable to algorithmic manipulations than our uncontractible graphs, as they would be, very often, endowed with infinitely many nodes.

It is outside the scope of this paper to analyze the algorithmic complexity of the problems involved in the notion of hyperset: e.g., establishing whether the equivalence $\mu \sim_G \nu$ holds between two nodes μ, ν of a graph G, or establishing whether G is uncontractible.

3 Properties of sets and hypersets

This section abstracts from the properties of our concrete sets and hypersets an axiomatic view of entities of both kinds. Two parallel axiomatic theories will result. In the rest of the paper we will work informally within the rails of these theories; accordingly, our reasonings about concrete entities will be transferable to any structure fulfilling the axioms.

Why do we not simply rely on [1]? Actually, the axioms **ZFC⁻ + AFA** therein have been our starting point, but using their full strenght would call into play infinite sets and hypersets, entities about whose existence we prefer to take no particular commitment here. To be coherent we are not willing, either, to adopt axioms like those studied in [4], that deliberately confine one's arguments to a realm of finite (hyper)sets. Our target is, in fact, an axiomatization very tightly tailored to meet the following goal: the axioms must have the power to either validate or disprove any sentence about sets/hypersets whose syntactic form matches one of a few pre-established patterns (cf. Sections 5, 6 and 7).

To describe properties of sets and hypersets, we will use the language which involves the infix predicate symbol $\in$ in addition to the standard first-order endowment (individual variables, identity predicate symbol $=$, propositional connectives, universal and existential quantifiers). Any *structure* $\mathcal{M} = (\mathcal{D}, \varepsilon)$ consisting of a non-empty domain $\mathcal{D}$ together with a binary relation ε over $\mathcal{D}$ can be exploited to *interpret* this language: quantified variables are then supposed to range over $\mathcal{D}$, free variables are arbitrarily replaced by elements of $\mathcal{D}$, $\in$ is interpreted as ε, and $=$ is interpreted as the identity relation $\{[\xi, \xi] : \xi \text{ in } \mathcal{D}\}$. In the privileged cases when $\mathcal{D}$ is either the collection of sets or the collection of hypersets and ε is the binary relation introduced by Def.5, ε is denoted as $\in$, at the price of a little but inoffensive confusion between the language and the meta-language.

The following series of lemmas makes use of these logical instruments:

Lemma 3. *The* extensionality *formula*

(E) $$\left(\forall x \left(x \in y_0 \leftrightarrow x \in y_1 \right) \right) \rightarrow y_0 = y_1$$

is valid about hypersets. As a consequence, **(E)** *holds about sets too.* □

Lemma 4. *The formulae*

(Z) $$\exists z \forall x\, x \notin z,$$
(W) $$\exists z \forall x \left(x \in z \leftrightarrow \left(x \in y_1 \lor x = y_2 \right) \right),$$
(L) $$\exists z \forall x \left(x \in z \leftrightarrow \left(x \in y_1 \,\&\, x \neq y_2 \right) \right),$$

stating that the interpreting structure contains an empty set *and is closed with respect to the* element insertion *and* element removal *operations, are valid both about sets and about hypersets.* □

To reflect **(W)**, **(L)** and **(Z)**, we introduce convenient Skolem functors:

Definition 6. *The operations of element insertion and removal are denoted by the left-associative operators* with *and* less; *the empty set is denoted* $\emptyset$, *and* $\{y_1, \ldots, y_n\}$ *stands for* $\emptyset$ with y_1 with $\cdots$ with y_n, *which denotes the hyperset whose only members are the hypersets denoted by* $y_1, \ldots, y_n$. □

Sets enjoy the so-called *regularity* property: $\exists z \forall x (x \in y \rightarrow (z \in y \,\&\, x \notin z))$, but we content ourselves with the statement of a simpler fact here:

Lemma 5. *The* acyclicity *formulae*

(A⁽ⁿ⁾) $$\left(\&_{0 \leq i < n}\, y_i \in y_{i+1} \right) \rightarrow y_n \notin y_0,$$

$n = 0, 1, 2, \ldots$, *are valid about sets. (N.B.: these formulae will be collectively referred to by the identifier* **(A)***).* □

The following lemma will be exploited in Section 5 for proving a proposition named *diagonalization lemma*, which will in turn play a crucial role in our treatment of $\exists^*\forall$-sentences.

Lemma 6. *The* anti-diagonal *formula*

$(\mathbf{D}_{\notin})$ $\qquad\qquad \exists z\,(\,z \notin z\ \&\ \forall x\,(\,x \in y\ \rightarrow\ z \notin x\,)\,)$

is valid about hypersets. In showing this, one can instantiate z as a set; hence $(\mathbf{D}_{\notin})$ is also valid about sets. $\qquad\square$

In the following we set forth the rationale behind the notion of bisimulation, which, as we have seen, lies deep in the characterization of hypersets. One wants from the outset (cf. [1] and [2]) the variety of hypersets to be rich enough that every graph G (or at least every finite one) admits a *decoration*, viz. a function $\nu \mapsto \xi_\nu$ from nodes to hypersets such that the members of each ξ_ν are the hypersets ξ_μ with $[\mu, \nu]$ an arc entering ν in G. Secondly, since ξ_{μ_0} might be a member of ξ_ν without $[\mu_0, \nu]$ being an arc of G (for that, it suffices that G has an arc $[\mu_1, \nu]$ with $\xi_{\mu_0} = \xi_{\mu_1}$), one wants to be able to recognize situations where ξ_{μ_0} could equal some ξ_{μ_1} with $\mu_0 \neq \mu_1$.

In order that $\xi_{\mu_0} = \xi_{\mu_1}$ can hold in a decoration $\nu \mapsto \xi_\nu$ of G, it is *a priori* necessary that $\mu_0\, B\, \mu_1$ for some bisimulation B on G —this will be the claim of Lemma 7. By adopting the latter as a sufficient condition too, one arrives at the characterization of hypersets we have given above; as an ensuing advantage, every graph will admit only one decoration (cf. Corollary 2 below).

Definition 7. *Let $x_0, \ldots, x_n$ be distinct variables associated with the nodes $0, \ldots, n$ of a graph G. By $\gamma_G = \gamma_G(x_0, \ldots, x_n)$ and $\Sigma_G = \Sigma_G(x_0, \ldots, x_n)$ we denote the following two formulae*

$$\gamma_G =_{\mathrm{Def}} \&_{0 \leq i \leq n}\, \&_{0 \leq j \leq n}\,(\,x_i \in x_j\ \leftrightarrow\ \bigvee_{\substack{0 \leq g \leq n \\ [g, j]\,\mathrm{arc\ of\ }G}}\, x_i = x_g\,),$$

$$\Sigma_G =_{\mathrm{Def}} \gamma_G\ \&\ \&_{0 \leq i \leq n}\neg\exists z\,(z \in x_i\ \&\ \&_{0 \leq j \leq n} z \neq x_j)\,.$$

With respect to a structure $\mathcal{M} = (\mathcal{D}, \varepsilon)$ for the language to which these formulae belong, one defines an M-decoration *to be an interpretation $x_0 \mapsto \xi_0, \ldots, x_n \mapsto \xi_n$ of the free variables by elements of the support domain $\mathcal{D}$ of $\mathcal{M}$ under which Σ_G is true.* $\qquad\square$

We now come to our first result about decorations, which has an obvious but nonetheless important corollary:

Lemma 7. *For any $\mathcal{M} = (\mathcal{D}, \varepsilon)$, if $x_i \mapsto \xi_i$ $(i = 0, \ldots, n)$ is an $\mathcal{M}$-decoration with $\xi_{j_0} = \xi_{j_1}$ of a graph G, then a bisimulation B on G exists with $j_0\, B\, j_1$.* $\quad\square$

Corollary 1. *An uncontractible graph G can be decorated with hypersets in a unique manner, namely by the injective function $\nu \mapsto G\lceil\nu$.* $\qquad\square$

The converse of Lemma 7 holds for hypersets. Hence we will draw various useful consequences:

Lemma 8. *If B is a bisimulation on a graph G, j_0 and j_1 are nodes of G with $j_0\, B\, j_1$, and $x_i \mapsto \xi_i$ $(i = 0, \ldots, n)$ is a decoration of G by means of hypersets, then $\xi_{j_0} = \xi_{j_1}$.* $\qquad\square$

Corollary 2. *The function $\nu \mapsto (G/\sim_G)\lceil\nu^{\sim_G}$ defined earlier, —injective if and only if G is uncontractible—, is the only decoration of a graph G by means of hypersets.* □

Corollary 3. *The following* <u>anti-regularity</u> *scheme, where G is a graph with nodes $0,\ldots,n$, is valid about hypersets for any n:*
(**Я**) $\qquad\qquad \exists x_0 \cdots \exists x_n \, \Sigma_G(x_0,\ldots,x_n)\,.$ □

The following analogue of Lemma 6 will, like it, help in the proof of the diagonalization lemma (Lemma 13 below):

Corollary 4. *The* <u>self-loop</u> *formula*
(**D$_\in$**) $\qquad\qquad \exists z \forall x \, (\, x \in z \, \leftrightarrow \, (\, x = z \vee \bigvee_{0<i\leq n} x = y_i\,)\,)$
is valid about hypersets, for $n = 0,1,2,\ldots$. □

Corollary 5. *The following* <u>hyperextensionality</u> *scheme, where G is a graph with nodes $0,\ldots,n$, is valid about hypersets for any n:*

$$
\begin{aligned}
&\big(\quad \gamma_G(y_0^0,\ldots,y_n^0)\ \&\ \gamma_G(y_0^1,\ldots,y_n^1)\ \&\\
\textbf{(H)}\qquad &\&_{0\leq j\leq n}\,\forall z\,(\,(\,z\in y_j^0\ \&\ \&_{i=0}^n z\neq y_i^0\,)\ \leftrightarrow\ (\,z\in y_j^1\ \&\ \&_{i=0}^n z\neq y_i^1\,)\,)\,\big)\\
&\longrightarrow\ \&_{j=0}^n y_j^0 = y_j^1\,.
\end{aligned}
$$

Proof. Let us assume against the thesis that there are hypersets $\xi_0^0,\ldots,\xi_n^0,\xi_0^1,\ldots,$ ξ_n^1 such that $\xi_{j_*}^0 \neq \xi_{j_*}^1$ for some j_* and that the antecedent of (**H**) holds under the interpretation $y_i^b \mapsto \xi_i^b$ ($b = 0,1$ and $i = 0,\ldots,n$). Let G_i' be an isomorphic copy of G_i, where $(G_i,0) = \xi_i^0$, with $G_1',\ldots,G_n'$ disjoint from one another. Let G be obtained by linking the G_i' graphs together by means of new arcs $[\nu_0,\nu_1]$ one for each pair ν_0,ν_1 of nodes such that: ν_0 belongs to some G_{i_0}', ν_1 to a G_{i_1}' with $i_0 \neq i_1$, and $G_{i_0}'\lceil\nu_0 \in G_{i_1}'\lceil\nu_1$. Two decorations of G are obtained as follows:

- first decoration: each node ν is sent to $G_i'\lceil\nu$, where G_i' is the graph to which ν belongs;
- second decoration: the image of each ν is as before, unless there exist j's with $G_i'\lceil\nu = \xi_j^0$, in which case one such j is selected (choosing in particular $j = i$ when ν is the designated node of G_i'), and the corresponding ξ_j^1 is taken as image of ν in the decoration.

Since the designated node of G_{j_*}' is sent to $\xi_{j_*}^0$ by the first decoration and to $\xi_{j_*}^1$ by the second, the two decorations are distinct, which leads to a conflict with Corollary 2. □

We have reached the focal point of this section:

Definition 8. *As* <u>axioms</u> *about sets and hypersets, we are adopting formulae or formula schemes introduced above: those about sets are* (**Z**), (**W**), (**L**), (**E**) *and* (**A**)*; those about hypersets are* (**Z**), (**W**), (**L**), (**E**), (**D$_\notin$**), (**D$_\in$**), (**H**) *and* (**Я**)*.* □

Remarks 1.

(1) Although we are not including (**D$_\notin$**) and (**H**) among the set axioms, these are formally provable about sets too: one easily deduces (**D$_\notin$**) from (**A$^{(1)}$**);

moreover, **(H)** about sets holds vacuously, thanks to **(A)**, if G is cyclic, and can otherwise be proved by induction on the height of G, exploiting **(E)**.

(2) **(E)** is deducible from **(H)**: this is why we are not including it among the hyperset axioms. To see this assume, under **(H)**, that x_0 and x_1 have the same members. If $x_0 \in x_0$ and $x_1 \in x_1$, then $y_0^0 = x_0$, $y_1^0 = x_1$, $y_0^1 = x_1$, $y_1^1 = x_0$ fulfill the antecedent of **(H)** for $G = \{[0,0],[1,0],[0,1],[1,1]\}$; hence $x_0 = x_1$. Similarly, exploiting the graph with isolated nodes $0,1$, one gets $x_0 = x_1$ from $x_0 \notin x_0$, $x_1 \notin x_1$. If $x_b \in x_b$ and $x_{1-b} \notin x_{1-b}$ ($b = 0$ or $b = 1$), then the graph $\{[0,0],[0,1]\}$ can be exploited in conjunction with $y_0^0 = x_b$, $y_1^0 = x_{1-b}$, $y_0^1 = x_b$, $y_1^1 = x_b$ to get $x_0 = x_1$ from $x_0 \neq x_1$; thus one reaches an absurd. Hence $x_0 = x_1$ holds in all possible cases, which proves **(E)**.

(3) Through a plain inductive argument, **(E)** generalizes into the following *extensionality scheme*, holding both for sets and for hypersets:

$$\exists x_0 \cdots \exists x_n \ \&_{0 \le i < j \le n+1} \left(\left(\&_{0 \le h < j} \left(x_h \in y_i \leftrightarrow x_h \in y_j \right) \right) \rightarrow y_i = y_j \right).$$

Roughly speaking, this says that given any tuple $y_0, \ldots, y_{n+1}$ of distinct sets/hypersets, a number $g \le n+1$ of members $x_0, \ldots, x_g$ of $y_0 \cup \cdots \cup y_{n+1}$ suffices to differentiate $y_0, \ldots, y_{n+1}$ from one another, in the sense that for $i \neq j$ one has $y_i \cap \{x_0, \ldots, x_g\} \neq y_j \cap \{x_0, \ldots, x_g\}$.

(4) The instances of **(R)** with G acyclic can be proved from **(Z)** and **(W)** alone. This remark implies that to state **(R)** about hypersets as economically as possible, it would suffice to postulate it subject to the restriction that G be cyclic and uncontractible. Notice that these cyclic instances of **(R)** are openly against **(A)**, and hence are against regularity.

(5) Among our axioms about hypersets, the only ones whose proof requires some ingenuity in Aczel's theory $\mathbf{ZFC}^- + \mathbf{AFA}$ are $(\mathbf{D}_{\notin})$ and **(H)**; let us hence investigate how one could prove them in that context. Concerning $(\mathbf{D}_{\notin})$, notice that by assuming that no z exists with $z \notin \bigcup y$ and $z \notin z$, one would incur the Russell antinomy (cf. [18]): in fact $\{z \ : \ z \in \bigcup y \,|\, z \notin z\}$ would equal $\{z \,|\, z \notin z\}$ under the absurd hypothesis, and hence the latter would be a hyperset with $\{z \,|\, z \notin z\} \in \{z \,|\, z \notin z\} \leftrightarrow \{z \,|\, z \notin z\} \notin \{z \,|\, z \notin z\}$, which is a contradiction. Concerning **(H)**, notice that we have deduced it from Lemma 8. The latter lemma is proved by [1], inside $\mathbf{ZFC}^- + \mathbf{AFA}$, through a technique that we could mimic in order to prove Lemma 8⋆ below. Hence **(H)** holds in Aczel's theory too. $\square$

4 Flexes and their decorations

To set the ground for the proof methods of Sections 6 and 7, we need to generalize our discourse to structures somewhat richer than graphs, creating a link between this paper and [23, 24]:

Definition 9. *A* flex *is a directed graph G whose nodes are subdivided into two disjoint collections:* principal nodes *and* places, *and which fulfills the following properties:*

- $\mathcal{G}$ has no arcs $[\pi_0, \pi_1]$ with π_0, π_1 places and $\pi_0 \neq \pi_1$;
- for each place π, $\mathcal{G}$ has at least one arc $[\pi, \nu]$ issuing from π that leads to a principal node.

A <u>bisimulation</u> on a flex $\mathcal{G}$ is a relation B between principal nodes of $\mathcal{G}$ such that for every pair ν_0, ν_1 of principal nodes with $\nu_0 \, B \, \nu_1$

- $\{\pi : [\pi, \nu_0]$ arc of $\mathcal{G} \,|\, \pi$ is a place $\} = \{\pi : [\pi, \nu_1]$ arc of $\mathcal{G} \,|\, \pi$ is a place $\}$ and $\{\pi : [\nu_0, \pi]$ arc of $\mathcal{G} \,|\, \pi$ is a place$\} = \{\pi : [\nu_1, \pi]$ arc of $\mathcal{G} \,|\, \pi$ is a place$\}$;
- corresponding to each arc $[\mu_b, \nu_b]$, with μ_b principal node, that enters ν_0 or ν_1 in $\mathcal{G}$ there is at least one arc $[\mu_{1-b}, \nu_{1-b}]$ of $\mathcal{G}$ with $\mu_0 \, B \, \mu_1$.

The notions of <u>trivial bisimulation</u> and <u>uncontractibility</u> are analogous to those relating to graphs (cf. Def 3). □

Through *numberings* one can define the *canonical representative* of any given flex, in analogy with what we have already done for pointed graphs. Actually, it is canonical flexes that really interest us here, and we convene in this connection that in numbering the nodes of a flex one is to exploit consecutive integers $1, \ldots, n$ for the principal nodes and subsequent consecutive integers $n+1, \ldots, n+\ell$ for the places.

For any flex $\mathcal{G}$, the relation $\sim_{\mathcal{G}}$ and the uncontractible flex $\mathcal{G}/\sim_{\mathcal{G}}$ can be defined in analogy with what has been done, following Def.3, in connection with ordinary graphs:

Definition 10. $\mu \sim_{\mathcal{G}} \nu$ holds if and only if $\mu \, B \, \nu$ holds for some bisimulation B on $\mathcal{G}$.

The principal nodes of $\mathcal{G}/\sim_{\mathcal{G}}$ are the $\sim_{\mathcal{G}}$-classes (formed by principal nodes of $\mathcal{G}$); its places are the same as the places of $\mathcal{G}$; its arcs are: all pairs $[\mu^{\sim\mathfrak{o}}, \nu^{\sim\mathfrak{o}}]$ corresponding to arcs $[\mu, \nu]$ between principal nodes of $\mathcal{G}$, and additional arcs determined as follows. For all pair π, ν consisting of a place and a principal node of $\mathcal{G}$: $[\pi, \nu^{\sim\mathfrak{o}}]$ is included among the arcs of $\mathcal{G}/\sim_{\mathcal{G}}$ if and only if $[\pi, \nu]$ is an arc of $\mathcal{G}$; $[\nu^{\sim\mathfrak{o}}, \pi]$ is included among the arcs of $\mathcal{G}/\sim_{\mathcal{G}}$ if and only if $[\nu, \pi]$ is an arc of $\mathcal{G}$. □

Let $\mathcal{G}$ be a flex with principal nodes $1, \ldots, n$ and places $n+1, \ldots, n+\ell$, and let $x_1, \ldots, x_n, z_{n+1}, \ldots, z_{n+\ell}$ be distinct variables. We will regard $\mathcal{G}$ as the encoding of a first-order formula $\Phi_{\mathcal{G}}$ which entails another formula, $\gamma_{\mathcal{G}}$, to depend exclusively on the restriction G of $\mathcal{G}$ —viewed as a graph— to principal nodes. Both $\Phi_{\mathcal{G}}$ and $\gamma_{\mathcal{G}}$ will involve $x_1, \ldots, x_n$ as free variables.

It is useful to momentarily refer by y_i to the variable x_i if $i \leq n$ and to the variable z_i if $i > n$. Thus we can introduce $y_i \in_{\mathcal{G}} y_j$ as a piece of notation for the literal $y_i \in y_j$ or for the literal $y_i \notin y_j$ depending on whether or not $[i, j]$ is an arc of $\mathcal{G}$. Notice, however, that the absence of an arc $[i, j]$ is only meant to indicate that $y_i \notin y_j$ when either y_i or y_j is one of the z_k's.

The definition of $\Phi_{\mathcal{G}}$ will exploit an auxiliary formula $\sigma_k(z_k)$ involving z_k, in addition to $x_1, \ldots, x_n$, as a free variable: $\sigma_k(z_0)$ will result from $\sigma_k(z_k)$ by replacement of all occurrences of z_k by a new variable z_0.

Definition 11.

$$\gamma_{\mathcal{G}} \quad =_{\text{Def}} \&_{0<i\leq n} \&_{0<j\leq n} (x_i \in x_j \mapsto \bigvee_{\substack{0 < g \leq n \\ [g,j] \text{ arc of } \mathcal{G}}} x_i = x_g),$$

$$\sigma_k(z_k) =_{\text{Def}} z_k \in_{\mathcal{G}} z_k \& \&_{0<i\leq n} (z_k \neq x_i \& z_k \in_{\mathcal{G}} x_i \& x_i \in_{\mathcal{G}} z_k),$$

$$\Sigma_{\mathcal{G}} \quad =_{\text{Def}} \gamma_{\mathcal{G}} \& \forall z_0 (((\bigvee_{i=1}^{n} z_0 \in x_i) \& \&_{j=1}^{n} z_0 \neq x_j) \to \bigvee_{k=n+1}^{n+\ell} \sigma_k(z_0)),$$

$$\Phi_{\mathcal{G}} \quad =_{\text{Def}} \Sigma_{\mathcal{G}} \& \&_{n<k\leq n+\ell} \exists z_k \, \sigma_k(z_k). \qquad \square$$

With respect to a structure $\mathcal{M} = (\mathcal{D}, \varepsilon)$ for the language, one defines:

Definition 12. *An* $\underline{\mathcal{M}\text{-decoration}}$ *of a flex* $\mathcal{G}$ *with principal nodes* $1, \ldots, n$ *is an interpretation* $x_1 \mapsto \xi_1, \ldots, x_n \mapsto \xi_n$ *of the free variables by elements of the support domain* $\mathcal{D}$ *of* $\mathcal{M}$ *under which* $\Phi_{\mathcal{G}}$ *is true.*
(Since the free variables of $\Phi_{\mathcal{G}}$ correspond to the principal nodes of $\mathcal{G}$, we will also call $\mathcal{M}$-decoration a function directly defined on such nodes in the appropriate fashion). $\qquad \square$

Notice that for a flex $\mathcal{G}$ devoid of places (hence undistinguishable from an ordinary graph, and with bisimulations that are also bisimulations in the sense of Def.3), the disjunction of the $\sigma_k(z_0)$'s and the conjunction of the $\exists z_k \sigma_k(z_k)$'s become empty —that is, they become false and true respectively—. Consequently the new definition of $\Sigma_{\mathcal{G}}$ comes to agree with the old one (Def.7), $\Phi_{\mathcal{G}}$ becomes equivalent to $\Sigma_{\mathcal{G}}$, and decorations in the new sense coincide with decorations in the old sense.

The results Lemma 7 and Lemma 8 about decorations of graphs have analogues about decorations of flexes:

Lemma 7⋆. *If* $x_i \mapsto \xi_i$ *(* $i = 1, \ldots, n$ *) is a decoration with* $\xi_{j_0} = \xi_{j_1}$ *of a flex* $\mathcal{G}$, *then a bisimulation* B *on* $\mathcal{G}$ *exists with* $j_0 \, B \, j_1$. $\qquad \square$

Lemma 8⋆. *Suppose* $\mathcal{M}$ *fulfills the scheme* (H). *If* B *is a bisimulation on a flex* $\mathcal{G}$, j_0 *and* j_1 *are principal nodes of* $\mathcal{G}$ *with* $j_0 \, B \, j_1$, *and* $x_i \mapsto \xi_i$ *(* $i = 1, \ldots, n$ *) is an* $\mathcal{M}$-*decoration of* $\mathcal{G}$, *then* $\xi_{j_0} = \xi_{j_1}$. $\qquad \square$

From Lemmas 7⋆ and 8⋆ one obtains:

Corollary 6. *In a theory in which all instances of* (H) *are provable and a specific* $\Phi_{\mathcal{G}}$ *cannot be refuted,* $x_i = x_j$ *can be inferred from* $\Phi_{\mathcal{G}}$ *if and only if* $i \sim_{\mathcal{G}} j$ *holds between the subscribing principal nodes* i, j. $\qquad \square$

The rest of this section is devoted to propositions that prefigure the completeness of a decision algorithm to be encountered in Section 6.

Lemma 9. *Every flex* $\mathcal{G}$ *can be decorated by hypersets.*

Proof. W.l.o.g., let us assume that $\mathcal{G}$ is uncontractible. Let $1, \ldots, n$ and $n + 1, \ldots, n+\ell$ be the principal nodes and the places of $\mathcal{G}$, respectively, and let H be the maximum height of a node (=maximum length of a path with no repeated nodes) in $\mathcal{G}$. Also let $M = 1 + \max\{\ell, H\}$, and $\mathcal{G}'$ be obtained by inserting into $\mathcal{G}$ new arcs

$$[\mu_M, \mu_{M-1}], \ldots, [\mu_2, \mu_1]; [\mu_1, \nu_1], \ldots, [\mu_1, \nu_\ell];$$
$$[\mu_2, \nu_2], \ldots, [\mu_\ell, \nu_\ell]; \qquad [\nu_1, n+1], \ldots, [\nu_\ell, n+\ell],$$

with $0, \ldots, n + \ell, \nu_1, \ldots, \nu_\ell, \mu_1, \ldots, \mu_M$ pairwise distinct nodes.

By treating $\mathcal{G}'$ as an ordinary graph, we construct $\mathcal{G}' / \sim_{\mathcal{G}'}$ as in the discussion following Def.3. It is a matter of routine to show that the function $x_i \mapsto (\mathcal{G}' / \sim_{\mathcal{G}'})\lceil i^{\sim_{\mathcal{G}'}}$, i varying over the principal nodes of $\mathcal{G}$, is a decoration. This depends on a few facts that are easy to prove: $\mu_i \not\sim_{\mathcal{G}'} \mu_j$ for $0 < i < j \leq M$; $\mu_i \not\sim_{\mathcal{G}'} \nu_g$ for $0 < i \leq M$ and $0 < g \leq \ell$; $n + g \not\sim_{\mathcal{G}'} n + h$ for $0 < g < h \leq \ell$; $i \not\sim_{\mathcal{G}'} n + g$ for $0 < i \leq n < n + g \leq n + \ell$; $i \not\sim_{\mathcal{G}'} j$ for $0 < i < j \leq n$. □

An analogous result can be drawn about sets (the proof is very much the same):

Lemma 10. *A flex can be decorated by sets if and only if it is acyclic.* □

A minor variant of the latter proposition can be obtained through the following lemma, which deserves some interest on its own:

Lemma 11. (Extensionality) *An acyclic flex cannot be uncontractible unless two of its principal nodes have the same immediate predecessors.* □

Corollary 7. *An acyclic flex $\mathcal{G}$ can be decorated by sets, different values being assigned to distinct nodes, if no two nodes of $\mathcal{G}$ have the same immediate predecessors.* □

5 A technique for re-expressing a single universal quantifier through restricted universal quantification

Our goal in this section is to re-express any given sentence of the form $\exists x_1 \cdots \exists x_n \, \forall y \, p$, with p unquantified, using restricted universal quantifiers $\forall y \in x_j$ instead of the unrestricted $\forall y$. The postulates of an axiomatic theory of hypersets, respectively those of a theory of sets, will be exploited in order to rewrite the given sentence into equivalent form

$$\exists x_1 \cdots \exists x_n (p_1 \, \& \cdots \& \, p_g \, \& \, (\forall y \in x_1 \, p_{g+1}) \, \& \cdots \& \, (\forall y \in x_n \, p_{g+n})) .$$

There will be only minor differences in the details of the rewriting technique between the case of sets and the case of hypersets. It goes without saying that the initial sentence and the resulting sentence belong to the first-order language exploited so far (without the Skolem functors of Def.6). A formula $\forall y \in x \, r$ involving a restricted quantifier is always assumed to have the variable x distinct from y, and is regarded as an abbreviation for $\forall y (y \in x \rightarrow r)$; we will manage to have no quantifiers, and no variables but the initial ones, $x_1, \ldots, x_n, y$, in the subformulae $p_1, \ldots, p_{g+n}$; also, y will not occur in $p_1, \ldots, p_g$.

Of the axioms about sets and hypersets specified in Def.8, (Z), (W) and $(A^{(1)})$ and, respectively, (Z), (W), $(D_{\not\in})$, $(D_\in)$ suffice to account for the correctness of the rewriting technique we are about to describe. A closely related technique (applicable to sets only) was expounded in [23] with richer clarifications than those we will give below.

It will be useful to have the following notation available:

Definition 13. *By $p_{x_j}^y$ one denotes the formula that results from the replacement in p of every occurrence of y by the variable x_j.*

Let q be a conjunction of literals of the two kinds $u \in v$ and $u \notin v$, with u, v drawn from among the distinct variables $x_1, \ldots, x_n, y$ of p. Suppose that q contains no pair $u \in v$, $u \notin v$ of complementary conjuncts. By $q \hookrightarrow p$ we will denote the implication

$$q \to ((\&_{0 < j \le n} \, y \ne x_j) \to p'),$$

where p' results from the replacement in p of

- *every occurrence of a conjunct $u \in v$ of q by* **true***;*
- *every occurrence $u \in v$ such that $u \notin v$ is a conjunct of q by* **false***;*
- *each of the atoms $y = x_1, \ldots, y = x_n, x_1 = y, \ldots, x_n = y$ by* **false***;*
- *each of the atoms $y = y, x_1 = x_1 \ldots, x_n = x_n$ by* **true***.* □

The following holds trivially, no matter how $\in$ is interpreted:

Lemma 12. $\exists x_1 \cdots \exists x_n \, \forall y \, p$ *is logically equivalent to*

$$\exists x_1 \cdots \exists x_n \left((\&_{0 < j \le n} \, p_{x_j}^y) \, \& \, \forall y ((\&_{0 < i \le n} (y \in x_i \hookrightarrow p)) \, \& \right.$$

$$\left. ((y \notin y \, \& \, \&_{0 < i \le n} y \notin x_i) \hookrightarrow p) \, \& \, ((y \in y \, \& \, \&_{0 < i \le n} y \notin x_i) \hookrightarrow p)) \right). \quad □$$

Since $\forall y$ can be distributed over the $n + 2$ conjuncts it embraces; since the first n of these conjuncts restrict $\forall y$; and since $(\mathbf{A}^{(1)})$ entails that $(y \in y \, \& \, \&_{0 < i \le n} y \notin x_i) \hookrightarrow p$ holds about sets, and hence can be dropped; we are left with only the problem of treating $(y \in y \, \& \, \&_{0 < i \le n} y \notin x_i) \hookrightarrow p$ in the case of hypersets and of treating $(y \notin y \, \& \, \&_{0 < i \le n} y \notin x_i) \hookrightarrow p$ both in the case of sets and in the case of hypersets. The following lemma will be crucial for this missing part of our rewriting technique.

Lemma 13. (Diagonalization) *Sets, as well as hypersets, fulfill the formula*

$$(\mathbf{Y}_{\notin}) \qquad \begin{aligned} (\&_{0 < j \le m} \&_{m < g \le n} \, x_j \ne x_g) &\to \exists y (y \notin y \, \& \, (\&_{0 < i \le n} y \notin x_i) \, \& \\ (\&_{0 < i \le n} \, y \ne x_i) &\, \& \, (\&_{0 < j \le m} \, x_j \in y) \, \& \, (\&_{m < g \le n} \, x_g \notin y)) \end{aligned}$$

with n, m integers, $0 \le m \le n$. Hypersets fulfill also the analogous formula which has $y \in y$ in place of $y \notin y$:

$$(\mathbf{Y}_\in) \qquad \begin{aligned} (\&_{0 < j \le m} \&_{m < g \le n} \, x_j \ne x_g) &\to \exists y (y \in y \, \& \, (\&_{0 < i \le n} y \notin x_i) \, \& \\ (\&_{0 < i \le n} \, y \ne x_i) &\, \& \, (\&_{0 < j \le m} \, x_j \in y) \, \& \, (\&_{m < g \le n} \, x_g \notin y)). \end{aligned}$$

Proof. Given hypersets or sets $x_1, \ldots, x_n$ that fulfill the antecedent of $(\mathbf{Y}_{\notin})$, one can exploit $(\mathbf{D}_{\notin})$ (cf. Remark 1(1)) and $(\mathbf{Z})$, $(\mathbf{W})$ to choose $z_0, z_1, \ldots, z_n$ such that $z_0 \notin z_0, z_i \notin z_i$, and

$$\forall x (x \in \{x_1, \ldots, x_n, \{x_1\}, \ldots, \{x_n\}\} \to z_0 \notin x),$$
$$\forall x (x \in x_i \text{ with } \{x_1\} \text{ with} \cdots \text{with} \{x_n\} \to z_i \notin x),$$

for $i = 1, \ldots, n$. Exploiting $(\mathbf{Z})$ and $(\mathbf{W})$ again to put $y = \{x_1, \ldots, x_m, z_0, \ldots, z_n\}$, one has that: y differs from each x_i (because $z_0 \in y, z_0 \notin x_i$); $y \notin y$ (as the opposite would imply $y = z_h$ for some $h \geq 0$, whereas $z_h \notin z_h$); $y \notin x_i$ for any x_i (because $z_i \in y$); $x_1, \ldots, x_m$ belong to y whereas $x_{m+1}, \ldots, x_n$ (each of which is distinct from $x_1, \ldots, x_m$ and from $z_0, \ldots, z_n$) do not. Thus y is as required by the consequent of $(\mathbf{Y}_\notin)$, and $(\mathbf{Y}_\notin)$ follows.

The proof of $(\mathbf{Y}_\in)$ is similar. One chooses $z_0, \ldots, z_n$ by the same criteria adopted above, and then exploits $(\mathbf{D}_\in)$ to choose a y fulfilling

$$\forall x \left(x \in y \leftrightarrow \left(x = y \vee (\bigvee_{0 < j \leq m} x = x_j) \vee (\bigvee_{0 \leq h \leq n} x = z_h) \right) \right). \qquad \square$$

Corollary 8. $\forall y \left((y \in^\pm y \ \& \ \&_{0 < i \leq n} y \notin x_i) \hookrightarrow p \right)$, *where* $\in^\pm$ *stands for* $\in$ *or for* $\notin$, *can be rewritten as an unquantified formula in the variables* $x_1, \ldots, x_n$, *both in the case of sets and of hypersets.*

Proof. Let us leave out the trivial case $\in^\pm = \in$ about sets. Inside the implication quantified by $\forall y$, the consequent of the consequent (named p' in Def.13) can be brought to conjunctive normal form $D_1 \ \& \cdots \& \ D_d$. Every D_h will consist of two sub-disjunctions, F_h and F'_h, with y occurring in no literal of F_h but occurring in all literals of F'_h. One can then rewrite the initial formula $\forall y \left((y \in^\pm y \ \& \ \&_{0 < i \leq n} y \notin x_i) \hookrightarrow p \right)$ as the unquantified conjunctive normal form $(F_1 \vee E_1) \& \cdots \& (F_d \vee E_d)$, where each E_h is the disjunction of all equalities $x_{j_0} = x_{j_1}$ such that both $x_{j_0} \in y$ and $x_{j_1} \notin y$ are disjuncts of F'_h. The equivalence is guaranteed by the preceding lemma. $\qquad \square$

Remark 2. Notice that $(\mathbf{Y}_\notin)$ and $(\mathbf{Y}_\in)$ bring to light the rationale behind the condition, imposed on flexes by Def.9, that no place π is destitute of principal successors. In order to keep the exposition close to [24] we should have <u>not</u> required that, but should have defined (slightly in contrast with Def.11 above) $\Phi_{\mathcal{G}}$ to be the formula

$$\Phi_{\mathcal{G}} =_{\mathrm{Def}} \gamma_{\mathcal{G}} \ \& \ \left(\forall z_0 \left((\&_{j=1}^n z_0 \neq x_j) \rightarrow \bigvee_{k=n+1}^{n+\ell} \sigma_k(z_0) \right) \right)$$
$$\& \ \&_{k=n+1}^{n+\ell} \exists z_k \ \sigma_k(z_k) .$$

With this alternative approach, however, $\Phi_{\mathcal{G}}$ would have turned out to be satisfiable when and only when $\mathcal{G}$ results from a flex $\overset{\circ}{\mathcal{G}}$ like those in this paper —$\overset{\circ}{\mathcal{G}}$ acyclic in the case of sets— via the following completion procedure:

For every family $Q_1, \ldots, Q_q$ of distinct $\sim_{\overset{\circ}{\mathcal{G}}}$-classes, a new place π is added to $\overset{\circ}{\mathcal{G}}$, that has immediate predecessors $Q_1 \cup \cdots \cup Q_q$ and has no successors.

In the case of hypersets, a place π' is also introduced into $\overset{\circ}{\mathcal{G}}$ that has immediate predecessors $Q_1 \cup \cdots \cup Q_q \cup \{\pi'\}$ and has no successors other than π' itself.

This other approach would have cluttered the exposition at various points, forcing us even to a minor revision of the definition of bisimulation on a flex. $\qquad \square$

6 A satisfiability decision test for a class of formulae involving restricted universal quantifiers

This section provides a satisfiability decision test applicable to formulae about hypersets that are conjunctions of conjuncts $\forall y_1 \in w_1 \cdots \forall y_m \in w_m \, p$, with $m \geq 0$, $y_1, \ldots, y_m$ distinct from one another and distinct from the variables w_j, and p a propositional combination of atoms of the two kinds $u = v$, $u \in v$ (u, v variables).

Our decision method is easily adapted to the case when formulae are to be interpreted over sets instead of over hypersets; in either case, a *solution* can be produced for the input formula r if the latter is indeed satisfiable: this is to say, a concrete substitution can be found of hypersets (respectively, sets) for the free variables of r that makes r true. One can even obtain, at least when $m \leq 1$, a detailed exhaustive specification of all possible solutions for r (cf. Theorem 2 below).

Although in showing the correctness of our decision method we will be explicitly referring to hypersets and sets (actually, we will invoke from Section 4 various propositions about such entities), the proposed method can be justified on purely proof-theoretic grounds: a careful analysis shows in fact that the axiomatization of sets/hypersets given at the end of Section 3 suffices to account for the correctness of the method. The satisfiability test can hence be used to prove or refute the existential closure $r^{\exists}$ of the given r under the axioms. The third possibility, $r^{\exists}$ neither provable nor refutable, is ruled out by the syntactic peculiarities of r.

The method to be proposed inherits from [5] (cf. also [7], Chapter 7), but is closer to a much improved decision algorithm given in [9]; both of those ancestors referred uniquely to sets, though. Notice that *the correctness of the methods crucially depends on the extensonality axiom* (**E**); actually, in the case of sets the axioms that enter into play are (**E**), (**A**), (**Z**) and (**W**), in the case of hypersets they are (**H**) and (**R**).

Let hence r be a finite (non-empty) conjunction of restrictedly quantified formulae of the kind described at the beginning. Fundamental to our analysis of r is the notion of *implicant*:

Definition 14. *Let $\mathcal{G}$ be a flex, whose principal nodes we view as free variables, and let r be of the form specified above; suppose that all free variables of r are principal nodes of $\mathcal{G}$. Then $\mathcal{G}$ is said to be an* implicant *of r if and only if r is entailed, without the help of any proper set/hyperset axioms, by the following amplification $\Phi'_{\mathcal{G}}$ of $\Phi_{\mathcal{G}}$: $\Phi'_{\mathcal{G}} =_{\text{Def}} \Phi_{\mathcal{G}} \,\&\, \&_{z \sim_{\mathcal{G}} x} z = x$.*

The number of principal nodes of $\mathcal{G}$ that have no free occurrences in r is called the exceeding size *of $\mathcal{G}$ with respect to r.* □

Checking whether $\mathcal{G}$ is an implicant of r is absolutely straightforward:

Lemma 14. *$\mathcal{G}$ is an implicant of r if and only if the finite graph structure $\mathcal{M} = (\mathcal{D}, \varepsilon)$ underlying $\mathcal{G}/\sim_{\mathcal{G}}$ makes r true under the interpretation $x \mapsto x^{\sim_{\mathcal{G}}}$ of the free variables of r.* □

We are now ready for a method to test r for *satisfiability*, i.e., a method to establish whether or not there is a substitution of sets/hypersets for the variables that makes r true:

Theorem 1. (Decidability) *r is satisfiable by hypersets if and only if there is an implicant $\mathcal{G}$ of r having:*

- *no arcs entering any place,*
- *an exceeding size g equal to the number of places and smaller than the number of free variables of r.*

The condition for the satisfiability of r by sets is entirely analogous, but with $\mathcal{G}$ acyclic. □

Proof. If part. By Lemma 9 (or by Lemma 10, when dealing with sets), every implicant (respectively, every acyclic implicant) $\mathcal{G}$ of r admits a decoration, clearly inducing an interpretation that satisfies r.

For the converse, let $x_1, \ldots, x_f$ be the free variables of r and let $x_i \mapsto \xi_i$ be an interpretation of them under which r is true. By generalized extensionality (cf. Remark 1(3)), fewer than f members of $\xi_1 \cup \cdots \cup \xi_f$, say $\eta_1, \ldots, \eta_g$ with $g < f$, suffice to discriminate $\xi_1, \ldots, \xi_f$ from one another. This is to say: every η_h belongs to some ξ_i and differs from any ξ_j and from any η_e with $e \neq h$; moreover, for all pair i, j with $\xi_i \neq \xi_j$, there is at least one λ among $\xi_1, \ldots, \xi_f, \eta_1, \ldots, \eta_g$ such that $\lambda \in \xi_i$ if and only if $\lambda \notin \xi_j$. We take the ξ's and η's as the principal nodes of a flex $\mathcal{G}$, which also has g places $\zeta_1, \ldots \zeta_g$. We introduce an arc $[\mu, \nu]$ between two principal nodes μ, ν of $\mathcal{G}$ if and only if $\mu \in \nu$; also, we introduce an arc from each ζ_h to the corresponding η_h.

Showing that all this characterizes an implicant of r requires well-established techniques (cf. [5, 7, 9]). □

Remark 3. Optimizations of the search method implicit in this theorem are at hand. The size of the search space can in fact be cut down by requiring that the sought $\mathcal{G}$ has:

- exceeding size g smaller than the number of variables that occur on the right of $\in$ in restricted quantifiers of r —these are named *critical variables*;
- a number $N = \lceil log_2(g + 1) \rceil$ of places (with no entering arcs).

In order to adapt the proof of the theorem to these changes, it suffices to

- choose $\eta_1, \ldots, \eta_g$ so that they discriminate from one another just those ξ's that are values of critical variables;
- introduce arcs from ζ's to η's in such a way that no two η's have the same immediate predecessors and that at least one arc $[\zeta_k, \eta_h]$ enters each η_h. □

Let us now consider the special sub-problem of the satisfiability problem under discussion in which the maximum value of m, taken over all conjuncts $\forall y_1 \in w_1 \cdots \forall y_m \in w_m\ p$ of r, is ≤ 1. In this case one has:

Theorem 2. (Syllogistic decomposition) *When $m \leq 1$, r can be rewritten as a finite disjunction of $\Phi_{\mathcal{G}}$'s, with $\mathcal{G}$ implicant of r and $\mathcal{G}$ acyclic in the case of sets.* (Notice that each $\Phi_{\mathcal{G}}$ in this decomposition is satisfiable).

Unless this decomposition of r is empty, at least one of its $\mathcal{G}$'s has fewer places than r has variables.

Proof. We are to explain how to associate with every interpretation $x_i \mapsto \xi_i$ ($i = 1, \dots, f$) satisfying r an implicant $\mathcal{G}$ of r whose size is elementarily bound to the number f of free variables in r, with $\Phi_{\mathcal{G}}$ holding in the interpretation. Thus, clearly, r will be equivalent to the disjunction of all $\Phi_{\mathcal{G}}$'s obtainable in this fashion.

The construction of $\mathcal{G}$ resembles very much the one we have just seen, but we will no longer demand, as we have done in Theorem 1, that no arcs enter into places; we will simply manage to have $\sigma_{k_0}(z) \neq \sigma_{k_1}(z)$ (cf. Def.11) for any two distinct places k_0, k_1 of $\mathcal{G}$.

For each triple P, Q, C with $P, Q \subseteq \{1, \dots, f\}$, $Q \neq \emptyset$ and $C \in \{\text{false}, \text{true}\}$, we select an $\eta_{P,Q,C}$ —if any— such that
$\xi_i \in \eta_{P,Q,C}$ if and only if $i \in P(i = 1, \dots, f)$;
$\eta_{P,Q,C} \in \xi_i$ if and only if $i \in Q(i = 1, \dots, f)$; $\eta_{P,Q,C} \in \eta_{P,Q,C}$ if and only if C.
Unlike in the proof of Theorem 1, the η's are directly selected as places this time, additional places ζ being not needed (of course the ξ's are again taken as principal nodes). The arcs of $\mathcal{G}$ will be all pairs $[\mu, \nu]$ of nodes with $\mu \in \nu$, save the pairs $[\eta', \eta'']$ of places with $\eta' \neq \eta''$, because the latter are forbidden as arcs by Def.9. An argument quite similar to the one exploitable for the above proof shows that $\mathcal{G}$ is an implicant of r and that $\Phi_{\mathcal{G}}$ is satisfied by the interpretation $x_i \mapsto \xi_i$ with which we have started.

Concerning the second part of the hypothesis, observe that if the η's had been chosen by the more restrictive criteria adopted in the proof of Theorem 1, then we would have obtained an implicant $\overset{\circ}{\mathcal{G}}$ of r endowed with fewer than f places. There is no guarantee that $\Phi_{\overset{\circ}{\mathcal{G}}}$ is satisfied by the interpretation $x_i \mapsto \xi_i$; nevertheless, since $\Phi_{\overset{\circ}{\mathcal{G}}}$ is satisfiable, it holds in some interpretation that clearly makes r true too. Hence $\Phi_{\overset{\circ}{\mathcal{G}}}$ contributes to the decomposition of r. □

It is a simple task to verify that a similar disjunctive decomposition can be carried out using $\Sigma_{\mathcal{G}}$'s (see Def. 11) instead of $\Phi_{\mathcal{G}}$'s. An advantage of proceeding in terms of the $\Sigma_{\mathcal{G}}$'s is that it brings to a decomposition with many fewer disjuncts; however, while the $\Phi_{\mathcal{G}}$'s logically exclude each other, this very seldom happens with the $\Sigma_{\mathcal{G}}$'s.

7 Unification of set/hyperset terms

In this section we narrow down the satisfiability problem solved in the preceding section to the collection of formulae r whose conjuncts are of the five kinds: $u = v$, $u \neq v$, $u \in v$, $u \notin v$ (u, v variables), and $\forall y \in wp$ with y *never occurring to the immediate right of* $\in$ (or of $\notin$) *inside* p —as before, y is distinct from w, and p is a propositional combination of atoms $u = v$, $u \in v$.

Taking advantage of these restraints on the structure of r, we will succeed in simplifying the satisfiability decision method, as well as the format of the exhaustive disjunctive decomposition of r. Under still stronger restraints, we will arrive at the rather important subcase which is the *unification problem* treated in this paper. The format of the syllogistic decomposition can then be made particularly appealing: it can be represented as a finite family of syntactic substitutions generating the space of all solutions to r. Borrowing the terminology from the field of equational unification theory (E-unification, cf. e.g. [26, 19]), we might say that the collection of formulae in question has a *finitary* unification problem.

Recall that by Theorem 2 implicants of r of exceeding size ≥ 1 are of no use when there is at most one quantifier per conjunct. This suggests one way in which the satisfiability conditions in Theorem 1 can be specialized to the case at hand:

Theorem 3. (Decidability) *A necessary and sufficient condition for the satisfiability of r by hypersets is the existence of an implicant $\mathcal{G}$ of r, of exceeding size 0, endowed with fewer places than r has variables, in which no arcs enter any place. The condition for the satisfiability of r by sets is entirely analogous, but with $\mathcal{G}$ acyclic.* □

Clearly, one's disregard for the principal predecessors of places in the proof of this theorem, is justified by the restraints imposed above on the occurrences of bound variables in r. Exploiting the same idea again, one can group the disjuncts of the syllogistic decomposition of r (cf. Theorem 2) into 'clusters', as follows: $\Phi_{\mathcal{G}_0}$ and $\Phi_{\mathcal{G}_1}$ *go into the same cluster if and only if* $\mathcal{G}_0, \mathcal{G}_1$ *get transformed into the same implicant $\mathcal{G}$ by the simplification process described in the preceding proof*. Referring by $\Phi_{\mathcal{G}}^{\vee}$ to the disjunction of all Φ's that reside in the cluster identified by $\mathcal{G}$, one easily obtains:

Lemma 15. $\Phi_{\mathcal{G}}^{\vee}$ *is logically equivalent to the formula that results from $\Phi_{\mathcal{G}}$ by replacement of each $\sigma_k(z_k)$ (cf. Def.11) by the weaker formula $\&_{0 < i \leq n}(z_k \neq x_i \& z_k \in_{\mathcal{G}} x_i)$, accordingly modifying $\sigma_k(z_0)$ too.* □

Then, straightforwardly:

Theorem 4. (Syllogistic decomposition) r *is equivalent to the disjunction of the (finitely many) $\Phi_{\mathcal{G}}^{\vee}$ formulae associated with implicants $\mathcal{G}$ of r that enjoy the following property: no arcs enter any place of $\mathcal{G}$, and no two places have the same immediate successors in $\mathcal{G}$.* □

Corollary 9. *It is possible to automatically produce an exhaustive characterization of the solutions, by sets/hypersets, of a conjunction of literals of the forms* $u = v$, $u \neq v$, $u \in v$, $u \notin v$, $u \subseteq v$, $u \not\subseteq v$, $u = \emptyset$, $u = v \cap z$, $u = v \setminus z$, $u = v \cup z$, $u = v$ *with* z, $u = v$ *less* z *(u, v, z variables), given in terms of $\Phi_{\mathcal{G}}^{\vee}$ formulae.*

Proof. All literals that involve constructs not officially present in our first-order language, can easily be rendered in terms of restricted quantifiers. To make an example, $(u = v \setminus z) =_{\text{Def}} \forall y \in u(y \in v \& y \notin z) \& \forall y \in v(y \in z \lor y \in u)$. Details are left to the reader. $\qquad\square$

Remark 4. The contents of the latter corollary, relative to sets, was first published in [8]. Much earlier and using a quite different terminology, in [16], it had been proved that a suitable $\sim_{\mathcal{G}}$ can be determined before $\mathcal{G}$ itself when there are no literals involving **with** and **less** and one is simply interested in the satisfiability —not in the decomposition— of the given conjunction. In such case, one can in fact fix $\sim_{\mathcal{G}}$ to be the relation holding between two variables x and y if and only if $x = y$ is satisfied by every interpretation { variables } $\longrightarrow \{ \underline{\emptyset}, \{ \underline{\emptyset} \} \}$ that satisfies all literals of the forms $u = v$, $u \subseteq v$, $u = v \cap z$, $u = v \setminus z$ and $u = v \cup z$ in the given conjunction. $\qquad\square$

We now restrict our discourse to formulae r whose conjuncts are of the four kinds $u = v$, $u \in v$, $u = \underline{\emptyset}$, $u = v \text{ with } z$. This restriction will remarkably simplify the way one can shape the syllogistic decomposition of r.

To obtain the slenderer decomposition we are aiming at, we could again use a disjunct clustering technique alike to the one above. However, since an approach of that kind might obscure the details of unification taken on its own, we opt for a direct approach this time. The decision algorithm is presented first, and later on we will prove its correctness, completeness and exhaustiveness.

Algorithm 1. (Unification) *A formula r of the form just described is to be tested for satisfiability over sets/hypersets. We denote by $\mathcal{X}$ the collection of all free variables of r. Also, we denote by $x \mapsto x_*$ a function from $\mathcal{X}$ to 'new' variables, i.e. to variables that do not occur in r. Arbitrarily impose a total ordering $<$ on $\mathcal{X}$, and convene that $\underline{\emptyset} < x$ is to hold for all x in $\mathcal{X}$, where $\underline{\emptyset}$ is the special constant introduced by Def.6.*

Non-deterministically construct a flex $\mathcal{G}$ as follows:

- *As family of the principal nodes of $\mathcal{G}$, take the whole $\mathcal{X}$.*
- *Guess the restriction G of $\mathcal{G}$ —viewed as a graph— to principal nodes. If r is being tested over sets, then G must be acyclic.*
- *Guess a function $x \mapsto x^*$ from $\mathcal{X}$ to $\mathcal{X} \cup \{ \underline{\emptyset} \}$, with $x^* \leq x$ for all x in $\mathcal{X}$.*
- *As places of $\mathcal{G}$ take the collection of all x_*^*'s, with x varying in $\mathcal{X}$ and x^* different from $\underline{\emptyset}$ (clearly x_*^* stands for $(x^*)_*$). As arcs leaving the places of $\mathcal{G}$, take all pairs $[x_*^*, x]$ with x^* different from $\underline{\emptyset}$. Let no arcs enter any place in $\mathcal{G}$.*

If $\mathcal{G}$ is not a implicant of r, then fail, else declare that r is satisfiable and output $\mathcal{G}$. $\qquad\square$

Of course a single failure in this algorithm is not meant to signal that r is unsatisfiable, but simply that wrong choices have been made: before r can securely be declared unsatisfiable, all possible choices for G and $x \mapsto x^*$ must have been tried unsuccessfully. When such choices lead to a success, the resulting

$\mathcal{G}$ can be produced in output in the following form, that makes the meaning of $\mathcal{G}$ explicit (notice that in our current context $\mathcal{G}$ represents a whole cluster of implicants of r, and hence it encodes more than the isolated formula $\Phi_{\mathcal{G}}$):

Algorithm 2. (Substitution output) *An implicant $\mathcal{G}$ of r, detected by the preceding algorithm, is given. Arbitrarily choose a representative variable from each $\sim_{\mathcal{G}}$-class. Let $x_1, \ldots, x_n$ be all such representative variables. For each x_i, let $x_{i1}, \ldots, x_{im_i}$ be the variables that have been elected to represent the immediate principal predecessors of $x_i^{\sim o}$ in $\mathcal{G}/\sim_{\mathcal{G}}$. For every x_i, output the equation $x_i = \{x_{i1}, \ldots, x_{im_i}\}$ if $x_i^\star$ is $\underline{\emptyset}$, else output the equation $x_i = (x_i)_\star^\star \mathtt{with} x_{i1} \mathtt{with} \cdots \mathtt{with} x_{im_i}$; in either case, also produce all equations $z = x_i$, with z —syntactically distinct from x_i— belonging to $x_i^{\sim o}$.* ☐

The correctness of the proposed unification algorithm is contained in the following two lemmas. It is interesting to observe that the proof of the former of these *is the only place where the element removal axiom* (**L**) *is ever exploited in this paper in connection with the satisfiability problems treated.*

Lemma 16. *Let r be of the form admitted as input to Algorithm 1. If r is satisfied by an interpretation $x \mapsto \xi_x$ (x in $\mathcal{X}$) of variables by sets/hypersets, then convenient choices can be performed by Algorithm 1 so that an implicant $\mathcal{G}$ of r is obtained whose corresponding conjunction $E_{\mathcal{G}}$ of equalities, produced by Algorithm 2, is satisfied by a suitable extension of the said interpretation.*

Proof. For all x in $\mathcal{X}$, let $\xi_x^\star =_{\mathrm{Def}} \xi_x \mathtt{less} \xi_{w_1} \mathtt{less} \cdots \mathtt{less} \xi_{w_g}$, where $w_1, \ldots, w_g$ are the variables forming $\mathcal{X}$. As arcs of G, Algorithm 1 must take all pairs $[u, v]$ with u, v in $\mathcal{X}$ and $\xi_u \in \xi_v$. As $x^\star$ it must take $\underline{\emptyset}$ if $\xi_x^\star$ is empty, the first z (w.r.t. $<$) in $\mathcal{X}$ for which $\xi_z^\star = \xi_x^\star$ otherwise. A line of reasoning parallelling the proof of Lemma 8⋆ shows that $u \sim_{\mathcal{G}} v$ holds between two variables if and only if u, v have the same immediate predecessors, if and only if $\xi_u = \xi_v$. Completing the proof is trivial. ☐

Lemma 17. *Let $\mathcal{G}$ be an implicant of r produced by Algorithm 1, and let $E_{\mathcal{G}}$ be the corresponding conjunction of equalities produced by Algorithm 2. Any interpretation of the variables by sets/hypersets that satisfies $E_{\mathcal{G}}$ satisfies r too.* ☐

We conclude that:

Corollary 10. *It is possible to automatically produce an exhaustive characterization of the solutions, by sets/hypersets, of a given conjunction of literals of the forms $u = v$, $u \in v$, $u = \underline{\emptyset}$, $u = v \mathtt{with} z$ (u, v, z variables), given in terms of syntactic substitutions involving, in addition to variables and $\underline{\emptyset}$, only the construct* $\mathtt{with}$. ☐

Remark 5. The latter lemma and corollary generalize under one salient respect, but restrain under others, what can be drawn about the Dovier–Pontelli unification algorithm described in [13] (cf. also [14]). Here unification can be performed not only relative to sets, but also relative to hypersets. On the other hand our Algorithm 1 —at least in its current form— does not treat at all uninterpreted Herbrand functors. Also, although conceptually very simple, it is less appealing from an implementer's point of view than a specification à la Robinson like the one given in [14], or (to cope with the new problems arising from the circularity of membership) à la Martelli-Montanari (cf. [22]). □

A result generalizing Corollary 10 can be proved in set/hyperset theories comprising, in addition to the axioms used so far (cf. Def.8), the _binary union_ and _difference_ axioms: $\exists z \, \forall x (x \in z \leftrightarrow (x \in y_1 \lor x \in y_2))$, $\exists z \, \forall x (x \in z \leftrightarrow (x \in y_1 \,\&\, x \notin y_2))$. The collection of restrictedly quantified formulae that then become treatable consists of those conjunctions r whose conjuncts are of the three kinds $u = v$, $u \in v$ and $\forall y \in wp$, where the literals ℓ_{ij} in the disjunctive normal form $\bigvee_{0 < i \le N} \&_{0 \le j \le M_i} \ell_{ij}$ of p are of two kinds: $x = z$ and $x \in t$, with t distinct from y. We are in this sense forbidding that negative literals appear in the matrix p of any quantified conjunct and that quantified variables occur to the right of $\in$.

Substitution pairs, in this case, will have the following form: $x_i = y_{i1} \cup \cdots \cup y_{in_i}$ with z_{i1} with $\cdots$ with z_{im_i}. Without proof, we state a new analogue of Corollary 9:

Lemma 18. _It is possible to automatically produce an exhaustive characterization of the solutions, by sets/hypersets, of a given conjunction of literals of the forms $u = v$, $u \in v$, $u \subseteq v$, $u \subseteq v \cap z$, $v \setminus z \subseteq u$, v less $z \subseteq u$, $u = \emptyset$, $u = v \cup z$, $u = v$ with z (u, v, z variables), given in terms of syntactic substitutions involving, in addition to variables and $\emptyset$, only the constructs_ with _and_ $\cup$. □

Acknowledgements. Funding came from MURST 40% projects (_Calcolo algebrico e simbolico, Logica matematica e applicazioni_), from Compulog 2 (Esprit project 6810), and from MURST 60% (_Programmazione logica con insiemi_).

References

1. P. Aczel. _Non-well-founded sets._ Vol 14, Lecture Notes, Center for the Study of Language and Information, Stanford, 1988.
2. J. Barwise, J. Etchemendy. _The liar: an essay on truth and circular propositions._ Oxford University Press, 1987.
3. M. Beeson. Towards a computation system based on set theory. Theoretical Computer Science 60,297–340, 1988.
4. S. Baratella, R. Ferro. A theory of sets with the negation of the axiom of infinity. AILA preprints n.8, 1991.
5. M. Breban, A. Ferro, E.G. Omodeo, J.T. Schwartz. Decision procedures for elementary sublanguages of set theory. II. Formulas involving restricted quantifiers, together with ordinal, integer, map, and domain notions. _Comm. on Pure and Appl. Mathematics,_ 34,177–195, 1981.
6. R. Boyer, E. Lusk, W. McCune, R. Overbeek, M. Stickel, L. Wos. Set theory in first-order logic: Clauses for Gödel's axioms. _J. of Automated Reasoning,_ 2,289–327, 1986.
7. D. Cantone, A. Ferro, E.G. Omodeo. _Computable set theory,_ 1st vol. Clarendon Press, Oxford, International Series of Monographs on Computer Science, 1989.

8. D. Cantone, S. Ghelfo, E.G. Omodeo. The automation of syllogistic. I. Syllogistic normal forms. *J. of Symbolic Computation*, 6(1),83–98, 1988.

9. D. Cantone, E.G. Omodeo, A. Policriti. The automation of syllogistic. II. Optimization and complexity issues. *J. of Automated Reasoning*, Kluwer Academic Publishers, 6(2),173–187, 1990.

10. B. Dreben, W. D. Goldfarb. *The decision problem: solvable classes of quantificational formulas.* Addison-Wesley, 1979.

11. R. Dionne, E. Mays, F.J. Oles. A non-well-founded approach to terminological cycles. Proceedings of AAAI-92, San Jose, Cal.

12. R. Dionne, E. Mays, F.J. Oles. The equivalence of model-theoretic and structural subsumption in description logics. To appear in the proceedings of IJCAI-93.

13. A. Dovier, E.G. Omodeo, E. Pontelli, G. Rossi. {log}: a logic programming language with finite sets. In K. Furukawa ed., *Logic Programming, Proceedings of the Eighth International Conference* (ICLP'91, Paris), The MIT Press, 111–124, 1991.

14. A. Dovier, E.G. Omodeo, E. Pontelli, G.-F. Rossi. Embedding finite sets in a logic programming language. In E. Lamma, P. Mello eds, *Extensions of Logic Programming*, Volume 660 of *Lecture Notes in Artificial Intelligence*, 150–167, Springer-Verlag, 1993.

15. M. Forti, F. Honsell. Set theory with free construction principles. *Annali Scuola Normale Superiore di Pisa*, Cl. Sc. (IV)10, 493–522, 1983.

16. A. Ferro, E.G. Omodeo, J.T. Schwartz. Decision procedures for elementary sublanguages of set theory. I. Multi-level syllogistic and some extensions. *Communications on Pure and Applied Mathematics*, 33,599–608, 1980.

17. M.R. Garey, D.S. Johnson. *Computers and intractability* - A Guide to the theory of NP-Completeness. W.H. Freeman and Co., New York, A series of books in the mathematical sciences, 1979.

18. J. v. Heijenoort, editor. *From Frege to Gödel* - A source book in mathematical logic, 1879–1931. Harvard University Press, Source books in the history of the sciences, 3rd printing, 1977.

19. J.P. Jouannaud, C. Kirchner. Solving equations in abstract algebras: a rule-based survey of unification. In J.-L. Lassez, G. Plotkin eds., *Alan Robinson's anniversary book*, 1991.

20. J.W. Lloyd. *Foundations of Logic Programming*. Springer-Verlag series *Symbolic Computation - Artificial Intelligence*, 2nd edition, 1987.

21. Yu.I. Manin. *A course in mathematical logic.* Springer-Verlag, New York, 1977.

22. A. Martelli, U. Montanari. An efficient unification algorithm. *ACM Transactions on Programming Languages and Systems* (TOPLAS), 4(2), 258–282, 1982.

23. E.G. Omodeo, F. Parlamento, A. Policriti. Decidability of $\exists^*\forall$-sentences in membership theories. University of Udine Research Report Nr.6, May 1992.

24. E.G. Omodeo, F. Parlamento, A. Policriti. A derived algorithm for evaluating ε-expressions over sets. Univ. of Rome "La Sapienza"–D.I.S., Rep.07.92 May 1992. (To appear on the *J. of Symbolic Computation*, special issue on automatic programming.)

25. J.T. Schwartz, R.K.B. Dewar, E. Dubinsky, E. Schonberg. *Programming with sets* - An introduction to SETL. Texts and Monographs in Computer Science, Springer-Verlag, 1986.

26. J.H. Siekmann. Unification Theory. A Survey. *J. of Simbolic Computation*, 1989. Special Issue on Unification. 7(3,4) 8(1-5).

Reasoning with Contexts[*]

William M. Farmer, Joshua D. Guttman, F. Javier Thayer

The MITRE Corporation, 202 Burlington Rd, Bedford, MA 01730-1420, USA

Telephone: 617-271-2749; Fax: 617-271-3816

Email: `farmer, guttman, jt@mitre.org`

Abstract. Contexts are sets of formulas used to manage the assumptions that arise in the course of a mathematical deduction or calculation. This paper describes some techniques for symbolic computation that are dependent on using contexts, and are implemented in IMPS, an Interactive Mathematical Proof System.

1 Introduction

A *context* is a set of formulas $\Gamma = \{\varphi_1, \ldots, \varphi_n\}$. The formulas of a context ordinarily serve as background assumptions. They may contain both closed formulas, such as the axioms of a theory, and open formulas, such as:

- $x \neq 0$,
- $0 < y \wedge f$ is differentiable at y, and
- G is a finite group.

A formula ψ is true in the context Γ if the members of Γ logically imply ψ. Contexts are a natural device for managing the assumptions that arise in the course of a mathematical deduction or calculation. Contexts also provide a mechanism for keeping track of what can be legitimately assumed at various places in a formula. Thus, in the formula $0 < x^2 \supset x/x = 1$, one can assume that the variable x in the subexpression $x/x = 1$ is nonzero. A "context at a place in a formula" is called a *local context* [10].

The following examples illustrate the need for contexts in rigorous mathematical reasoning.

- If G is a set with a binary operation, then a theorem about finite groups can be applied to G only when "G is a finite group" is assumed.
- The term x/x can be simplified to 1 only if $x \neq 0$ is known to be true.
- The equation

$$\int_a^b x^\alpha dx = \frac{b^{\alpha+1} - a^{\alpha+1}}{\alpha + 1}$$

holds only when $\alpha \neq -1$.

---

[*] Supported by the MITRE-Sponsored Research program.

− The power series expansion

$$\frac{1}{1-x} = \sum_{i=1}^{\infty} x^i$$

is legitimate only for $|x| < 1$.

Although symbolic computation can be as highly context-dependent as any other form of mathematical reasoning, existing symbolic manipulation systems usually are not sensitive to contexts. As a consequence, these systems may do such things as simplifying x/x to 1 even when there is no assumption that $x \neq 0$. Computations in these systems typically behave as though they were applying universally valid equalities without auxiliary conditions.

This simple approach is problematic because many indispensable equalities, like $x/x = 1$, *do* have auxiliary conditions. Keeping track of such auxiliary conditions within complex expressions can become extremely difficult, and without precautions, can very easily lead to falsehoods. One might hope to eliminate side conditions by using generic quantification, that is, validity "almost everywhere" of a formula. For instance, if $P(x,y)$ is the formula $y - x^2 \neq 0$, then P holds for all values of the parameter $(x,y) \in \mathbf{R}^2 \setminus K$, where the set K is the parabola given by the equation $y = x^2$. K is a small subset of $\mathbf{R}^2$ (in measure), so that it is customary to say "P is true for almost all values of (x,y)." However, this kind of quantification is fraught with difficulties in that the most basic logical operations, such as instantiation of universals, fail. Thus it is not true that for an arbitrary polynomial function $\varphi : \mathbf{R} \to \mathbf{R}^2$, $P(\varphi(t))$ holds outside some small set in $\mathbf{R}$. Indeed, if ran $\varphi \subseteq K$, then $P(\varphi(t))$ is always false.

Logically, assumptions can be incorporated into an expression by turning the expression into a conditional. For example, $|x + |3 - x||$ has the same value as the conditional

$$\begin{cases} 2x - 3 \text{ if } x \geq 3 \\ 3 \qquad \text{otherwise} \end{cases}$$

However, this reduction is usually undesirable, firstly because the conditional is much harder to understand than the original expression, and secondly because there are no simple algebraic rules for reducing conditionals. Alternatively, one could consider the original expression in two different contexts: one context Γ_1 contains the assumption $x \geq 3$ and the other Γ_2 contains the assumption $x < 3$. In the context Γ_1, the expression $|x + |3 - x||$ simplifies to $P_1 = 2x - 3$, while in Γ_2, it simplifies to $P_2 = 3$. The original expression is thus replaced with two polynomial expressions P_i in the variable x.

Reasoning with contexts is one of the main design principles of the IMPS Interactive Mathematical Proof System [5, 6]. IMPS is intended to support the traditional techniques of mathematical reasoning. In particular, the logic of IMPS [3, 4, 8] allows functions to be partial and terms to be undefined. The system consists of a data base of mathematics and a collection of tools for exploring, applying, extending, and communicating the mathematics in the data base. One of the chief tools is a facility for developing formal proofs. In contrast to the formal proofs described in logic textbooks, IMPS proofs are a blend of computation and high-level inference. Consequently, they resemble intelligible informal proofs, but unlike informal proofs, all details of an IMPS proof are machine checked.

Reasoning in IMPS is always performed with respect to some context. In the process of constructing a proof, many different contexts are usually needed. IMPS represents each context with a separate data structure that holds the set of assumptions of the context as well as various kinds of derived information. The entries in an IMPS proof are compound formulas, called *sequents*, consisting of a single formula called the *assertion* together with a context. Low-level reasoning is performed by two kinds of context-dependent computation: automatic expression simplification and semi-automatic theorem application.

The rest of the paper describes the mechanisms in IMPS for expression simplification and theorem application. These mechanisms illustrate that managing symbolic computation with contexts is feasible as well as desirable. An example of an IMPS proof involving expression simplification and theorem application is given in Section 5.

Each context in IMPS belongs to a *theory*, which is a formal language plus a set of closed formulas of the language called *axioms*. All the contexts created in course of a proof belong to the same theory. The assumptions of a context, by definition, form a superset of the axioms of its theory. Thus there are two classes of context assumptions: the theory axioms are closed formulas used as permanent assumptions, and the remaining assumptions are possibly open formulas used as temporary assumptions. Section 5 contains statistics about the set of contexts generated by the example proof.

IMPS supports the "little theories" version of the axiomatic method [7]. In this approach, a number of theories are used in the course of developing a portion of mathematics. Different theorems are proved in different theories, depending on the amount and kind of mathematics that is required. Theories are linked together by interpretations which serve as conduits to pass results from one theory to another. We argue in [7] that this way of organizing mathematics across a network of linked theories is advantageous for managing complex mathematics by means of abstraction and reuse. IMPS also supports the "big theory" version of the axiomatic method in which all reasoning is performed within a single powerful and highly expressive axiomatic theory, such as Zermelo-Fraenkel set theory. Both versions of the axiomatic method are well established in modern mathematical practice.

2 The Simplifier

The IMPS simplifier is responsible for substantial size steps in computation and inference. It is always invoked on an expression relative to a context Γ, and serves three primary purposes:

- To invoke a variety of theory-specific transformations on expressions, such as rewrite rules and simplification of polynomials (given that the theory has suitable algebraic structure, such as that of a ring);

- To make simplifications based on the logical structure of an expression, often at locations deeply nested within it;
- To discharge the great majority of the assertions about definedness needed to apply many inferences and lemmas.

This last goal, and many other aspects of the simplifier, are required by the relatively permissive IMPS logic. Since in the IMPS logic functions may be partial and terms may be undefined, term simplification must involve a considerable amount of definedness checking. For example, simplifying expressions naively may cancel undefined terms, reducing a possibly undefined expression such as $1/x - 1/x$ to 0, which is certainly defined. In this example, the previous replacement is valid if the context Γ can be seen to entail the definedness or "convergence" of $1/x$. In general, algebraic reductions of this kind produce intermediate definedness formulas to which the simplifier is applied recursively. These formulas are called *convergence requirements*.

The IMPS logic also allows subtypes. By a type we mean a basic domain of objects, such as the points of a space, the real numbers, or the partial real functions $\mathbf{R} \rightarrow \mathbf{R}$. A type may have several distinguished subtypes, which denote nonempty subsets of the whole domain of the type. A *sort* is a type or a subtype. A variable always ranges over some particular sort. The logic supplies a logical operator **defined-in**, written $\downarrow$, which can be used to assert that a term has a value in a particular sort, for instance $6.2/3.1 \downarrow \mathbf{Z}$. There is also a logical operator **is-defined**, written as $\downarrow$ without a trailing sort, which asserts only that the term takes some value. This is equivalent to asserting that the term is defined in the sort that its syntax predicts. For instance, $6.2/3.1 \downarrow$ is equivalent to $6.2/3.1 \downarrow \mathbf{R}$. Thus by a convergence requirement we mean a formula of the form $t \downarrow \alpha$ or $t \downarrow$ which must be true to justify a manipulation.

The notion of quasi-equality serves as the correctness requirement for the simplifier. We say that s and t are quasi-equal, written $s \simeq t$, if

$$(s \downarrow \vee t \downarrow) \supset s = t.$$

Since in the IMPS logic $s = t$ can hold only if both terms are defined, $s \simeq t$ holds just in case the terms are either both undefined, or both defined with the same value. In IMPS, quasi-equality justifies substituting t in place of s at any occurrence at which t is free for s.

The simplifier is designed to transform an expression e to e' relative to the assumptions of a context Γ (in a theory $\mathcal{T}$), only if Γ entails $e \simeq e'$.

The algorithm traverses the expression recursively; as it traverses propositional connectives it does simplification with respect to a richer context. Thus, for instance, in simplifying an implication $A \supset B$, A may be assumed true in the "local context" relative to which B is simplified. Similarly, in simplifying the last conjunct C of a ternary conjunction $A \wedge B \wedge C$, A and B may be assumed in the "local context." On the other hand, when a variable-binding operator is traversed, and there are context assumptions in which the bound variable occurs free, then the simplifier must either rename the bound variable or discard the offending assumptions. The strategy of exploiting local contexts is justified in [10].

At any stage in this recursive descent, if a theory-specific procedure may successfully be used to transform the expression, it is applied. These procedures currently include:

- Algebraic simplification of polynomials, relative to a range of algebraic theories (see Section 3);
- A decision procedure for linear inequalities, based on the variable elimination method used in many other theorem provers, for instance by Boyer and Moore [1];
- Rewrite rules for the current theory T, or for certain theories T_0 for which IMPS can find interpretations of T_0 in T.

The procedures for algebraic simplification have been coded so as to record the convergence requirements for all the transformations they make. When the procedure returns, the simplified value is used only if the requirements can be discharged by a recursive invocation of the simplifier.

Rewrite rules also generate convergence requirements. Suppose that we have a theorem such as for instance

$$\forall x, y : \mathbf{Z} \, . \, x < y \Leftrightarrow x + 1 \leq y$$

which is being used as a rewrite rule from left to right. If a portion of an expression being simplified is of the form $s < t$, then we would like it to be rewritten to $s + 1 \leq t$, but only if s and t both take values in $\mathbf{Z}$. If one of the terms is undefined, or if it has a value in $\mathbf{R}$ but not in $\mathbf{Z}$, then the change is not justified as an instance of the theorem.

If no transform is applicable, then a simplification routine determined by the top-most constructor or quasi-constructor of the expression to be simplified is applied. These routines normally invoke the simplifier recursively on subexpressions, with different contexts. The routines for a few constructors, especially the definedness constructors (Section 2.1), use special routines exploiting information extracted from the axioms and theorems of the context's theory.

The emphasis on a powerful simplification procedure to allow large inference steps in the course of interactive proof development is shared with Eves and its predecessor m-Eves [2], as well as the more recent PVS [11].

2.1 Reasoning about Definedness

Because simplification and theorem application involve large numbers of convergence requirements, it is important to automate, to the greatest extent possible, the process of checking that expressions are well-defined or defined with a value in a particular sort. This kind of reasoning must rely heavily on axioms and theorems of the axiomatic theory at issue. The algorithm for simplifying definedness assertions is separated into two layers, according to whether recursive calls to the simplifier are involved.

The Lower Level of Definedness Checking. In the lower level, there are no recursive calls to the simplifier; two kinds of information are used:

- *Totality* theorems of the form $\forall x_1 : \alpha_1, \ldots, x_n : \alpha_n . f(x_1, \ldots, x_n) \downarrow \alpha$.
- *Unconditional sort coercions* of the form $\forall x : \alpha . x \downarrow \beta$.

A set S of unconditional sort coercion theorems determine a pre-order $\ll_S$ on sorts. $\alpha \ll_S \beta$ if and only if in every model of S, the denotation of α is included in the denotation of β. The relation $\ll_S$ is a pre-order rather than a partial order because for two different syntactic sorts α and β, we may have $\alpha \ll_S \beta$ and $\beta \ll_S \alpha$; in this case α and β have the same denotation in every model of S. Fix some collection S of axioms and theorems of T, with respect to which definedness-checking is being carried out.

The relation $\ll_S$ together with the totality theorems are used in IMPS by an algorithm for checking definedness. We use totality information and unconditional sort coercions to extract "critical pairs of subterms and sorts," or simply critical pairs, from t. By a set of critical pairs, we mean a set of pairs $\langle s_i, \beta_i \rangle$ such that:

- Each s_i is a subterm of t, and
- If $s_i \downarrow \beta_i$ holds for each i, then $t \downarrow \alpha$.

In particular, if the null set is a set of critical pairs for t and α, then $t \downarrow \alpha$ is true. Naturally, $\{\langle t, \alpha \rangle\}$ is always a set of critical pairs for t and α. More useful sets of critical pairs may be computed for many expressions using two main principles:

- Suppose that $C \cup \{\langle s_i, \beta_i \rangle\}$ is a set of critical pairs, where s_i is a variable, constant, or λ-expression, and γ is its syntactically declared sort. If $\gamma \ll_S \beta_i$, then $s_i \downarrow \beta_i$ is patently true, so C is also a set of critical pairs.
- Suppose that $\gamma \ll_S \alpha$, t is an application $f(a_1, \ldots, a_n)$, and S contains

$$\forall x_1 : \beta_1, \ldots, x_n : \beta_n . f(x_1, \ldots, x_n) \downarrow \gamma.$$

If C_i is a set of critical pairs for a_i and β_i, then $\bigcup_i C_i$ is a set of critical pairs for t and α.

These principles mechanize definedness checking for a fragment of the IMPS logic that corresponds to order sorted theories in higher order logic [9].

Frequently, a set of critical pairs will be relatively small, even if it is nonnull. Moreover, the terms it contains may be far smaller than t. For instance, consider the term t:

$$(i + j - k) \cdot (i - j + k) \cdot (i - k + j/2)$$

where k, j, i range over the integers $\mathbf{Z}$, and all of the function symbols denote the usual binary functions on the reals. The only critical pairs for t to be defined among the rationals $\mathbf{Q}$ is $\langle j/2, \mathbf{Q} \rangle$. In this case, we would like to combine the results of the lower level with the fact that

$$\forall p, q : \mathbf{Q} . q \neq 0 \supset p/q \downarrow \mathbf{Q}.$$

For this reason, the results of the lower level of definedness-checking are passed to the upper layer, which uses this sort of conditional information.

The Upper Level of Definedness Checking In this layer, conditional information about definedness is consulted. The simplifier is invoked on the resulting assertions, in an attempt to reduce them to truth. This layer uses two primary kinds of information about the domain and range of functions, and the relations between sorts, in the theory.

- *Definedness conditions* of the form

$$\forall x_1 : \alpha_1, \ldots, x_n : \alpha_n . \psi(x_1, \ldots, x_n) \supset f(x_1, \ldots, x_n) \downarrow \alpha.$$

- *Conditional sort coercions* of the form $\forall x : \beta . \varphi(x) \supset x \downarrow \alpha.$

To check the definedness of a term $f(t_1, \ldots, t_n)$ in sort α, we look for a definedness condition for f and α. If successful, we simplify the new goal $\psi(t_1, \ldots, t_n)$.

Alternatively, we look next for a sort coercion condition $\forall x : \beta . \varphi(x) \supset x \downarrow \alpha$, where β is the syntactic sort of $f(t_1, \ldots, t_n)$ (i.e., the declared range of f). If successful, we call the simplifier on the assertion $(\lambda x : \beta . \varphi)(f(t_1, \ldots, t_n))$. This is logically equivalent to the conjunction $f(t_1, \ldots, t_n) \downarrow \beta \wedge \varphi[f(t_1, \ldots, t_n)/x]$. The first conjunct may require a recursive call for definedness checking.

The assertions that, in IMPS, are expressed using partial functions and subtypes can also be expressed, more cumbrously, in ordinary simple type theory. Nevertheless, the machinery of subtypes and definedness assertions helps to guide IMPS's automated support. It provides syntactic cues that the reasoning embodied in these algorithms is likely to be useful. One might initially think that partial functions and subtyping would introduce restrictions that would make sound, effective simplification infeasible. Yet the IMPS simplifier is able to work effectively, partly because the necessary checking is guided by the syntactically explicit definedness operators of the logic.

3   Algebraic Processors

The IMPS simplifier makes repeated calls to procedures called transforms. For this discussion, a transform can be thought of as a function T on pairs (Γ, e), where Γ is a context and e is an expression, which returns two values: an expression $T(\Gamma, e)$ and a set of formulas $\{c_1(\Gamma, e), \ldots, c_n(\Gamma, e)\}$ called convergence requirements. The significance of this set of formulas is explained below.

The two kinds of transforms used by the IMPS simplifier are rewrite rule application and algebraic manipulation. Since any universally quantified equality can be used as a rewrite rule, the framework for applying rewrite rules is entirely general, and uses pattern matching and substitution in a familiar way. By contrast, the transforms that perform algebraic manipulation use specially coded procedures, and they are applied to expressions in a way that may not be easily expressed as patterns. Nevertheless, their validity, like the validity of rewrites, depends on theorems—many of which are universal, unconditional equalities (associative and commutative laws, for instance).

Two issues must be balanced in providing algebraic manipulation using specially coded procedures. On the one hand, ad-hoc implementations have the benefit of allowing optimizations not possible with more generic procedures, and they may be coded in a rigorously correct way for particular kinds of structure. However, they are limited in scope: to provide similar computational facilities for a new kind of algebraic structure such as fields or vector spaces would require adding more special-purpose code to the reasoning machinery. With this approach, users who dealt with theories of an algebraic nature would either write new low-level code for algebraic simplification themselves (for example, using source code provided with the system as a model) or do without algebraic simplification altogether.

We have attempted to maintain the advantages of special-purpose algebraic manipulation while providing easy extensibility to a wide variety of algebraic structures. Instead of providing a fixed set transforms for manipulating expressions in a limited class of algebraic structures, we have implemented a facility for automatically generating and installing such transforms for general classes of algebraic structures. This is possible, since algorithmically the transforms are the same in many cases; only the names have to be changed, so to speak. The algebraic manipulation transform is one component of a data structure called an *algebraic processor*.

An algebraic processor has either two or three associated transforms. There is one transform for handling terms of the form $f(a, b)$ where f is one of the algebraic operations in the processor definition. A separate transform handles equalities of algebraic expressions. Frequently, when the structure has an ordering relation, there is a third transform for handling inequalities of algebraic expressions.

A transform T is valid if and only if for every context Γ and expression e the following holds:

- If all the convergence requirements $c_1(\Gamma, e), \cdots, c_n(\Gamma, e)$ are valid, then $T(\Gamma, e) \simeq e$.
- If any of the convergence requirements is false, then $T(\Gamma, e)$ is undefined.

When an algebraic processor is created, the system generates a set of formulas which must hold in order for its manipulations to be valid. The processor's transforms are installed in a theory T only if the system can ascertain that each of these formulas is a theorem of T.

The user builds an algebraic processor using a specification such as

```
(def-algebraic-processor FIELD-ALGEBRAIC-PROCESSOR
  (language fields)
  (base ((operations
          (+ +_kk)
          (* *_kk)
          (- -_kk)
          (zero o_kk)
          (unit i_kk))
         commutes)))
```

This specification has the effect of building a processor with a transform for handling terms of the form $f(a, b)$, where f is one the arithmetic functions $+_K$, $*_K$, $-_K$. For example, the entry (+ +_kk) tells the transform to treat the function symbol $+_K$ as addition. The transforms are installed in the theory *Fields* by the specification

```
(def-theory-processors FIELDS
  (algebraic-simplifier
    (field-algebraic-processor *_kk +_kk -_kk))
  (algebraic-term-comparator field-algebraic-processor))
```

When this form is evaluated, IMPS checks that the conditions for doing algebraic manipulation do indeed hold in the theory *Fields*.

4 Computing with Theorems

In addition to its simplifier, IMPS also provides a separate mechanism for manipulating formal expressions through applying theorems. It is used when simplification does not go far enough or does the wrong thing. Unlike simplification, it is under the direct or indirect control of the user. The basic idea is that expression manipulations rules—for instance, conditional rewrite rules—are represented as theorems. The theory/context mechanism of IMPS provides a way of making sure that rules represented as theorems are applied correctly in as wide a scope as possible.

Application. How a theorem is applied is determined by its syntactic form. For example, consider the theorem $\forall x : \mathbf{R} \ . \ 0 \leq x \supset |x| = x$ which is proved in a theory $\mathcal{R}$ of the real numbers. This theorem has the form of a conditional rewrite rule; the rule replaces the absolute value of a nonnegative number with itself. Suppose we would like to apply this theorem to a subexpression e' of an expression e in a context in a theory $\mathcal{T}$. This can done in two ways.

The first and simplest way can be used when the home theory $\mathcal{R}$ of the theorem is a subtheory of $\mathcal{T}$; it is based on ordinary expression matching. First, e' is matched to the left hand side $|x|$ of the equation. If there is a substitution, say matching t with x, the simplifier is called to check whether $t \downarrow \mathbf{R}$. If it succeeds, then it is applied to the condition $0 \leq t$. If this also succeeds, e' is then replaced by t in this case, or generally by the instance of the right hand side of the equation under the substitution. If any one of the tests fails, e' is left unchanged.

Alternatively, if $\mathcal{R}$ is not a subtheory of $\mathcal{T}$, "translation matching" is used instead of ordinary matching. An expression e_1 in a theory $\mathcal{T}_1$ *translation matches* an expression e_2 in a theory $\mathcal{T}_2$ if there is a theory interpretation Φ of $\mathcal{T}_1$ in $\mathcal{T}_2$ and a substitution σ such that σ matches the translation of e_1 via Φ to e_2. Translation matching allows theorems to be applied outside of their home theory.

For theorems of different syntactic form, the behavior of theorem application differs. For instance, a form of the triangle inequality for metric spaces,

$$\forall x, y, z : \mathbf{P}, r : \mathbf{R} \ . \ \mathrm{dist}(x, z) + \mathrm{dist}(y, z) \leq r \supset \mathrm{dist}(x, y) \leq r,$$

causes a formula of the form $\text{dist}(x,y) \leq r$ to be replaced by the corresponding existential assertion $\exists z : \mathbf{P} \cdot \text{dist}(x,z) + \text{dist}(y,z) \leq r$. In a case such as this, in which an implication (rather than an equality or a biconditional) is being used, the theorem provides a form of backchaining. Thus, it is legitimate only when the target formula occurs positively in the goal being proved.[2]

Installation. In IMPS theorems are usually installed by the user (but in some cases theorems are installed automatically by the system). To install a theorem in a theory T, the user has to supply a proof of the theorem in T. Once the theorem is installed, it can be used in any theory containing T as well as any theory which contains the structure of T via a theory interpretation. Thus the user can add new expression manipulation rules to the system by simply proving and installing theorems.

Programming. Theorems are actually applied in IMPS by means of procedures called *macetes*. When a theorem is installed, the IMPS system automatically makes an *elementary macete* which will apply the theorem via ordinary matching. If requested by the user, the system will also make a *transportable macete* which will apply the theorem via translation matching.

IMPS contains a rudimentary programming language for building more complex macetes. A *compound macete* is ultimately constructed from elementary and transportable macetes, together with a few special macetes such as beta-reduction and simplification, using a small number of *macete constructors*, which are just functions from macetes to macetes. Compound macetes provide a simple mechanism to apply a collection of theorems in an organized manner. Macetes are usually invoked directly by the user; the user decides what macete is applied on what subexpressions of the goal formula.

As an illustration, consider the compound macete specified by

```
(def-compound-macete REMOVE-ABSOLUTE-VALUES
  (repeat
    absolute-value-non-negative
    absolute-value-non-positive))
```

where absolute-value-non-negative and absolute-value-non-positive are, respectively, the names of the elementary macetes created from the theorem mentioned above and the theorem $\forall x : \mathbf{R} \cdot x \leq 0 \supset |x| = -x$. This macete removes as many absolute value signs as possible from an expression.

---

[2] To define the notion of a positive occurrence, consider the ordering of the truth values in which $\mathbf{F}$ is less than $\mathbf{T}$. A location in a formula is positive if the formula is a monotonically nondecreasing function of the truth value of the subformula at that location. It is negative if it is monotonically non-increasing. A location may also be neither positive nor negative. A syntactic approximation is easily defined by induction on the structure of formulas; for instance, if A occurs positively [negatively; neither] in B, then it occurs negatively [positively; neither] in $\neg B$ or $B \supset C$.

Retrieval. Since theorem application via macetes is under the direct control of the user, it is important that the system have a facility for finding the macetes that are applicable to a particular goal. In IMPS this is done by applying each available macete to the goal using fast matching algorithms. In situations where there are over 500 macetes, the system rarely presents the user with more than 10 macetes as options. Any macete that is not presented is certainly not applicable. However, because the fast matching algorithms are not guaranteed to be sound, some of the presented macetes may not do anything when they are applied with the normal sound matching algorithms.

5 Example

To illustrate some of the preceding ideas, we present in this section an example of an IMPS proof.

Each theorem specification has three important components:

- The statement of the theorem, which is a formula in a formal language.
- The theory in which the theorem is valid. A theorem is *valid in* a theory if it is a semantic consequence of the set of assumptions encapsulated in the theory.
- The proof of the theorem. The proof of a theorem is a sequence of proof commands and control statements called a *script*. This script is executed prior to installing the theorem.

For example, the following entry installs the combinatorial identity as a theorem (called comb-ident).

```
(def-theorem comb-ident
  "forall(k,m:zz,
     1<=k and k<=m implies comb(1+m,k)=comb(m,k-1)+comb(m,k))"
  (theory h-o-real-arithmetic)
  (proof
   ((unfold-single-defined-constant-globally comb)
    (apply-macete-with-minor-premises
      fractional-expression-manipulation)
    (label-node compound)
    direct-and-antecedent-inference-strategy
    (jump-to-node compound)
    (for-nodes
     (unsupported-descendents)
     (if (matches? "with(t:rr, #(t^[-1]))")
         (apply-macete-with-minor-premises
           definedness-manipulations)
         (block
          (apply-macete-with-minor-premises
           factorial-reduction)
          simplify))))))))
```

The proof of the combinatorial identity has five main steps:

1. Expand the definition of **comb**, which in IMPS is given in terms of the factorial function. The resulting sequent consists of a context containing the formulas $k \leq m$ and $1 \leq k$ and the assertion

$$\frac{(1+m)!}{k! \cdot (1+m-k)!} = \frac{m!}{(k-1)!\,(m-(k-1))!} + \frac{m!}{k!\,(m-k)!}.$$

2. Apply the macete fractional-expression-manipulation to the sequent above. Among other things, this multiplies out denominators in equalities and inequalities. This expansion is not done by the simplifier since the necessary computations require extensive branching on cases (for instance, depending on the signs of denominators.) Since the simplifier is called at all levels of the IMPS system, making these reductions universally would be computationally too expensive.

3. Apply the command direct-and-antecedent-inference-strategy to the result of previous step. This reduces the proof to proving three sequents: two concerning definedness and one whose assertion is an equation.

4. For the sequents concerning definedness, use the macete definedness-manipulations. This combines a number of facts about the definedness of the arithmetic operations and the factorial function. One of these two sequents, for example, consists of a context containing $k \leq m$ and $1 \leq k$ and an assertion which says $(k! \cdot (1+m+(-1) \cdot k)!)^{-1}$ is defined.

5. For the remaining sequent, use the macete factorial-reduction which combines the following two theorems:
   - $\forall m : \mathbf{Z} \,.\, 1 \leq m \supseteq m! = (m-1)!\,m$
   - $0! = 1$

The finished proof has the following characteristics:

- Number of sequents created by the proof: 14.
- Number of contexts created during the proof: 79.
- Average number of assumptions (other than axioms) per context: 5.61.

The current IMPS theory library contains approximately 1100 theorems with proofs. The context information for these proofs can be summarized as follows:

- Average number of sequents created by a proof: 24.83.
- Average number of contexts created during a proof: 22.63.
- Average number of assumptions (other than axioms) per context: 2.25.

6   Conclusion

Systematic use of contexts is desirable for symbolic computation, and the IMPS system demonstrates that they are feasible to implement. Indeed, a context determines a class of identities that may be validly applied, and the constrained use of such theorems is in effect the heart of symbolic computation. IMPS uses these ideas in order to unify some of the strengths of theorem provers with symbolic computation.

References

1. R. S. Boyer and J Strother Moore. Integrating decision procedures into heuristic theorem provers: A case study of linear arithmetic. Technical Report ICSCA-CMP-44, Institute for Computing Science, University of Texas at Austin, January 1985.
2. D. Craigen, S. Kromodimoeljo, I. Meisels, B. Pase, and M. Saaltink. Eves system description. In D. Kapur, editor, *Automated Deduction—CADE-11*, volume 607 of *Lecture Notes in Computer Science*, pages 771–775. Springer-Verlag, 1992.
3. W. M. Farmer. A simple type theory with partial functions and subtypes. *Annals of Pure and Applied Logic*. Forthcoming.
4. W. M. Farmer. A partial functions version of Church's simple theory of types. *Journal of Symbolic Logic*, 55:1269–91, 1990.
5. W. M. Farmer, J. D. Guttman, and F. J. Thayer. IMPS: an Interactive Mathematical Proof System. *Journal of Automated Reasoning*. Forthcoming.
6. W. M. Farmer, J. D. Guttman, and F. J. Thayer. IMPS: System description. In D. Kapur, editor, *Automated Deduction—CADE-11*, volume 607 of *Lecture Notes in Computer Science*, pages 701–705. Springer-Verlag, 1992.
7. W. M. Farmer, J. D. Guttman, and F. J. Thayer. Little theories. In D. Kapur, editor, *Automated Deduction—CADE-11*, volume 607 of *Lecture Notes in Computer Science*, pages 567–581. Springer-Verlag, 1992.
8. J. D. Guttman. A proposed interface logic for verification environments. Technical Report M91-19, The MITRE Corporation, 1991.
9. M. Kohlhase. Unification in order-sorted type theory. In A. Voronkov, editor, *Logic Programming and Automated Reasoning*, volume 624 of *Lecture Notes in Computer Science*, pages 421–432. Springer-Verlag, July 1992.
10. L. G. Monk. Inference rules using local contexts. *Journal of Automated Reasoning*, 4:445–462, 1988.
11. S. Owre, J. M. Rushby, and N. Shankar. PVS: A prototype verification system. In D. Kapur, editor, *Automated Deduction—CADE-11*, volume 607 of *Lecture Notes in Computer Science*, pages 748–752. Springer-Verlag, 1992.

GLEF$_{\text{ATINF}}$:A Graphic Framework for Combining Theorem Provers and Editing Proofs for Different Logics

Ricardo Caferra   Michel Herment
LIFIA-IMAG

46 Av. Félix Viallet, 38031 Grenoble Cedex, FRANCE
{caferra | herment}@lifia.imag.fr
Phone: (33) 76 57-46 59 or -48 05 FAX: (33) 76 57 46 02

Keywords. Graphic Proof Presentation, Proof Edition, Logical Frameworks.

Abstract. A running system for combining theorem provers, editing and checking proofs in different logics is presented. It is based on a formalism similar to and developed from the Calculus of Constructions (Coquand and Huet).

Its abilities are independent of the logic and the calculus used. It is therefore possible to present and combine proofs coming from any theorem prover. The user may define the full graphic presentation of proofs, formulas,... he wishes.

Some other original features are: proofs can be viewed and browsed in different ways, (parts of) proofs can be memorized and reused, the size of inference steps for proof presentation can be fixed arbitrarily, symmetries can be handled in a natural manner, equational and non equational proofs are treated in an homogeneous way,...

The principles of the underlying formalism and the several languages used are given. The system has been implemented on Macintosh (and will soon be transported on SUN workstations).

The system should also allow an easy integration of theorem provers with symbolic computation systems.

Several running examples in classical and non-classical logics show evidence of the capabilities of the system. In particular, we present an OTTER [McC90] proof in which we handle symmetries in the proved theorem. Finally, the main lines of current and future work are given.

1   Introduction

As a natural consequence of the progress of Automated Deduction, two important streams of research have been developed in the last years (though the basic ideas they are based on are not necessarily new) simultaneously with work in more traditional style emphasizing efficiency. These ways of research can be characterized as aiming for (1) *generality* and (2) *ease of interaction*.

(1) Search for generality is easily understood if one considers the proliferation of ad hoc theorem provers for different logics and calculi. Notions such as that of *generic theorem provers* [Pau89] converge with approaches coming from pure logics and fundamental computer science, such as those of logical frameworks for defining logics (see [AHMP92] [HHP89] and also [CH88] and [Avr91]).

(2) One of the causes of search for ease of interaction is surely the increasing of power of theorem provers allowing to cope with "difficult theorems". It is possible to classify the works in the domain in two classes (for a synthetic account of the state of the art see [And91]):

(a) Works about *translation* of proofs between different calculi in the *same* logic (see for ex. [And91], [Mil84]). This translation has been also attempted between logic and natural language ([Che76], [Hua90]).

(b) Works about *presentation* of proofs in a structured way (see for ex. [Lin90], [LP90], [CHZ91]).

The present work can be considered -from a conceptual point of view- as a combination of (part of) (1) and (2b).

More precisely, we describe an extensible (and running) system able to present proofs *independently* of the logic and the calculus in which the proof was found (i.e. *independently of the theorem prover* used to obtain the proof). The system allows to define (a large class of) logics. For these logics it is a proof editor and a proof checker, that is to say, it allows handling of proofs with the corresponding verification steps, and in particular to build proofs step by step. It is possible to call different theorem provers from it. It has been designed in order to accept extensions of the edition capabilities.

This paper is structured in 8 sections. In section 2 the underlying philosophy and design bases are given. Section 3 explains the architecture of the system, its definition formalism and the different languages used. Section 4 describes the main principles of the graphic user interface. Section 5 gives a detailed example of a proof construction with GLEF$_{ATINF}$. Section 6 contains different examples of proof presentation in different logic with different calculi. Section 7 presents an OTTER proof emphasizing symmetries. Conclusion and some lines of present and future work are given in section 8.

2 The Underlying Principles and Design Bases

We present first the ideas and then the design bases leading to the specification of our system

2.1 The Underlying Principles

The general design philosophy is guided by practical and simple principles: the user may need to handle different logics and calculi and to present proofs coming from any theorem prover and this in *a human-oriented way*, that is to say for example that (s)he would like to dispose of different ways of structuring proofs, to be able to name subproofs, to handle symmetries, to identify inference steps of arbitrary size,... on acceptable running time.

An easy way to convince the reader of the interest of a pragmatic approach is to consider, for example, the very nice and general work in [AHMP92], [HHP89], [CH88]. These approaches are much more concerned with foundations than with theorem proving and human oriented proof handling.

We have combined the formalism of the Calculus of Constructions (abbreviated CC) [CH88] with other languages, in order to build a system with the desired abilities.

2.2 The Design Bases

GLEF$_{ATINF}$ has been developed in the framework of our inference laboratory ATINF ([CHZ91], [CH93]) and fulfills the following informal specification. It must:
- require as few as possible "heavy" work to the user (who should not be supposed to be an expert programmer). Proofs must be presented in the customary user's way. In particular the system must incorporate graphic facilities such as a box discipline.
- allow to define as much logics and calculi as possible (for classical-equational and non-equational and non-classical logics, geometry...).
- use a formalism allowing "incremental" definitions of logics, that is to say, common features between two logics must be defined only once and can be reused.
- allow to present proofs coming from theorem provers *not* in ATINF (such as OTTER, see section 7)
- allow *different hierarchical and graphic presentations* of a proof for the *same* proof system.
- be able to exhibit and hide *any part of a proof* and not only a local neighbourhood in the proof tree of a formula at hand.
- offer to the user the possibility of applying tools such as matching and unification independently of a particular theorem prover
- be able to *memorize* and to *modify* any part of a proof.
- offer to the user the possibility of naming and handling in an homogeneous way any syntactical unit: (sub)proofs, formulas, symbols,...

- offer to the user the possibility of gathering an arbitrary number of elementary steps in an inference rule of any size (this feature is very useful in hierarchical presentation of proofs).
- be able to edit and, when possible, check proofs in defined logics.

The second point in the list is the strongest requirement and implies the use of a language for handling proofs independent of any internal representation in a theorem prover.

In order to satisfy this informal specification we have adopted a formalism issued from CC. CC is a higher-order formalism for constructive proofs in natural deduction style [CH88].

We give a taste of this formalism, pointing out some interesting features and syntactical differences with CC.

3 Architecture

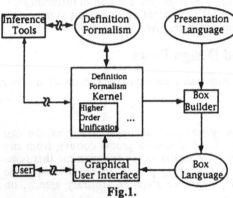

Fig.1.

GLEF$_{ATINF}$ is written in C++ and running on Macintosh. Most of its components are portable.

It is structured around different languages (fig.1): A *definition* formalism is used to define logics and to store proofs. A *presentation* language is used to specify presentation of logics.

Formulae and proofs are presented to the user via a *graphic* language: a *box* language. A *box builder* uses the presentation language specification to translate the *internal representation* of definition formalism terms in definition formalism kernel into boxes.

The user can *handle* the *box structure* directly with the *mouse*. This handling is reflected into the definition formalism kernel. To each box representing a term is associated a formal representation of the term, therefore, there is no missing information.

Communication with inference tools does occur at definition formalism and definition formalism kernel levels. Obviously, communication with the user occurs via the graphical user interface.

Let us briefly explain the languages used in the system.

3.1 Definition Formalism

The definition formalism is useful to GLEF$_{ATINF}$ for *communication* with inference tools. It is also used for *verification* of user's handling and proofs obtained from inference tools.

It is a Π-*typed* λ-*calculus* similar to CC (Calculus of Constructions) [CH88]. His production rules are those of CC extended by notions for multisets, multiset types and multi types. Though the study of its formal properties has not been deepened, its use is safe and powerful enough in the scope of the present work.

*	Universe
<x:M>N	Abstraction
[x:M]N	Dependent product
(M N)	Application
x	Variable
{M$_1$..., M$_n$}	Multiset
M+	Multiset type
{M$_1$..., M$_n$}	Type[1]

---

[1]In concrete syntax, symbols '{' and '}' are written '{' and '}'. Distinction between types and multisets is done *syntactically*. For the sake of compatibility with CC, one can write T for the singleton {T}.

A formal and complete description of the definition language is given by the inference rules for type judgement. The form for type judgement is $\Gamma \vdash t : T$, it means that in the context Γ, t is a well-typed term of type T.

[] is the empty context. Γ is a context. T, T_i are types, their form are: $\{t_1...,t_n\}$. M, M_i are terms. t, t_i are terms which are neither abstractions nor multisets.

universe: $\qquad [] \vdash *:\{\}$

valid context: $\qquad \dfrac{\Gamma \vdash t_1:T_1 \ ... \ \Gamma \vdash t_n:T_n}{\Gamma[x:\{t_1...,t_n\}] \vdash *:\{\}}$

variable: $\qquad \dfrac{\Gamma \vdash *:\{\}}{\Gamma \vdash x:type\_get(\Gamma,x)}$

product formation: $\qquad \dfrac{\Gamma[x:T_1] \vdash t:T_2}{\Gamma \vdash [x:T_1]t:\{*\}}$

list formation: $\qquad \dfrac{\Gamma \vdash t_1:T_1 \ ... \ \Gamma \vdash t_n:T_n}{\Gamma \vdash \{t_1...,t_n\}+:\{*\}}$

abstraction: $\qquad \dfrac{\Gamma[x:T] \vdash M:\{t_1...,t_n\}}{\Gamma \vdash <x:T>M:\{[x:T]t_1...,[x:T]t_n\}}$

multiset: $\qquad \dfrac{\Gamma \vdash M_1:T_1 \ ... \ \Gamma \vdash M_n:T_n}{\Gamma \vdash \{M_1...,M_n\}:\{type\_union(T_1...,T_n)+\}}$

application: $\qquad \dfrac{\Gamma \vdash M_1:T_1 \ \ \Gamma \vdash M_2:T_2 \ type\_apply(T_1,T_2)\neq\{\}}{\Gamma \vdash (M_1 \ M_2):type\_apply(T_1,T_2)}$

Where (with the usual set theory notions for types):

$type\_union(T_1...,T_n) = \bigcup_i^n T_i$

$type\_apply(T_1,T_2) = \{t_3 \mid \exists t_1 \in T_1 \ t_1=[x:T_3]t_3 \ tc(T_3,T_2)\}$

$tc(T_1,T_2) = (\forall t_1 \in T_1 \ \exists t_2 \in T_2 \ to(t_1,t_2))$
$t'c(T_1,T_2) = (\forall t_2 \in T_2 \ \exists t_1 \in T_1 \ to(t_1,t_2))$
$to(t_1,t_2) = (t_1=t_2 \vee t_1=[x:T_1]t'_1 \wedge t_2=[x:T_2]t'_2 \wedge to(t'_1,t'_2) \wedge tc(T_2,T_1)$
$\qquad\qquad\qquad\qquad\qquad\qquad\qquad \vee t_1=T_1+ \wedge t_2=T_2+ \wedge t'c(T_1,T_2)$

And type_get is recursively defined by:

$type\_get(\Gamma[y:T],x) =$ if x=y then T else $type\_get(\Gamma,x)$

A full logic definition cannot be given here (it would take a big place in the paper, it is similar to those in [CH88] or [HHP89], see also [CH93]). Below we give some examples of the new product rules:
- [and:[formula+]formula] defines a function with *undefined arity*. For instance, in the context: [P,Q,R:formula], it is possible to write: (and {P,Q,R}).
- A *polymorphic* function can be defined like:

$\qquad$ [rule:{
$\qquad\qquad$ [A,B:formula][(equiv A B)](implies_right A B),
$\qquad\qquad$ [A,B:formula][(equiv A B)](implies_left A B)}]

- One can assume *multi-typed arguments* (= denotes the syntactic naming facility):

$\qquad$ [apply= <f:{[type1]formula, [type2]formula}>
$\qquad\qquad$ <A:type1><B:type2>(and {(f A),(f B)}]

3.2 Box Language

The box language is similar to languages used in *structured document editing* [Qui87]. It allows natural and graphical presentation of structured objects.

Fig.2.

A box is an *invisible* rectangle whose sides are either horizontal or vertical. The rectangle allows to place and size each elementary or composed graphic unit (fig.2).

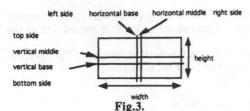

Fig.3.

A box is defined by four sides, four axis and two sizes, used to place and size it (fig.3). Some relations link these fields to those of surrounding boxes. A set of box instances is calculated automatically by an incremental attribute evaluator.

Moreover, the box language allows natural and homogeneous handling of terms (terms, formulae, proofs) using the mouse.

3.3 Presentation Language

The presentation language allows to present terms of the definition formalism.

A default presentation is assumed for each constructor (abstraction, product...) of the definition formalism. This default presentations are described in a separate presentation file.

The presentation of each defined syntactical unit may be specified. The presentation algorithm uses those presentations rather than the default ones when presenting variables and applications.

This algorithm is directed by the current term to present and the current presentation, in order to build the current box.

Some primitives of the language allow to specify *boxes contents*, *positions* and *sizes*. A box can be "filled" with text, any symbol, picture, external presentation of a term, or recursively with other boxes. Positions and sizes are specified using linear equation on fields of the current box and its surrounding boxes (a box containing it, contained in it, previous, next...).

In order to control boxes building, some other primitives allow to traverse the term being presented. It is not necessary to describe the full term presentation, since the presentation algorithm can be called recursively to present subterms. *This is useful when presenting a full proof: it suffices to describe the presentation of logical connectives and inference rules.*

For example, in the context:

```
[term,formula: *]
[forall:        [[term]formula]formula]
[or:            [formula][formula]formula]
[P,Q:           [term]formula]
```

the term: `(forall <x:term> (or (P x) (Q x)))` can be presented as $\forall x\, P(x) \lor Q(x)$, describing only the presentation of syntactical units or and forall.

For instance, the following is the common presentation specification of exists and forall logical connectives:

```
exists,forall:BEGIN
    PRESENT:box;
    TEXT '(':BEGIN
        PRESENT:inbox;
    END;
    SYMBOL NAME:BEGIN
        PRESENT:line_t;
    END;
    PLACE A.1.REFERENCE:BEGIN
        PRESENT:line_t;
    END;
    PLACE A.1.BLOC:BEGIN
        PRESENT:line_t_space;
    END;
    TEXT ')':BEGIN
```

```
                        PRESENT:line_t;
            END;
    END;
```

The language is well adapted to *code reusability*. box, inbox, line_t, line_t_space are not keywords but references to auxiliary presentation specifications:

```
            line_t_space:BEGIN
                LEFT:=PREVIOUS.RIGHT+5;
                TOP:=PREVIOUS.TOP;
            END;
```

4 User Interface

Only the main principles of the graphic user interface will be described in this section. Each edited proof is presented in a window topped by buttons and menus. A trash and a "copy machine" appear in its bottom-right corner.

4.1 Visualization

Selection. The user can select *term boxes*. A term box is a box which corresponds to the full presentation of a term.

Buttons. Frequently, the user doesn't want to see the full proof. Thus a proof is often presented with only its conclusion and hypothesis at first. In fact, the presentation specification defines what must be hidden. If deriving conclusion form hypothesis doesn't seem obvious to the user, (s)he can use five buttons to examine the desired proof parts.
- The *Show* button allows to visualize the hidden boxes of highest level, enclosed in the selected boxes. All hidden boxes can be shown with successive clicks.
- The *Show all* button is a short-cut to show all boxes in the selected boxes.
- The *Hide* button hides all the boxes that can be hidden in the selected boxes.
- The *Catch* button hides all graphic stuff surrounding the selected boxes. This command can be called recursively.
- The *Release* button undoes Catch command effects, step by step.

Namings. A definition, a proof for instance, can contain *names*. One can bind a presentation specification to a name, it is used for the presentation of its instances, instead of the named term presentation.

However, the named term presentation is shown for the instance presented in *top boxes*. A top box is a box without enclosing box. Thus, when a naming instance is *caught* with the Catch button, the named term is shown.

Note that the named term can be an abstraction of which instances are applications.

4.2 Edition

Dragging. A term box can be selected and dragged into another term box. The term bound to the top box enclosing the destination box (which can contain the source box) is *instanciated* in order to obtain an instance of the source box at the destination box occurrence (see the example of proof construction in section 5).

A dragging is not always possible. Thus, GLEF$_{ATINF}$ tells if it is possible by highlighting the current destination box under the mouse pointer. Telling if a dragging is possible and instanciating the top box enclosing it are done using *unification*.

Note that this method is not intended to replace tactics (a la LCF), which allows the user to build many inference steps with few input (and will soon be added to GLEF$_{ATINF}$). But it is natural for step by step proof construction.

Menus. Menus are mainly used to add syntactical units defined in the definition formalism or definition formalism constructors.

Definition formalism constructors are not very used. In particular, they are useless to build proofs in logics without variables, with fixed arity logical connectives and inference rules. Nevertheless, abstraction is needed for logics with variables and multiset is needed for variable arity functions. Note that to build a logic definition, one may need all constructors.

Other menus allow to add syntactical units. If these units are functions (logical connectives, inference rules…) they are applied to a maximum of variables, with correct types, before being added to the proof. This application makes proof construction easier (See section 5).

Selecting a destination box and calling a menu is a *useful short-cut* for calling a menu and dragging the new box into the destination box (See section 5).

Scrapbook and other facilities. Classical commands, like *Undo, Cut, Copy, Paste, Duplicate, Erase*, are available. In particular, the scrapbook can contain any term. Note that Erase and Cut commands can delete a subterm. It should be observed that this deletion is easy in λ-calculus, but tricky in typed λ-calculus.

The *trash* and the *duplicator* are short cuts for erase and duplicate commands. One just have to drag selected boxes into them.

Keyboard. For presentation purpose, a variable can be renamed. It is done by selecting it and editing its new name with the keyboard (An example is given in next section).

5   A Simple Example of Proof Construction

Let us describe the construction of a proof of the theorem $A \Rightarrow A$ in a well known Hilbert like system for propositional logic. First, the following is the definition of the calculus:

Axioms:
$$a1: \quad A \Rightarrow (B \Rightarrow A)$$
$$a2: \quad (A \Rightarrow (B \Rightarrow C)) \Rightarrow ((A \Rightarrow B) \Rightarrow (A \Rightarrow C))$$
$$a3: \quad (\neg B \Rightarrow \neg A) \Rightarrow ((\neg B \Rightarrow A) \Rightarrow B)$$

Inference rule: $\qquad$ modus-ponens : $\dfrac{B \quad B \Rightarrow A}{A}$

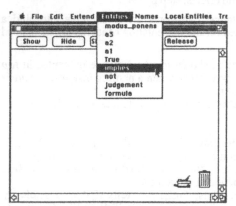

Fig.4.

Fig.4 shows the screen, just before building the proof, the definition and the presentation of the logic have already been parsed (parsing is done automatically by GLEF$_{\text{ATINF}}$).

To start building the proof of $A \Rightarrow A$ we can first call the $\Rightarrow$ constructor by selection in the Entities menu (logical connectives and inference rules are handled homogeneously). It will appear on the screen with its arguments filled with different variables.

One of these variables can be selected and dragged into the other . While such a dragging, the term under the mouse pointer is highlighted if it is *unifiable* with the source term. As the user releases the mouse button, if the destination term is unifiable with the source term, the substitution is applied to the term bound to the top box containing the destination box .

To continue building the proof one can call the modus-ponens rule and drag the previous term into its conclusion (fig.5).

Fig.5.

Fig.6.

The "Show" button allows to show the hidden part of the inference rule, its premises in this case (fig.6). These premises are variables whose types can be obtained with the *Get type* command (fig.7).

After determining premises types, one would like to apply axiom a1 to the first premise (figs.7,8). Rather than calling the axiom with the Entities menu and then dragging its box in the premise's one, one can first select the premise. As the Entities menu is used, all the menu entries are *unified* with the selected term. Not unifiable menu entries are dimmed. The substitution corresponding to the user's choice is finally applied to the term bound to the top box containing the selected box (fig.9).

Figs.7,8,9.

After dragging the useless premises types in the trash, one can continue the proof building using both methods (figs.8,9,10,11,12).

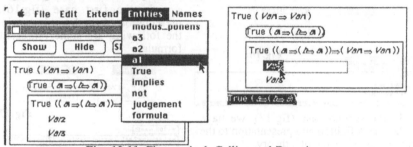

Figs.10,11. First method: Calling and Dragging.

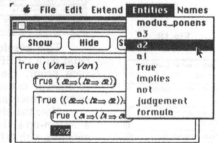

Fig. 12. Second method: Selecting and Calling.

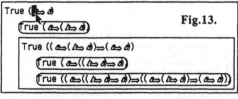

Fig.13.

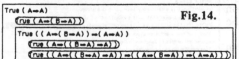

Fig.14.

The proof (in a top-down style) is now complete, we can rename free variables by selecting them and editing their names (figs.13,14).

Note that to build a proof, one can try to call inference rules before logical connectives. This method allows to avoid useless variable bindings. Before using GLEF$_{ATINF}$ we knew a proof of $A \Rightarrow A$ with the same inference rules, but where B was replaced by A (fig.14). Using this method showed that B could be free.

6 Some Examples of GLEF$_{ATINF}$ Capabilities

In this section we shall show via examples some of the main abilities of GLEF$_{ATINF}$.

6.1 A Gentzen System

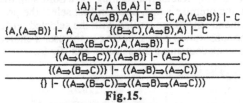

Fig.15.

the one that creates an empty multiset (fig.15).

The definition of this calculus is based on that of propositional language.

In order to create the contexts in the left part of the sequents, we have used the multiset notion of the definition formalism. The proof presented here has been built by GLEF$_{ATINF}$ using only one constructor in the definition formalism: the one that creates an empty multiset (fig.15).

6.2 Many Sorted First Order Language

The following two examples illustrate possibilities of realization and overloading of the definition formalism. They are based on MSFOL ("Many Sorted First Order Logic"). no calculus has been defined for it.

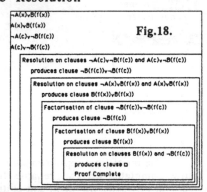

Fig.16.

In the first case (fig.16), the presentation of the logic is overloaded, the following are some representative formulas.

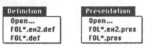

Fig.17.

In the second case (fig.17), we have extended the definition and presentation to the language and axiomatic of a theory.

6.3 Resolution

¬A(x)∨B(f(x))
A(x)∨B(f(x))
¬A(c)∨¬B(f(c))
A(c)∨¬B(f(c))

Fig.18.

Resolution on clauses ¬A(c)∨¬B(f(c)) and A(c)∨¬B(f(c))
produces clause ¬B(f(c))∨¬B(f(c))

Resolution on clauses ¬A(x)∨B(f(x)) and A(x)∨B(f(x))
produces clause B(f(x))∨B(f(x))

Factorisation of clause ¬B(f(c))∨¬B(f(c))
produces clause ¬B(f(c))

Factorisation of clause B(f(x))∨B(f(x))
produces clause B(f(x))

Resolution on clauses B(f(x)) and ¬B(f(c))
produces clause □
Proof Complete

This resolution proof has been build with GLEF$_{ATINF}$ (fig.18). To build such proof, the only two definition formalism constructors used are the empty multiset and abstraction.

Note that this proof is presented in a bottom-up style. The resolution proof obtained with OTTER will be presented in the top-down style (see section 7).

6.4 Natural Language

This "natural language" proof (fig.19) is presented in [Hua90] and is a translation of a formal proof into pseudo-natural language. There is no difference between the definition of this pseudo-natural language and the definition of any logic with GLEF$_{ATINF}$.

The selection shows that any term can be selected.

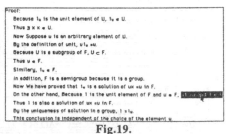

Fig.19.

6.5 Tableaux in Modal Logic S4

This tableaux proof in modal logic S4 has been obtained with the tableaux-based theorem prover of ATINF [CHZ91], which is linked to GLEF$_{ATINF}$ (fig.20).

Note that semantic tableaux in modal logic S4 specification is the last layer of a multi-layer specification:

- First-order logic
- Semantic tableaux in first order logic[2]
- Semantic tableaux in modal logic
- Semantic tableaux in modal logic S4

The very little box in the right (right part of the β-rule) corresponds to an undeveloped part of the proof. Of course it is also possible to present the entire proof.

6.6 Equational Logic

The homogeneous treatment of logics and calculi (inherited from CC) includes equational logic as a particular case. The following proof has been obtained with a theorem prover in ATINF. This theorem prover is linked to GLEF$_{ATINF}$ (fig.21).

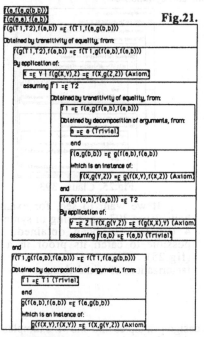

Fig.21.

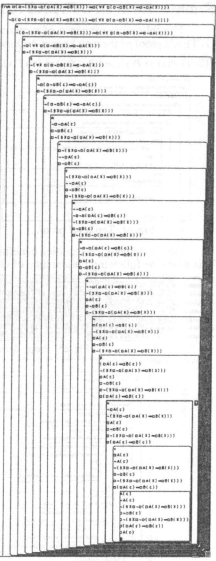

Fig.20.

Note that terms T_1 and T_2 are names, their definition appears at the top of the proof.

---

[2]For this calculus there exists a theorem prover in ATINF which is linked to GLEF$_{ATINF}$.

7 Presenting Proofs of Existing Theorem Provers: Handling Symmetries in an OTTER Proof

The following sequence of screens is the presentation our system can give of the OTTER's proof of the well known Andrew's challenge theorem:

$$((\exists x \forall y\, P(x) \Leftrightarrow P(y)) \Leftrightarrow ((\exists u\, Q(u)) \Leftrightarrow (\forall v\, P(v)))) \Leftrightarrow ((\exists w \forall z\, Q(w) \Leftrightarrow Q(z)) \Leftrightarrow ((\exists s\, P(s)) \Leftrightarrow (\forall t\, Q(t))))$$

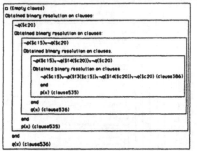

Fig.22. Top level proof.

From the statement of the theorem, it seems clear that several partial subproofs should be the same or have, at least, the same type (i.e. conclusion and hypothesis). It is also clear that a *human-oriented* presentation of the proof should take account of these symmetries.

The symmetries are discovered and filtered (on the finished proof) by GLEF$_{\text{ATINF}}$. A naming is used to factorize each symmetry. Note that the named term is an abstraction, unless the symmetry is an equality.

Thus, symmetry presentation benefits from optimal hierarchical naming presentation. This gives the following presentation of OTTER's proof (figs.22,23,24,25).

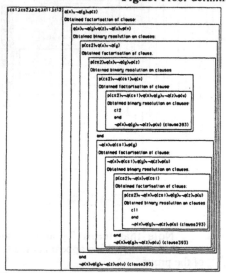

Fig.23. Proof definitions and symmetries.

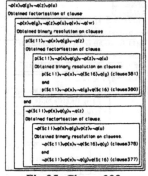

Fig.24. The biggest symmetry is symmetry 4.

Fig.25. Clause 393.

If we want to see, for example, how clause 393, appearing in symmetry 4 (fig.24) has been obtained, it is possible to catch its proof in a box (fig.25). Other clauses are just factorizations of initial clauses.

8 Conclusion and Future Work

We have described the theoretical bases, design principles and shown examples of a running extensible graphic tool for editing proofs and combining theorem provers with several original capabilities: ease to use, logic-independent, different ways of structuring the same proof,... Obviously, it can also be used for *teaching logic* purposes.

We are trying to increase the capabilities of GLEF$_{ATINF}$ in order to use it as a help in:
- handling symmetries (similar to section 7)
- handling analogies (to be combined with the approach in [BC87]).
- proof planning.
- translating formal proofs into natural language. GLEF$_{ATINF}$ can be profitably used as a flexible tool for presenting and structuring proofs in natural language in order to extend the approaches in [Hua90] and [Dyb82].
- proof presentation in geometry. We experiment presently with GEOTHER, a geometric theorem prover being incorporated to ATINF [Wang93].
- Presently we extend the edition capabilities. We experiment with higher-order unification and matching.
- GLEF$_{ATINF}$ should allow an easy integration of symbolic computation systems such as Mathematica (similarly to [CZ92]). We are presently studying this point.

9 Bibliography

[And91] P.B. Andrews: *More on the problem of finding a mapping between clause representation and natural deduction representation*, Journal of Automated Reasoning 7 (1991), pp. 285-286.

[AHMP92] A. Avron, F.A. Honsell, I.A. Mason, R. Pollack: *Using typed lambda calculus to implement formal systems on a machine*, Journal of Automated Reasoning 9 (1992), pp. 309-354.

[Avr91] A. Avron: *Simple consequence relations*, Information and Computation 92 (1991), pp. 105-139.

[BC87] T. Boy de la Tour, R. Caferra: *Proof analogy in interactive theorem proving: a method to use and express it via second order pattern matching*, Proc. AAAI-87, Morgan & Kaufmann 1987, pp. 95-99.

[CH88] T. Coquand, G. Huet: *The calculus of constructions*, Information and Computation 76 (1988), pp. 95-120.

[CH93] R. Caferra, M. Herment: *GLEF$_{ATINF}$ A graphic framework for combining provers and editing proofs for different logics*, (long version). In preparation.

[Che76] D. Chester: *The translation of formal proofs into English*, Artificial Intelligence 7 (1976), pp. 261-278.

[CHZ91] R. Caferra, M. Herment, N. Zabel: *User-oriented theorem proving with the ATINF graphic proof editor*, Proc. FAIR'91, LNAI 535, Springer-Verlag 1991, pp. 2-10.

[CZ92] E. Clarke, X. Zhao: *Analytica-An experiment in combinig theorem proving and symbolic computation*, RR CMU-CS-92-117, September 1992.

[Dyb82] P. Dybjer: *Mathematical proofs in natural language*, Report 5, Programming Methodology Group, University of Goteborg, April 1982.

[HHP89] R. Harper, F. Honsell, G. Plotkin: *A framework for defining logics*, RR CMU-CS-89-173, Carnegie Mellon University, January 1989.

[Hua90] X. Huang: *Reference choices in mathematical proofs*, Proc. ECAI'90, Pitman 1990, pp. 720-725.

[Lin90] C. Lingenfelder: *Transformation and structuring of computer generated proofs*, SEKI Report SR-90-26, University of Kaiserslautern 1990.

[LP90] C. Lingenfelder, A. Präcklein: *Presentation of proofs in equational calculus*, SEKI Report SR-90-15 (SFB), University of Kaiserslautern 1990.

[McC90] W.W. McCune: *OTTER 2.0 users guide*, Technical Report ANL-90/9, Argonne National Laboratory, Illinois 1990.

[Mil84] D.A. Miller: *Expansion tree proofs and their conversion to natural deduction*, Proc. CADE-7, LNCS 170, Springer-Verlag 1984, pp. 375-393.

[Pau89] L.C. Paulson: *The foundation of a generic theorem prover*, Journal of Automated Reasoning 5 (1989), pp. 363-397.

[Qui87] V. Quint: *Une approche de l'Edition Structurée des documents*, Ph.D Thesis, Université Sc. Tech. et Méd. de Grenoble (1987).

[Wang93] D.M. Wang: *An elimination method based on Seidenberg's theory and its applications in computational algebraic geometry* In Progress in Mathematics 109, pp. 301-328, Birkhäuser, 1993.

Extending RISC-CLP(*Real*) to Handle Symbolic Functions

Olga Caprotti*

Research Institute for Symbolic Computation
Johannes Kepler University
A-4040 Linz, Austria
e-mail: ocaprott@risc.uni-linz.ac.at

Abstract. In the previous version of the constraint logic programming language RISC-CLP(*Real*), the domain *Real* of real numbers was the intended domain of computation. In this paper, we extend it to the domain *TreeReal* of finite symbolic trees with real numbers as leaves, that is the integration of the domain of reals with the domain of finite Herbrand trees. In the extended language, a system of constraints over the new domain is decided by first decomposing equations into a tree-solved form produced by an adapted unification algorithm. Then, polynomial real constraints are decided by the partial cylindrical algebraic decomposition method and a solution to the original system is constructed.

1 Introduction

In this paper we describe how the language RISC-CLP(*Real*) [12] has been extended by the introduction of symbolic function symbols. This language is an instance of the Constraint Logic Programming (CLP) scheme that was defined by Jaffar and Lassez in [16] and describes a class of logic programming languages combining logic programming with constraint solving. The CLP scheme relaxes the restriction on the Herbrand universe by allowing a variety of domains of computation, such as trees, booleans and reals, and it relaxes the restriction on unification by employing domain-specific decision algorithms. Therefore, there are two key implementation issues:

1. a clear definition of what is the intended domain of computation,
2. the existence of a decision procedure to solve constraints over such a domain.

We dealt with both these issues while extending the language RISC-CLP(*Real*).

Currently, among the main CLP representatives are the languages: Prolog III, CLP(*R*), and CAL. Colmerauer's Prolog III [7, 6] combines the domains of rational numbers, lists and booleans into the domain of infinite rational trees. It supports construction, boolean, arithmetical operations and binary relations

---

* In the frame of the ACCLAIM project sponsored by
European Community ESPRIT BRA 7195
Austrian Science Foundation P9374 – PHY.

such as $\{\neq, >, <, \leq, \geq\}$ together with unary relations that declare the domain to which a variable belongs.

Jaffar and his group's CLP($\mathcal{R}$) [9, 14] computes on the domain of trees over reals and supports decision of linear constraints among arithmetical terms [15, 19]. An analysis of the main differences among CLP($\mathcal{R}$) and RISC-CLP is presented in [12].

Sakai and Aiba's CAL (Contrainte avec Logique) [23] offers nonlinear constraints over the complex numbers by employing Buchberger algorithm as decision algorithm [2, 3]. RISC-CLP($\mathcal{R}eal$) incorporates Gröbner bases method to check unsolvability before resorting to the more expensive cylindrical algebraic decomposition (CAD) method. Monfroy in [20] also describes a CLP language that deals with nonlinearity over the complex numbers using Gröbner bases method. Examples of other CLP languages are CHIP over finite domains [1], CLP($\varSigma^*$) over regular sets [25], Echidna over discrete sets [8], BNR-Prolog over real intervals [22].

Whereas most of the well known implementation of the CLP scheme are concerned with finite domains or only with linear constraints, RISC-CLP($\mathcal{R}eal$) is able to deal with nonlinearity over the real numbers by employing partial cylindrical algebraic decomposition as quantifier elimination procedure for real closed fields [4, 10]. On the other hand, RISC-CLP($\mathcal{R}eal$) offered so far no possibility of data structuring via non arithmetical terms because symbolic functions and lists were not supported. We extended the domain of computation of RISC-CLP($\mathcal{R}eal$) by adding symbolic function identifiers to the set of arithmetical function symbols, $\{+, -, *\}$, and we modified the solver module in order to decide the new type of constraints arising. The new version of RISC-CLP($\mathcal{R}eal$) has as domain of computation the domain of *finite symbolic trees with real numbers* which is the usual Herbrand universe equipped with real numbers. Because the decision procedure on the Herbrand universe is unification, it is natural that a unification algorithm, inspired by the one described in [17] and presented as a rule based algorithm like in [18], is the major change we have integrated into the solver.

The paper is structured as follows. Section 2 describes the domain of symbolic trees with real numbers as a combination of two well-known domains: real numbers and finite trees. Section 3 shows how a system of constraints over $\mathcal{T}ree\mathcal{R}eal$ is decided by the solver that combines a decision algorithm for the real numbers with the "adapted unification" algorithm. For the details of how the solver decides nonlinear constraints over the reals, the interested reader is referred to [12, 10]. Section 4 shows an example program the RISC-CLP language that can be solve problems in chemical reaction systems.

2 Domain of Computation and Language

In the following Section, we recall the two domains: $\mathcal{R}eal$ of real numbers and $\mathcal{T}ree$ of finite trees, and we describe the combined domain: $\mathcal{T}ree\mathcal{R}eal$.

Real constant symbols:	fractions, e.g. 2/3, -1, 3.75
Real function symbols:	arithmetical, i. e. $\{+, -, *\}$.
Real relation symbols:	arithmetical, i. e. $\{\neq, >, <, \leq, \geq\}$, real equality $=_R$.

| Real-term: | is a real constant, or is a variable or has the form '$t_1 \ \alpha \ t_2$' where α is a real function symbol and t_1, t_2 are real-terms. |
| Real-constraint: | is of the form '$t_1 \ \rho \ t_2$' where t_1, t_2 are real-terms and ρ is a real relation symbol. |

Fig. 1. Language and syntax of the domain $\mathcal{R}eal$.

Tree constant symbols:	symbolic.
Tree function symbols:	symbolic tree constructors.
Tree relation symbols:	tree equality $=_T$.

| Tree-term: | is a tree constant, or is a variable or has the form: '$f(t_1, \ldots, t_n)$' where f is a tree function symbol of arity $n > 0$ and $t_1, \ldots, t_n$ are tree-terms. |
| Tree-constraint: | has the form '$t_1 =_T t_2$' where t_1, t_2 are tree-terms. |

Fig. 2. Language and syntax of the domain $\mathcal{T}ree$.

Constant symbols:	real and tree constant symbols.
Function symbols:	real and tree function symbols.
Relation symbols:	arithmetical, i.e. $\{\neq, >, <, \leq, \geq\}$, equality $=$.

| Term: | is a variable, a real-term, a tree-term or has the form: '$f(t_1, \ldots, t_n)$' where $t_1, \ldots, t_n$ are terms and f is a tree function symbol of arity $n > 0$. |
| Constraint: | is of the form '$t_1 \ \rho \ t_2$' where t_1, t_2 are terms and ρ is a relation symbol. |

Fig. 3. Language and syntax of the domain $\mathcal{T}ree\mathcal{R}eal$.

The domain $\mathcal{R}eal$ is obviously the set of all real numbers, including all real algebraic and transcendental numbers. A careful reader might wonder how it is possible to compute on the domain $\mathcal{R}eal$ when such a domain is known to be uncountable. The answer is provided by Tarski's theorem [24] which states that the validity of any first order sentence over real numbers is the same as the validity over algebraic numbers, and these are indeed countable. Hence, it is convenient to assume that the domain $\mathcal{R}eal$ is the set of only real algebraic numbers. The language and the syntax of the domain of $\mathcal{R}eal$ are recalled in Figure 1. Real-terms and real-constraints are interpreted in the conventional way of arithmetic. In this domain, all functions are total and conjunction of real-constraints is decided by the quantifier elimination algorithm based on partial CAD [4, 10, 5].

The domain $\mathcal{T}ree$ consists in the set of finite trees labelled by identifiers. Language and syntax of the domain of $\mathcal{T}ree$ are recalled in Figure 2. In this domain, tree terms are interpreted as labelled n-ary trees. Like in the domain $\mathcal{R}eal$, all the functions are total. Syntactical unification is the well known decision procedure for solving a conjunction (set) of tree-constraints.

Now we describe the combined domain $\mathcal{T}ree\mathcal{R}eal$ that is similar to the domain of CLP($\mathcal{R}$). $\mathcal{T}ree\mathcal{R}eal$ is the set of finite symbolic trees having possibly real numbers as leaves. It can be recursively defined as follows.

Leaves: Nodes labelled by real numbers in $\mathcal{R}eal$, and by identifiers in $\mathcal{T}ree$, are in $\mathcal{T}ree\mathcal{R}eal$.

Recursion: For all tree function symbols f of arity $n > 0$, if $d_1, \ldots, d_n$ are in $\mathcal{T}ree\mathcal{R}eal$, then the tree, whose root is labelled by the identifier "f" and whose children are the trees $d_1, \ldots, d_n$, is in $\mathcal{T}ree\mathcal{R}eal$.

Its language consists in the merging of the languages of the two domains $\mathcal{R}eal$ and $\mathcal{T}ree$, and the syntax is as described in Figure 3. Note that we do not keep the equalities $=_R$ and $=_T$, but we have instead a unique symbol $=$. The (logical or declarative) semantics interprets arithmetical symbols as the usual operations and relations over $\mathcal{R}eal$, while tree functions are interpreted over $\mathcal{T}ree\mathcal{R}eal$.

Terms: Real constants are interpreted as nodes labelled by real numbers in $\mathcal{R}eal$.

Tree constants are interpreted as nodes labelled by identifiers in $\mathcal{T}ree$.

Real-terms are interpreted as nodes labelled by the real numbers that are their interpretation over $\mathcal{R}eal$.

A term of the form '$f(t_1, \ldots, t_n)$' is the tree in $\mathcal{T}ree\mathcal{R}eal$ whose root is labelled by the identifier "f" and whose children are the the interpretations of $t_1, \ldots, t_n$ over $\mathcal{T}ree\mathcal{R}eal$.

Constraints: '$t_1 \rho t_2$', where ρ is among $\{\neq, >, <, \leq, \geq\}$ is true iff it is true when t_1 and t_2 are interpreted over $\mathcal{R}eal$.

'$t_1 = t_2$' is true if the trees in $\mathcal{T}ree\mathcal{R}eal$ corresponding to t_1 and to t_2 are the same, more precisely:

nodes labelled by real numbers are equal as in $\mathcal{R}eal$,
nodes labelled by identifiers are equal as in $\mathcal{T}ree$,
and trees corresponding to $'f(t_1, \ldots, t_n)'$ and to $'g(s_1, \ldots, s_m)'$
are equal iff the identifier "f" is equal to "g" and m is equal
to n.

We now recall the notions of solution of a system of constraints and fix some notation for the rest of the paper.

Let V be the set of all variables, let W be the subset $\{X_0, \ldots, X_n\} \subset V$, and let t and s be terms. A *system of constraints* S is a finite set of constraints. An *assignment* α *of* W *on a domain* D is a set having the form: $\{X_0 := d_0, \ldots, X_n := d_n\}$ in which $d_0, \ldots, d_n$ are in D. An assignment of W on the domain D satisfies a system of constraints S in variables W iff, interpreting the variables by the assigned element of D, each constraint in the system evaluates to true on D. Such an assignment is called a *solution* of S. We will not specify the domain unless it is needed to avoid confusion. If W' is a subset of the variables W occurring in a system of constraints S, then an assignment α' of W' is a solution of S on W' iff it is a subset of at least one solution of S.

Two constraints or two systems of constraints are *equivalent* iff the sets of their solutions are equal. A constraint or a system of constraints admitting at least one solution is called *solvable*.

3   The Solver

A solver for a domain is an algorithm which decides the solvability of a given system of constraints over the domain and produces a "simplified" form in the positive case.

For instance, the well-known unification is a solver for the domain $\mathcal{T}ree$. Also the partial cylindrical algebraic decomposition method (used in RISC-CLP($\mathcal{R}eal$)) is a solver for the domain $\mathcal{R}eal$.

In this section, we describe a solver for the combined domain $\mathcal{T}ree\mathcal{R}eal$. The solver is essentially an integration of the unification and the partial cylindrical algebraic decomposition method.

It first applies an "adapted unification" on the given system of constraints which decomposes the system of equations into an equivalent but simpler system of equations called "a tree-solved form". Then, it passes some part of the tree-solved form and the other constraints to the partial cylindrical algebraic decomposition method. In the following we describe the details.

First, we begin by defining the notion of "tree-solved form". We define the *depth* of a term t, $\delta(t)$, as a function from the terms in the language of $\mathcal{T}ree\mathcal{R}eal$ to the natural numbers as follows.

$$\delta(t) := \begin{cases} 1 & \text{if } t \text{ is a variable, a constant or a real-term} \\ 1 + max\{\delta(t_1), \ldots, \delta(t_n)\} & \text{if } t \text{ is of the form } f(t_1, \ldots, t_n) \end{cases}$$

The definition is extended to equations and to finite systems of equations.

$$\delta_e(t = s) := min\{ \delta(t), \delta(s) \},$$

$$\delta_s(\{\ e_1,\ldots,\ e_n\ \}) := max\{\ \delta_e(e_i)\ |\ 1 \le i \le n\ \}.$$

Definition 1. A finite system S of equations in variables W is in *tree-solved form* iff there exists a subset of variables $\{X_1, \ldots, X_h\} \subseteq W$ such that S has the form:

$$S = \{X_1 = t_1, \ldots, X_h = t_h,\ s_{h+1} = t_{h+1}, \ldots, s_n = t_n\}$$

and the following holds :
(i) S has depth 1 and contains no identity.
(ii) X_i does not occur in t_i nor anywhere else in S, $1 \le i \le h \le n$.

The reason behind defining such form is that, once the equality constraints are in tree-solved form, we can easily decide the whole system of constraints by using a decision procedure for the domain $\mathcal{R}eal$, as will be shown now.

Assuming the availability of an algorithm for transforming equality constraints into the tree-solved form, we give a solver for the domain $TreeReal$ in Figure 4.

Let C be solvable system of constraints in variables $W' \subseteq W$, and let S be a new system of constraints in variables W. The solver decides if the system $C \cup S$ has a solution on W over $TreeReal$. In case it does, then the simplified output is called a "tree-real partition" of $C \cup S$ of W. Let C be a solvable system of constraints over $TreeReal$ in variables W. $C_e \uplus C_r$ is a *tree-real partition of C of W* if, for every solution σ on $\mathcal{R}eal$ of C on the variables W_r occurring in C_r, there exists a unique assignment α of $W \setminus W_r$ on $TreeReal$ such that $\sigma \cup \alpha$ is a solution of C over $TreeReal$.

We assume that the input to the solver is a tree-real partition of system $C = C_e \uplus C_r$ and a new system of constraints S. The algorithm is structured mainly in three parts: firstly, it checks whether an empty system of new constraints is given. Then, it constructs the equation part of the tree-real partition of $C \cup S$ from the tree-solved form of the system of equations $C_e \cup S_e$. If this construction succeeds without causing any type error, the solver decides, in the last step, whether the real part of the tree-real partition of $C \cup S$ has a solution over $\mathcal{R}eal$ on the variables therein. If it does, then we are done. If it does not, then the system $C \cup S$ is not solvable because the real-constraints are not.

The correctness is based on the observation that the system obtained from the tree-solved form C'_e in step (2.c) of the algorithm is a "solved system" in the usual sense of unification: each equation contains a variable that occurs nowhere else, in particular (after type checking) it does not occur in any arithmetical term. Therefore, it is possible to extract an associated substitution representing the required assignment α for the tree-real partition $C' = C'_e \uplus C'_r$ that combined with a solution σ of C'_r is a solution of C'.

As promised earlier, now we give in Figure 5 an algorithm for computing a tree solved form of a given system of equations. The algorithm is a unification algorithm adapted from [17, 18].

In the algorithm description, the following (standard) notion of "substitution" is used. A *substitution* σ of W is a set: $\{X_0 \leftarrow t_0, \ldots, X_n \leftarrow t_n\}$,

$$SOLVER\ (C, S;\ u, C')$$

Input: $C = C_e \uplus C_r$ is a tree-real partition of C of $W' \subseteq W$.
              S is a system of new constraints in variables W.

Output: $C' = C'_e \uplus C'_r$ is a tree-real partition of $C \cup S$ of W.
              u is 'solvable' if a partition of $C \cup S$ of W is found, 'unsolvable' otherwise.

(1) [Trivial.] If the new set of constraints S is empty, set $C' \leftarrow C, u \leftarrow$ 'solvable' and return.

(2) [Tree-solve.]
    (a) Set $S \leftarrow S_e \cup S_r$, where equations are in S_e, and constraints involving arithmetical relation symbols are in S_r.
    (b) Apply *TREE-SOLVE* $(S_e \cup C_e;\ C'_e)$. If the tree-solved form C'_e is FALSE, then $C \cup S$ is not solvable. Set $u \leftarrow$ 'unsolvable' and return.
    (c) Set $C'_e \leftarrow C'_e \setminus S'_r$, where S'_r is the set of equations '$t = s$' in which at least one among t or s is a real-term or is a variable occurring in a real-term or in $C_r \cup S_r$.
    (d) Type-check each variable X appearing in equations '$X = t$' of C'_e where t is not a variable. If X occurs in $C_r \cup S_r \cup S'_r$, then there is a type-error and $C \cup S$ is not solvable. Set $u \leftarrow$ 'unsolvable' and return.

(3) [Decide.] Decide the truth of the sentence $(\exists \vec{W})(C_r \cup S_r \cup S'_r)$ by the solver over *Real*. If it is FALSE, then $C \cup S$ is not solvable. Set $u \leftarrow$ 'unsolvable' and return. Otherwise, set $C' \leftarrow C'_e \uplus C'_r$, where C'_r is the simplified system returned by the solver over *Real*. Set $u \leftarrow$ 'solvable' and return.

Fig. 4. Algorithm *SOLVER*.

where $t_0, \ldots, t_n$ are terms such that $X_i \neq X_j$, and X_i does not occur in t_i for $1 \leq i,\ j \leq n$. By $t\sigma$ we denote the term obtained by simultaneously replacing in t each occurrence of the variable X_i with the term t_i, for all $1 \leq i \leq n$. We denote by $c\sigma$ and by $S\sigma$ the natural extension of the application of a substitution to a constraint c and to a system of constraints S. By FALSE we mean that the system is not solvable.

To prove the correctness of *TREE-SOLVE* we must show that the algorithm terminates according to a control strategy and that it conforms to the specification. Let the control strategy choose always the rule that applies to an equation in S of greater depth. Termination is based on the fact that after each application of a rule, either S is FALSE, or S is "simpler" in the following sense:

- S contains an identity less than before;
- S contains less occurrences for the variable substituted by rule (SV), and the rule cannot be applied anymore;
- S has a depth less than it had before the decomposition performed by rule (ET).

TREE-SOLVE $(S; S')$

Input: S is a system of equations in variables W.

Output: S' is a tree-solved form equivalent to S or FALSE.

(1) Transform S by applying one of the following rules until no change is possible or S is FALSE.

-RI- [Remove Identities]

   $S \cup \{s = s\} \Longrightarrow S$.

-ET- [Equate Terms]

   $S \cup \{f(s_1, \ldots, s_n) = f(t_1, \ldots, t_n)\} \Longrightarrow S \cup \{s_1 = t_1, \ldots, s_n = t_n\}$.

   $S \cup \{f(s_1, \ldots, s_n) = g(t_1, \ldots, t_m)\} \Longrightarrow$ FALSE,

$\qquad\qquad\qquad\qquad\qquad\qquad$ if "f" is different from "g"

$\qquad\qquad\qquad\qquad\qquad\qquad$ or m is not equal to n.

-EC- [Equate Constants]

   $S \cup \{m = n\} \Longrightarrow$ FALSE,

$\qquad\qquad\qquad$ if m and n are not equivalent constants.

-SV- [Substitute Variables]

   $S \cup \{X = s\} \Longrightarrow S\{X \leftarrow s\} \cup \{X = s\}$,

$\qquad\qquad\qquad$ if X is a variable in W occurring in S,

$\qquad\qquad\qquad$ s is not a variable and X does not occur in s.

   $S \cup \{X = Y\} \Longrightarrow S\{X \leftarrow Y\} \cup \{X = Y\}$,

$\qquad\qquad\qquad$ if X and Y are distinct variables in W and

$\qquad\qquad\qquad$ both occur in S.

-TC- [Type Check]

   $S \cup \{s = t\} \Longrightarrow$ FALSE,

$\qquad\qquad\qquad$ if s is a real-term and t is not a real-term.

-OC- [Occur Check]

   $S \cup \{X = s\} \Longrightarrow$ FALSE,

$\qquad\qquad\qquad$ if X is a variable occurring in s and s is not a real-term.

(2) Assign S to S' and return.

Fig. 5. Unification algorithm:*TREE-SOLVE*.

Therefore, because the initial set is finite and infinite trees are not allowed, *TREE-SOLVE* terminates.

Next, we show that each rule preserves the set of solutions of S. Rules (RI), (ET), (EC), (SV) and (OC) are clear, rule (TC) is a consequence of having defined arithmetical functions partial.

Finally, it remains to show that, given a system of equations S in variables W, *TREE-SOLVE* produces an equivalent tree-solved form S' or FALSE. Let S' be different from FALSE and be obtained from S by application of the rules until no change is possible. S' does not contain any trivial identity nor it contains equations among constants, because they are removed by rules (RI) and (EC).

Rule (ET) decreases the depth of S, because no more rule can be applied to S', it is clear that S' must have depth equal to 1. Let the equation '$X = s$' be in S'. If X is a variable not occurring in the term s, then by the application of rule (SV), X does not appear anywhere else in S'. This shows that S' is a tree-solved form.

4 Example: Chemical Systems

This last section is devoted to showing by an example a possible application for the extended RISC-CLP($\mathcal{R}eal$) language. We have decided to give one big example instead of many "toy" examples, because we believe that this will better demonstrate the new language and its additional possibilities.

The scenario of our example is that of systems of chemical reactions. In a container we place some substances, say H_2O, CO, Cl, ..., and so on. Now, these substances will interact with each other, new substances will be produced and some will be used up, until possibly an equilibrium state is reached among them. At equilibrium, reactions still continue in the system but the net amounts of each substance remain the same.

In the usual model of a chemical system of reactions, we can denote the substances by $A_i, A_j, \ldots$ and their concentrations (atoms per unit of volume, measured in $moles \cdot liter^{-1}$) by $[A_i], [A_j], \ldots$, respectively. Each system is also described by reactions laws among substances that tell us what is the kinetics of the system. As an example, consider the system of reversible reactions:

$$A_i + A_j \xrightarrow{k_1} A_k$$
$$A_k \xrightarrow{k_{-1}} A_i + A_j$$

where k_1 and k_{-1} are the reaction constants (in sec^{-1}) measuring the "rate" by which the substances on the left transform into substances on the right. For each particular system, we can associate a system of differential equations that tells us about the dynamic of the reactions. In case of the system above, the differential equations are as follows:

$$\frac{d[A_i]}{dt} = \frac{d[A_j]}{dt} = -k_1[A_i][A_j] + k_{-1}[A_k]$$
$$\frac{d[A_k]}{dt} = k_1[A_i][A_j] - k_{-1}[A_k]$$

Note that each equation is not a "stand alone" description of each single reaction, but takes into account the entire system. The system described above reaches an equilibrium state that is characterized by the equations:

$$0 = -k_1[A_i][A_j] + k_{-1}[A_k]$$
$$0 = k_1[A_i][A_j] - k_{-1}[A_k]$$

One can ask various questions about this system, for instance about the relative concentrations of the substances that made it possible to reach the equilibrium state.

The modelling process can be translated into a RISC-CLP($\mathcal{R}eal$) program by using the added capabilities provided by the extended language. In Figure 6, we show a RISC-CLP($\mathcal{R}eal$) example program that captures the scenario of reaction chains in its generality.

We now show how the model works by some examples, beginning with the description of water:

$$H_2O \xrightarrow{k_1} H^+ + OH^-$$
$$H^+ + OH^- \xrightarrow{k_2} H_2O$$

At equilibrium, we can ask RISC-CLP($\mathcal{R}eal$) what is the system of equations describing water by the following query:

```
?-  system([h2o,h,oh], [ [ [[h2o],[h,oh]],[H2O],K1],
                         [ [[h,oh],[h2o]],[H,OH],K2]],
            [0,0,0]).
```

The answer of RISC-CLP($\mathcal{R}eal$) is the general equilibrium state equation for a reversible system, like that given above.

```
K1 > 0
H2O >= 0
K2 > 0
H >= 0
H OH K2 - H2O K1 = 0
```

Suppose we want to compute the pH (defined as $-log_{10}[H^+]$), given that the concentration of pure water is known to be $55.5 \cdot moles \cdot liter^{-1}$, the equilibrium constant is $K = k_2/k_1 = 55.5 \cdot 10^{14}$ and the concentrations of H^+ and of OH^- are the same because there is no other source outside of water. The query would look as follows. Note that we are able to specify, in curly braces, which variables should appear in the answer.

```
?- H2O = 55.5, K1 * 55.5 * 10^14 = K2, H = OH,
   system([h2o,h,oh], [ [ [[h2o],[h,oh]],[H2O],K1],
                        [ [[h,oh],[h2o]], [H,OH],K2]],
           [0,0,0]) {H}.
```

The answer produced by RISC-CLP($\mathcal{R}eal$) is correct, the pH of water is 7 ($[H^+] = 10^{-7}$):

```
H = 1e-07
```

```
%===========================
% chemical reactions laws
%===========================

% predicate system(A,L,R): A is a list of names of substances, L is a
% list of reactions among substances and R is a list of rates Rate such
% that d[Ar]/dt = Rate, for each name Ar in list A, wrt system L.
%
system([],L,[]).
system([Ar|T],L,[Rate|Tail]) :-
                              rate(Ar,L,Rate),
                              system(T,L,Tail).

%
% predicate rate(Ar,L,Rate):  Ar is a substance name,  L  is a list of
% reactions among substances and Rate is the rate d[Ar]/dt wrt system L.
% Each element of L is of the form [ T,C,K ]: T is the reaction template
% list made of two lists, reactants names and products names; C is the
% list of the concentrations of reactants and K is the reaction constant.
%
rate(Ar,[],0).

% part of the rate of Ar, when Ar is among the reactants of a reaction.
%
rate(Ar,[ [ [R,P],[R1|Rt],K] |L ],Rate) :-
                              K > 0, R1 >= 0,
                              Rate = -K*R1*Rest + Tr,
                              member(Ar,R),
                              product(Rt,Rest),
                              rate(Ar,L,Tr).

%
% part of the rate of Ar, when Ar is among the products of a reaction.
%
rate(Ar,[ [ [R,P],[R1|Rt],K ]|L ],Rate) :-
                              K > 0, P1 >= 0,
                              Rate = K*R1*Rest + Tr,
                              member(Ar,P),
                              product(Rt,Rest),
                              rate(Ar,L,Tr).

%
% part of the rate of Ar, when Ar does not appear in the reaction.
%
rate(Ar,[ [ L1,Rs,K ]|L ],Rate) :-
                              rate(Ar,L,Rate).

% utility predicates
%
product([],1).
product([H|T],H*Pt) :- product(T,Pt).

member(E,[E|T]).
member(E,[H,T]) :- member(E,T).
```

Fig. 6. An example program in the RISC-CLP($\mathcal{R}eal$) language.

A similar system of acid-base reactions involves the "Acetate buffer", described by the following equilibrium.

$$HOAc \xrightarrow{k_1} H^+ + OAc^-$$
$$H^+ + OAc^- \xrightarrow{k_2} HOAc$$

Buffer solutions are used to maintain constant pH in a system: they contain a base that reacts if acids are added and, conversely, an acid that reacts if bases are added.

The "buffering" effect that characterizes a buffer solution can be illustrated with the following example. First we compute the pH in a buffered solution for the initial concentrations of $[HOAc]_0 = [OAc]_0 = 1$ $moles \cdot liter^{-1}$ with the addition of no extra acid ($X = 0$).

```
?-  HOAc = HOAcO + X, OAc = OAcO - X, K2 * 18 = K1 * 10^6,
    X = 0, HOAcO = 1, OAcO = 1,
    system([hoac,h,oac], [ [ [[hoac],[h,oac]],[HOAc],K1],
                           [ [[h,oac],[hoac]], [H,OAc],K2]],
            [0,0,0]) {H}.
```

We obtain a pH of 4.75:

```
H = 1.8e-05
```

Normally, the addition of a large amount of acid from an outside source would change the pH dramatically. But in a buffered solution, even the addition of a relatively large amount of strong acid, changes the pH only slightly. Below we add .1 $moles \cdot liter^{-1}$ to the solution and ask RISC-CLP($\mathcal{R}eal$) about the new value of the pH.

```
?-  HOAc = HOAcO + X, OAc = OAcO - X, K2 * 18 = K1 * 10^6,
    X = 1/10, HOAcO = 1, OAcO = 1,
    system([hoac,h,oac], [ [ [[hoac],[h,oac]],[HOAc],K1],
                           [ [[h,oac],[hoac]], [H,OAc],K2]],
            [0,0,0]) {H}.
```

Surprisingly, we see that the pH changes only to 4.65:

```
H = 2.2e-05
```

We can also ask the reverse question: how much acid should be added to decrease the pH to 4.

```
?-  HOAc = HOAcO + X, OAc = OAcO - X, K2 * 18 = K1 * 10^6,
    H = 1/10^4, HOAcO = 1, OAcO = 1,
    system([hoac,h,oac], [ [ [[hoac],[h,oac]],[HOAc],K1],
                           [ [[h,oac],[hoac]], [H,OAc],K2]],
            [0,0,0]).
```

The answer is striking, we must add a lot of acid: $.69\ moles \cdot liter^{-1}$.

```
X = 0.694915
```

We have tried to solve all the above queries by using $CLP(\mathcal{R})$, but because they require solving non-linear equations, $CLP(\mathcal{R})$ cannot give the complete answer. Non-linearity arises very naturally here, in the context of chemical systems, but also in the majority of "real world" problems.

5 Conclusions

In this paper we have presented the work that we have been doing in extending the domain of computation of RISC-CLP from *Real* to *TreeReal*. The motivation for supporting the more general domain of trees with real numbers is that the possibility of integrating various distinct decidable domains enhances declarativeness.

Solving systems of constraints over the reals is a very difficult task and a lot of work has still to be done in order to attain efficiency. The number of free variables is a critical measure of the complexity of deciding the polynomial system of equations [11] and therefore, it should be minimized. One examples are the variables generated by the renaming mechanism because they do not occur, in general, in the set of already solved constraints as remarked in [17]. This suggests that one could try to eliminate them without resorting to the expensive solver. A more detailed study of the problem of elimination of variables from polynomial systems of equations and inequalities has yet to be done. A first step in this direction is represented by [13].

The crucial role of a notion of canonical form for a system non-linear constraints became also clear. Work has been pursued in these directions by developers of $CLP(\mathcal{R})$, who have been mainly concerned about an efficient treatment of linear systems of constraints. All the above problems are still open for the more general setting of polynomial systems which we intend to pursue further.

Acknowledgments

I sincerely thank my advisor Hoon Hong for discussions, criticism and for his implementation of the partial CAD method, without which this work could not have been done.

References

1. A. Aggoun, M. Dincbas, A. Herold, H. Simonis, and P. Van Hentenryck. The CHIP System. Technical Report TR-LP-24, European Computer Industry Research Centre (ECRC), Munich, Germany, June 1987.
2. B. Buchberger. *An Algorithm for Finding a Basis for the Residue Class Ring of a Zero-Dimensional Polynomial Ideal*. PhD thesis, Universitat Innsbruck, Institut fur Mathematik, 1965. German.

3. B. Buchberger. Groebner bases: An algorithmic method in polynomial ideal theory. In N. K. Bose, editor, *Recent Trends in Multidimensional Systems Theory*, chapter 6. D. Riedel Publ. Comp., 1983.

4. G. E. Collins. Quantifier elimination for the elementary theory of real closed fields by cylindrical algebraic decomposition. In *Lecture Notes In Computer Science*, pages 134–183. Springer-Verlag, Berlin, 1975. Vol. 33.

5. G. E. Collins and H. Hong. Partial cylindrical algebraic decomposition for quantifier elimination. *Journal of Symbolic Computation*, 12(3):299–328, September 1991.

6. A. Colmerauer. An introduction to Prolog III. *Communications of the ACM*, 33(7):69–90, July 1990.

7. Alain Colmerauer. Final specifications for PROLOG-III. Technical Report P1219(1106), ESPRIT, February 1988.

8. William Havens, Susan Sidebottom, Greg Sidebottom, John Jones, and Russel Ovans. Echidna: a constraint logic programming shell. In *Proceedings of the 1992 Pacific Rim International Conference on Artificial Intelligence*, volume I, pages 165–171, Seoul, Korea, 1992.

9. N. Heintze, J. Jaffar, S. Michaylov, P. Stuckey, and R. Yap. *The CLP(R) Programmer's Manual Version 1.1*, November 1991.

10. H. Hong. *Improvements in CAD-based Quantifier Elimination*. PhD thesis, The Ohio State University, 1990.

11. H. Hong. Comparison of several decision algorithms for the existential theory of the reals. Technical Report 91–41.0, Research Institute for Symbolic Computation, Johannes Kepler University A-4040 Linz, Austria, 1991. Submitted to *Journal of Symbolic Computation*.

12. H. Hong. Non-linear real constraints in constraint logic programming. In *International Conference on Algebraic and Logic Programming (LNCS 632)*, pages 201–212, 1992.

13. Hoon Hong. Quantifier elimination for formulas constrainted by quadratic equations via slope resultants. *The Computer Journal, Special issue on Quantifier Elimination*, 1993.

14. J. Jaffar and S. Michaylov. Methodology and implementation of a CLP system. In J.-L. Lassez, editor, Proceedings 4th ICLP, pages 196–218, Cambridge, MA, May 1987. The MIT Press.

15. J. Jaffar, S. Michaylov, and Roland H. C. Yap. A methodology for managing hard constraints in clp systems. In *Proceedings of the ACM-SIGPLAN Conference on Programming Language Design and Implementation*, pages 306–316, Toronto, June 1991.

16. Joxan Jaffar and Jean-Louis Lassez. Constraint logic programming. In *Proceedings of the 14th ACM Symposium on Principles of Programming Languages, Munich, Germany*, pages 111–119. ACM, January 1987.

17. Joxan Jaffar, Spiro Michaylov, Peter Stuckey, and Roland H. C. Yap. The CLP(R) Language and System. Research Report RC 16292, IBM Research Division, T. J. Watson Research Center, Yorktown Heights, NY 10598, November 1990.

18. Jean-Pierre Jouannaud and Claude Kirchner. Solving equations in abstract algebras: A rule-based survey of unification. In J.-L. Lassez and G. Plotkin, editors, *Computational Logic. Essays in honour of Alan Robinson*, chapter 8, pages 257–321. MIT Press, Cambridge, (MA) USA, 1991.

19. D.A. Kohler. *Projection of Convex Polyhedral Sets*. PhD thesis, University of California, Berkeley, 1967.

20. E. Monfroy. Specification of Geometrical Constraints. Technical Report ECRC-92-31, ECRC, Munich, Germany, 1992. presented at AAGR 92, Linz, Austria.
21. Jean Michel Nataf. Algorithm of simplification of nonlinear equations systems. *SIGSAM Bulletin*, 26(3):9–16, April 1992.
22. William Older and Andrè Vellino. Constraint arithmetic on real intervals. In Frèdèric Benhamou and Alain Colmerauer, editors, *Constraint Logic Programming: Selected Research*. The MIT Press, Cambridge, 1992.
23. K. Sakai and A. Aiba. CAL: A Theoretical Background of Constraint Logic Programming and its Applications. *Journal of Symbolic Computation*, 8:589–603, 1989.
24. A. Tarski. *A Decision Method for Elementary Algebra and Geometry*. Univ. of California Press, Berkeley, second edition, 1951.
25. Clifford Walinsky. Clp(Σ^*): Constraint logic programming with regular sets. In *Logic Programming: Proceedings 6th International Conference, Jerusalem, Israel*, pages 181–196. MIT Press, June 1989.

Dynamic Term Rewriting Calculus and Its Application to Inductive Equational Reasoning

Su Feng, Toshiki Sakabe and Yasuyoshi Inagaki

Department of Information Engineering, Faculty of Engineering, Nagoya University,
Furo-cho, Chikusa-ku, Nagoya, 464-01 JAPAN

Abstract. *Dynamic Term Rewriting Calculus (DTRC)* is a new computation model proposed by the authors for the purpose of formal description and verification of algorithms treating *Term Rewriting Systems (TRSs)*. The computation of DTRC is basically term rewriting. The characteristic features of DTRC are dynamic change of rewriting rules during computation and hierarchical declaration of not only function symbols and variables but also rewriting rules. These features allow us to program meta-computation of TRSs in DTRC, i.e., we can implement in DTRC in a natural way those algorithms which manipulate TRSs as well as those procedures which verify such algorithms. We show here that we can use DTRC to represent the proof of an inductive theorem of an equational axiom system, i.e., we can translate the statements of base and induction steps in the proof of the inductive theorem into a DTRC term. The translation reduces the proof of the statements into the evaluation of the DTRC term.

1 Introduction

Recent researches on TRSs[1] give rise to importance of meta-computation of TRSs, that is, formal description and verification of algorithms treating TRSs. For example, Knuth-Bendix (KB) algorithm [9] transforms TRSs to find a complete TRS for a given equational axiom system. The computation done by KB algorithm is considered as a meta-computation, since a TRS expresses the computation to find a normal form of a given input term, and KB algorithm repeats transformations from one TRS to another.

In this paper, we propose a new computation model called *Dynamic Term Rewriting Calculus (DTRC,* for short) for the purpose of formal treatment of meta-computation for TRSs.

Computation of DTRC is basically term rewriting. The characteristic features of DTRC are dynamic change of rewriting rules during computation and hierarchical declaration of not only function symbols and variables but also rewriting rules. These features distinguish DTRC from other computation models, especially other reduction calculus, and allow us to implement in DTRC in a natural way those algorithms which manipulate TRSs as well as those verification procedures for such algorithms.

---

[1] Here we suppose the reader is familiar with TRSs (see [6] for example).

Related works on meta-computation are found in [11] and [5]. The former proposed λProlog which uses λ-terms as data and Prolog as programming language. The latter discussed schematic rewriting systems which use λ-terms as data and rewriting systems as programming language. Comparing with them, in DTRC, data and programs are in the same frame work. One can even write interpreter of DTRC in DTRC. Further, if we consider [ ], ;, etc. as special function symbols and matching as a restricted form of that of TRSs, then matching of DTRC is regarded syntactically first order, while unification of λProlog and matching of schematic rewriting systems are both higher order.

After introducing the syntax and the operational semantics of DTRC, we show that DTRC is capable of inductive equational reasoning. That is, we show that we can translate the statements of base and induction steps to prove an inductive theorem of an equational axiom system into a DTRC term. This means that, we can reduce the proof of the statements to the evaluation of the DTRC term. This exemplifies the expressive power of DTRC.[2]

2 Syntax

A typical term of DTRC has the form

$$[F; V: R](t)$$

where F and V are declarations of function symbols and variables, respectively, R is a set of rules, and t is a term to be rewritten. We will call a term of this form a program.

Intuitively, the subexpression $[F; V: R]$ is a declaration which defines all functions in F in terms of rules of R, and $[F; V: R](t)$ is the value obtained by interpreting t under the definition of the functions. It should be emphasized that both sides of a rule and the term to be rewritten may contain subterms of the form of program. This feature allows us to write hierarchical programs.

Formal definitions of terms and related concepts are given as follows. Let $\mathcal{N}$ be the set of nonnegative integers.

Definition 1. For a set $\mathcal{F}$ of function symbols with a mapping $arity \colon \mathcal{F} \to \mathcal{N}$ and a set $\mathcal{V}$ of variables, the set of all *terms* over $\mathcal{F}$ and $\mathcal{V}$, $\mathbf{T}(\mathcal{F}, \mathcal{V})$, is the smallest set satisfying the following conditions:

1. If $x \in \mathcal{V}$ then

$$x \in \mathbf{T}(\mathcal{F}, \mathcal{V})$$

2. If $t_1, \ldots, t_n \in \mathbf{T}(\mathcal{F}, \mathcal{V})$, $f \in \mathcal{F}$, $arity(f) = n$ then

$$f(t_1, \ldots, t_n) \in \mathbf{T}(\mathcal{F}, \mathcal{V})$$

---

[2] DTRC was originally proposed in [1]. Intensive revision and extension has been done in this paper.

3. If $f_1, ..., f_l \in \mathcal{F}$, $x_1, ..., x_m \in \mathcal{V}$, $l_1, ..., l_n, r_1, ..., r_n, t \in \mathbf{T}(\mathcal{F}, \mathcal{V})$ then

$$[f_1, ..., f_l; x_1, ..., x_m : l_1 \to r_1, ..., l_n \to r_n](t) \in \mathbf{T}(\mathcal{F}, \mathcal{V})$$

Definition 2. Let $\Box$ be a special symbol which is not in $\mathcal{F} \cup \mathcal{V}$. A *context* is a term over $\mathcal{F}$ and $\mathcal{V} \cup \{\Box\}$ in which $\Box$ occurs exactly once at a place of the term outside variable declarations. A context is denoted as $C[\![\ ]\!]$ and the term obtained from $C[\ ]$ by replacing $\Box$ with a term t is written as $C[t]$.

Definition 3. Let t, s be terms over $\mathcal{F}$ and $\mathcal{V}$. If there exists a context $C[\![\ ]\!]$ such that $t = C[\![s]\!]$ then s is said to be a *subterm* of t.

Definition 4. For a term $t = [F; V : R](s)$, the *scope of the function declaration* F is the whole expression of t. The *scope of the variable declaration* V is the subexpression $V : R$.

The notion of scope induces free and bound occurrences of function symbols and variables.

Definition 5. An occurrence of a function symbol f in a term t is *bound* if it is within the scope of a function declaration in t which declares f, otherwise it is said to be *free*. A bound and free occurrence of a variable is defined in a same way.

Here we use "occurrence" with intuitive meaning without giving a formal definition of it.

In the sequel, a term of the form

$$[F; V : R](t)$$

is called a *program*. The subexpression

$$[F; V : R]$$

is called a *system* and F, V and R are called *function, variable* and *rule declarations*, respectively. A term not containing programs as subterms is called a *simple term*. The set of all simple terms over $\mathcal{F}$ and $\mathcal{V}$ is denoted as $\mathbf{S}(\mathcal{F}, \mathcal{V})$.

3 Operational Semantics

Roughly speaking, the subexpression $[F; V : R]$ of a program $[F; V : R](t)$ defines an environment in terms of rules of R, and the program itself means the value of t under the environment. That is, interpretation of the program is done by rewriting t with rules of R. But we should note that declarations are hierarchical and R and t may include other programs as subexpressions.

If a function symbol f which is not declared in F occurs freely in t or R, it is left uninterpreted. The function symbol f is interpreted when the program is placed in a context which declares f. This implies that rules of a subprogram can be changed by the main program during computation.

If a variable x which is not declared in V occurs freely in a rule of R or t, it is treated as a constant. The variable x is treated as a variable when the program is inserted into a context in which $\Box$ occurs in the scope of a variable declaration that declares x. Thus free variables can serve as meta-variables which are very important in inductive equational reasoning as will be shown later.

If R has a rule with a subprogram in its right hand side, application of the rule introduces a new environment in which some computation is done for evaluation of functions in the left hand side.

We define the operational semantics of DTRC as a reduction relation $\to$ on $\mathbf{T}(\mathcal{F}, V)$. We first define the notions of substitution and matching.

Definition 6. A *substitution* σ is a finite set $\{t_1/x_1, \ldots, t_n/x_n\}$ of pairs such that x_i are all distinct variables and t_i are terms. The *domain of* σ, denoted as $dom(\sigma)$, is defined to be $\{x_1, \ldots, x_n\}$. For a term t and a substitution σ, we define the term $t\sigma$ by induction on the structure of terms as follows:

1. If $t \in V$ then
$$t\sigma = \begin{cases} t_i & \text{if } t = x_i \\ t & \text{otherwise} \end{cases}$$

2. If $t = f(s_1, \ldots, s_m)$ then
$$t\sigma = f(s_1\sigma, \ldots, s_m\sigma)$$

3. If $t = [F; V : l_1 \to r_1, \ldots, l_m \to r_m](s)$ then
$$t\sigma = [F; V : l_1\sigma' \to r_1\sigma', \ldots, l_m\sigma' \to r_m\sigma'](s\sigma)$$

where $\sigma' = \{t/x \in \sigma \mid x \notin V\}$

For a substitution $\sigma = \{t_1/x_1, \ldots, t_n/x_n\}$ and a term t, $t\sigma$ is the term obtained from t by simultaneously replacing every free occurrence of x_i by the corresponding term t_i for each $i = 1, \ldots, n$.

Definition 7. Let $C[\![\ ]\!]$ be a context, s, t be terms, and σ be a substitution. Then s *matches* t by σ under $C[\![\ ]\!]$ if and only if $s\sigma = t$ and any free occurrence of function symbols in t does not become bound in $C[\![t]\!]$.

For instance, $base(x)$ does not match $base(0)$ under $[base; y : base(y) \to y](\Box)$, for the free occurrence of the function symbol $base$ in $base(0)$ is bound in $[base; y : base(y) \to y](base(0))$.

We define the *reduction relation* $\to$ on $\mathbf{T}(\mathcal{F}, V)$ as follows:

Definition 8. For any program $[F; V : R](C[\![t]\!])$, if there is a rule $l \to r \in R$ and a substitution σ such that $dom(\sigma) \subseteq V$ and l matches t by σ under $C[\![\ ]\!]$, then $[F; V : R](C[\![t]\!])$ *is rewritten to* $[F; V : R](C[\![r\sigma]\!])$ and we call the pair of these terms as a *rewrite*. For any context $B[\![\ ]\!]$ and terms t, t', if (t, t') is a rewrite then

$$B[\![t]\!] \to B[\![t']\!]$$

Definition 9. Let $\overset{*}{\rightarrow}$ stand for the reflexive and transitive closure of $\rightarrow$. A term s is a *normal form* of a term t if $t \overset{*}{\rightarrow} s$ and there is no term u such that $s \rightarrow u$. A term s is *weakly normalizing* if there is a normal form of s. A term s is *strongly normalizing* if all reduction sequences $s \rightarrow s_1 \rightarrow s_2 \rightarrow \ldots$ terminate.

4 Examples of DTRC programs

In this section, we show examples of DTRC terms, which make use of characteristic features of DTRC.

Let's consider the following program $Sys_1(t)$:

$Sys_1(t) \overset{\triangle}{=}$
    [recursive; x, y :
      recursive(0, y) → base(y),
      recursive(s(x), y) → rec(y, recursive(x, y))] (t)

The system Sys_1 defines **recursive** in a general form of recursive functions with two arguments. This definition is "general" in the sense that function symbols **base** and **rec** are not declared in Sys_1. To obtain the total value of $Sys_1(t)$, we place $Sys_1(t)$ in a context which provides $Sys_1(t)$ with definitions of **base** and **rec**.

Next program $Sys(t)$ is an application of $Sys_1(t)$:

$Sys(t) \overset{\triangle}{=}$
    [recur, add, 0, s; u, v:
      recur(u, v) → Sys_1(recursive(u, v)),
      add(u, v) → [base, rec; x, y:
      base(x) → x,
      rec(x, y) → s(y)](recur(u, v))] (t)

In Sys, **add** is defined to be **recursive** of Sys_1 in the environment where **base** is the identity function and **rec** is the function that returns the second argument plus one.

Following is one of the reduction sequences for the program Sys(add(s(0), s(0))), which intuitively means $1+1$. In the reduction sequence, we use *addition* for the abbreviation of the system

$addition \overset{\triangle}{=}$
    [base, rec; x, y:
      base(x) → x,
      rec(x, y) → s(y)]

$Sys(\underline{add(s(0), s(0))})$

$\rightarrow$

```
[ recur, add, 0, s; u, v:
  recur(u, v) → Sys₁(recursive(u, v)),
  add(u, v) → addition(recur(u, v)) ]
    (addition(recur(s(0), s(0)))))
```

$\rightarrow$

```
[ recur, add, 0, s; u, v:
  recur(u, v) → Sys₁(recursive(u, v)),
  add(u, v) → addition(recur(u, v)) ]
    (addition(Sys₁(recursive(s(0), s(0))))))
```

$\rightarrow$

```
[ recur, add, 0, s; u, v:
  recur(u, v) → Sys₁(recursive(u, v)),
  add(u, v) → addition(recur(u, v)) ]
    (addition(Sys₁(rec(s(0), recursive(0,s(0)))))))
```

$\rightarrow$

```
[ recur, add, 0, s; u, v:
  recur(u, v) → Sys₁(recursive(u, v)),
  add(u, v) → addition(recur(u, v)) ]
    (addition(Sys₁(rec(s(0), base(s(0))))))
```

$\rightarrow$

```
[ recur, add, 0, s; u, v:
  recur(u, v) → Sys₁(recursive(u, v)),
  add(u, v) → addition(recur(u, v)) ]
    (addition(Sys₁(rec(s(0), s(0)))))
```

$\rightarrow$

```
[ recur, add, 0, s; u, v:
  recur(u, v) → Sys₁(recursive(u, v)),
  add(u, v) → addition(recur(u, v)) ]
    (addition(Sys₁(s(s(0)))))
```

In the reduction sequence, the underlined part indicates the subterm rewritten. The last program of this section is an extension of $Sys(t)$:

```
[ recur, add, mult, 0, s; u, v:
  recur(u, v) → Sys₁(recursive(u, v)),
  add(u, v) → [base, rec; x, y:
    base(x) → x,
    rec(x, y) → s(y)](recur(u, v)),
  mult(u, v) → [base, rec; x, y:
    base(x) → 0,
    rec(x, y) → add(x, y)](recur(u, v)) ](t)
```

In this program, add and mult are defined in almost the same way by using Sys_1, but the definitions of **base** and **rec** are different. That is, **mult** is defined to be **recursive** of Sys_1 in the environment where **base** is zero and **rec** is addition.

In [3] we showed that for any simple term t programs above are strongly normalizing and confluent, which means, they have unique normal forms for any reduction sequences.

The examples show that DTRC can facilitate modularization: we can write programs containing other programs as subterms (e.g., $Sys_1(\mathtt{recursive(u,v)})$ in $Sys(t)$). DTRC also has the feature of parameterization in the sense that a term with no bound function symbols (e.g., **base(y)** in Sys_1) can be considered as a formal parameter and an actual parameter is passed dynamically during the computation. Such parameters may be terms containing programs as subterms and may also be changed to terms containing programs as subterms. We can reuse components in such terms naturally by changing some subterms of the components during computation, just as the reuse of $Sys_1(s)$ in the last program. This kind of modularization and parameterization is somehow stronger than the usual, since the parameters are terms instead of function symbols.

5 Inductive Equational Reasoning by DTRC

There are two main approaches on inductive equational reasoning based on TRSs: one applies explicit induction arguments on the structure of terms [13, 14] and the other uses inductionless induction method [4, 7, 12]; while [10] combines these two methods based on the notion of test-set.

In this section, it is shown that DTRC can be used to automate proof of inductive equational theorems by the two explicit induction methods: structural induction and cover set induction [13, 14].

When we prove by structural induction that an equation is an inductive theorem of an equational axiom system, we use meta-variables and equational axioms as rewrite rules. A *meta-variable* is a variable ranging over ground constructor terms w.r.t the equational axiom system, and is treated as a constant distinguished from that belonging to the symbol set of the equational language.

Cover set induction is an extension of structural induction. Sakai et al. [13] introduced a kind of cover set induction, which is different from that of Zhang et al. [14], for proving an inductive theorem of an equational axiom system. [13] also showed a method to mechanize the cover set induction as well as the usual structural induction, in which term rewriting systems and the concept of meta-variables played important roles.

We show that DTRC can treat meta-variables very easily and mechanize inductive equational reasoning based on the work of [13].

5.1 Inductive Equational Theorems

Definition 10.

1. *An equation over $\mathcal{F}$ and $\mathcal{V}$ is an equation $\alpha = \beta$ with $\alpha, \beta \in \mathbf{S}(\mathcal{F}, \mathcal{V})$. For a substitution σ, $\alpha\sigma = \beta\sigma$ is also denoted as $(\alpha = \beta)\sigma$.*

2. *An equational axiom system over $\mathcal{F}$ and $\mathcal{V}$ is a set of equations over $\mathcal{F}$ and $\mathcal{V}$.*

Definition 11. Let E be an equational axiom system over $\mathcal{F}$ and $\mathcal{V}$ and e be an equation $\alpha = \beta$ over $\mathcal{F}$ and $\mathcal{V}$. e is an *inductive theorem* of E, denoted as $E \vdash_{ind} e$, if and only if for any substitution $\sigma = \{t_1/x_1, ..., t_n/x_n\}$ with x_i occurring freely in α or β and $t_i \in \mathbf{S}(\mathcal{F}, \mathcal{V})$ for $i = 1, ..., n$, $\alpha\sigma = \beta\sigma$ is derived from E, denoted as $E \vdash e\sigma$, whenever $\alpha\sigma, \beta\sigma \in \mathbf{S}(\mathcal{F}, \emptyset)$, where $\mathbf{S}(\mathcal{F}, \emptyset)$ is the set of all *ground simple term* over $\mathcal{F}$, i.e., the set of all simple terms over $\mathcal{F}$ and $\mathcal{V}$ containing no variables.

All theorems of E are inductive theorems of E, but generally, there are inductive theorems of E which are not theorems of E.

Definition 12. Let E be an equational axiom system over $\mathcal{F}$ and $\mathcal{V}$. A set $C \subset \mathcal{F}$ is a *set of constructors* w.r.t. E if for any $t \in \mathbf{S}(\mathcal{F}, \emptyset)$ there exists a term $s \in \mathbf{S}(C, \emptyset)$ such that $E \vdash t = s$. An element of C is called a *constructor* and a term of $\mathbf{S}(C, \emptyset)$ is called a *ground constructor term*.

The following is a well-known result [4].

Theorem 13. *Let E be an equational axiom system over $\mathcal{F}$ and $\mathcal{V}$, C be a set of constructors w.r.t. E and e be an equation $\alpha = \beta$ over $\mathcal{F}$ and $\mathcal{V}$. $E \vdash_{ind} e$ if and only if for any substitution $\sigma = \{t_1/x_1, ..., t_n/x_n\}$ with x_i occurring freely in α or β and $t_i \in \mathbf{S}(C, \mathcal{V})$ for $i = 1, ..., n$, $E \vdash e\sigma$, whenever $\alpha\sigma, \beta\sigma \in \mathbf{S}(\mathcal{F}, \emptyset)$.*

In the sequel, we will omit phrase "over $\mathcal{F}$ and C" where no difficulty is caused.

5.2 Structural Induction

Let E be an equational axiom system and e be an equation in which a variable x occurs. To prove by *structural induction* on x that e is an inductive theorem of E (w.r.t. x), we prove the following two statements. We assume here that if there are other variables occurring in e then they are substituted by distinct meta-variables. Let p_i ($i = 1, 2, ...$) be meta-variables.

(1) Base step:
   For each constant c, $e\{c/x\}$ is derived from E.
(2) Induction step:
   For any positive integer n and any constructor f with arity n, if $e\{p_i/x\}(i = 1, ..., n)$ are derived form E, then $e\{f(p_1, ..., p_n)/x\}$ is derived from E.

As easily known, these two statements are merged into the following statement: For each constructor f with arity n (including constants with arity 0),

$$E \cup \{e\{p_i/x\} \mid i = 1, ..., n\} \vdash e\{f(p_1, ..., p_n)/x\}$$

Correctness of structural induction is assured by theorem 13.

The following example shows that the structural induction can be coded in DTRC.

Example 1. Let E be $\{0 + y = y,\ s(x) + y = s(x + y)\}$ and e be $(x + y) + z = x + (y + z)$. It is known that e is an inductive theorem of E w.r.t. x. The proof of this fact by structural induction w.r.t. x is coded in DTRC as follows where 0 and s are constructors:

$T \overset{\Delta}{=}$

```
[ 0, s, +, true, eq, and; t:
    eq(t,t) → true,
    and( true,[;:(p+q)+r → p+(q+r)](true) ) → true ]
( [ ; x, y:
    0+y → y,
    s(x)+y → s(x+y) ]
  (and(eq((0+q)+r,0+(q+r)),
       [;:(p+q)+r→p+(q+r)](eq((s(p)+q)+r,s(p)+(q+r)))))))
```

In this DTRC term T, the outermost declaration (abbreviated as Sys_{eq})

```
[ 0, s, +, true, eq, and; t:
      eq(t,t) → true,
      and( true,[;:(p+q)+r → p+(q+r)](true) ) → true ]
```

defines **eq** as the equality function and **and** as a kind of conjunction. The rules in the second level declaration (abbreviated as Sys_E) corresponds to E. It should be noted that variables **x, y** in the rules corresponding to E are bound so that equational inference is done by rewriting. The term

```
and(eq((0+q)+r,0+(q+r)),
    [;:(p+q)+r → p+(q+r)] (eq((s(p)+q)+r,s(p)+(q+r))))
```

represents both of the base step (**eq((0+q)+r,0+(q+r))**) and the induction step (**[;:(p+q)+r → p+(q+r)](eq((s(p)+q)+r,s(p)+(q+r)))**) where the system **[;:(p+q)+r → p+(q+r)]** corresponds to the induction hypothesis, in which **p,q,r** are free variables serving as meta-variables denoting any ground (simple) constructor terms, and **eq((s(p)+q)+r,s(p)+(q+r))** corresponds to the predicate to be proved from the induction hypothesis.

T
$\overset{*}{\rightarrow}$

$Sys_{eq}(Sys_E$
  (and(eq(q+r,q+r),
    [;:(p+q)+r → p+(q+r)](eq((s(p)+q)+r,s(p)+(q+r))))))

$\rightarrow$

$Sys_{eq}(Sys_E$
  (and(true,[;:(p+q)+r→p+(q+r)](eq((s(p)+q)+r,s(p)+(q+r))))))

$\overset{*}{\rightarrow}$

$Sys_{eq}(Sys_E$
  (and(true,[;:(p+q)+r→p+(q+r)](eq(s((p+q)+r),s(p+(q+r)))))))

$\rightarrow$

$Sys_{eq}(Sys_E$
  $(and(\ true, [;:(p+q)+r \to p+(q+r)] (eq(s(p+(q+r)),s(p+(q+r)))))\ )))$
$\to$
  $Sys_{eq}(Sys_E(and(\ true, [;:(p+q)+r\ \to\ p+(q+r)] (true)\ )))$
$\to$
  $Sys_{eq}(Sys_E\ (true))$

From the above sketch of reductions, it is easily seen that a process of proof by structural induction naturally corresponds to a reduction sequence starting with T, and vice versa. We have accomplished the proof that $(x+y)+z = x+(y+z)$ is an inductive theorem of $\{0+y=y,\ s(x)+y=s(x+y)\}$, for T has unique normal form $Sys_{eq}(Sys_E(true))$.

The term T above is strongly normalizing and has unique normal form since the innermost system representing induction hypothesis cannot be changed by the rules in the second level declaration. But this is not always true. To avoid such cases, we modify and then generalize the above example by exchanging the position of the last two systems to obtain the following two propositions.

Proposition 14. *Let E be an equational axiom system and e be an equation $\alpha = \beta$ with a variable x. If the following DTRC term T has a normal form like $[\ldots](true)$, then e is an inductive theorem of E w.r.t. x.*

```
[ F, true,equiv,eq,and; u,v,t:
    equiv(u,v) → [; V: R_E](eq(u,v)),
    eq(t,t) → true,
    [; V: R_E](true) → true,
    and(Sys₁(true),...,Sysₙ(true) ) → true ]
    ( and( Sys₁(t₁),...,Sysₙ(tₙ) ) )
```

where

- *F is the set of function symbols occurring in E and V is the set of variables occurring in E.*
- *R_E is E with equations directed.*
- *Sys_i stands for the system*

$$[;: \alpha\{p_1/x\} \to \beta\{p_1/x\},\ldots,\alpha\{p_{arity(f_i)}/x\} \to \beta\{p_{arity(f_i)}/x\}]$$

for each constructor f_i w.r.t. E.
- *t_i stands for the term*

$$equiv(\alpha\{f_i(p_1,\ldots,p_{arity(f_i)})/x\},\beta\{f_i(p_1,\ldots,p_{arity(f_i)})/x\})$$

Proof. (Sketch) Note that Sys_i represents the inductive hypothesis

$$\alpha\{p_1/x\} = \beta\{p_1/x\},\ldots,\alpha\{p_{arity(f_i)}/x\} = \beta\{p_{arity(f_i)}/x\}$$

and thus for each constructor f_i,

```
[ F, true,equiv,eq,and; u,v,t:
  equiv(u,v) → [; V: R_E](eq(u,v)),
  eq(t,t) → true,
  [; V: R_E](true) → true,
  and(Sys_1(true),...,Sys_n(true) ) → true ]
    ( Sys_i(t_i) )
→*
  [...] ( Sys_i(true) )
```

syntactically means

$$\alpha\{f_i(p_1,\ldots,p_{arity(f_i)})/x\} = \beta\{f_i(p_1,\ldots,p_{arity(f_i)})/x\}$$

under such hypothesis and E. Therefore the existence of a normal form of T like $[\ldots]$(**true**) just means

$$E \cup \{e\{p_i/x\} \mid i = 1,\ldots,n\} \vdash e\{f(p_1,\ldots,p_n)/x\}.$$

for each constructor f with arity n, i.e., e is an inductive theorem of E w.r.t. x.

This proposition does not say all inductive theorems are proved or disproved by evaluating the corresponding DTRC terms. If evaluation of the DTRC term does never terminate, we cannot say anything about the proof. Even if it has a normal form different from the expected form $[\ldots]$(**true**), we cannot conclude that e is disproved. Only when the DTRC term has the expected normal form $[\ldots]$(**true**), we can conclude that e is an inductive theorem of E. In the case that R_E is in the form of a complete TRS and the corresponding DTRC term is strongly normalizing, we can decide whether structural induction can prove that e is an inductive theorem of E in the sense that

Proposition 15. *Let E,e,T,x be as the same as proposition 14 where R_E is in the form of a complete TRS and T is strongly normalizing. Then structural induction can prove that e is an inductive theorem of E w.r.t. x if and only if T has a normal form like $[\ldots]$(true).*

Proof. (Sketch) This is obtained from the completeness of R_E and the description in the proof of proposition 14.

5.3 Cover Set Induction

We employ the cover set induction proposed by [13] as an extension of structural induction. Another cover set induction is found in [14].

An n-fold *cover set* for an equational language is a set of n-tuples of terms over constructors and meta-variables such that all n-tuples of ground constructor terms of the symbol set of the language are generated from the cover set by appropriately substituting ground constructor terms for meta-variables.

Proof by one-fold cover set induction proceeds as follows. Given an equational axiom system E, a cover set M and an equation e with a variable x, one can prove that e is an inductive theorem of E by showing that for every $\xi \in M$

$$E \cup \{e\{\xi'/x\} \mid \xi' \prec \xi\} \vdash e\{\xi/x\}.$$

where $\prec$ is a strict partial order such that $\eta \prec \xi$ means that η is obtained from a subterm ξ_1 of ξ by replacing some subterm of ξ_1 by a meta-variable occurring in the subterm. We assumed here if there are other variables occurring in e then they are substituted by distinct meta-variables. It is easy to generalize the one-fold case to the n-fold case by using the pointwise extension of $\prec$. Correctness of one-fold and n-fold cover set induction was shown in [13]. We can obtain the following proposition, which implies that process of n-fold cover set induction can be coded into a DTRC term.

Proposition 16. *Let E be a set of equations, e be an equation $\alpha = \beta$ with variables $\mathbf{x} = x_1, \ldots, x_n$, and $M = \{\boldsymbol{\xi_1}, \ldots, \boldsymbol{\xi_k}\}$ be an n-fold cover set.*

If the following DTRC term T has a normal form like $[\ldots](\mathbf{true})$, then e is an inductive theorem of E w.r.t. $\mathbf{x}$.

```
[ F, true,equiv,eq,and; u,v,t:
    equiv(u,v) → [ ; V:R_E](eq(u,v)),
    eq(t,t) → true,
    [ ; V:R_E](true) → true,
    and(Sys_1(true),..., Sys_k(true) ) → true]
    (and(Sys_1(t_1),..., Sys_k(t_k)))
```

where

- *F is the set of function symbols occurring in E and V is the set of variables occurring in E.*
- *R_E is E with equations directed.*
- *Sys_i stands for the system*

$$[;: \alpha\{\xi'_{i,1}/\mathbf{x}\} \to \beta\{\xi'_{i,1}/\mathbf{x}\}, \ldots, \alpha\{\xi'_{i,i_m}/\mathbf{x}\} \to \beta\{\xi'_{i,i_m}/\mathbf{x}\}]$$

where $\xi'_{i,j}$ ($j = 1, \ldots, i_m$) are all n-tuples smaller than ξ_i and $\{\{t_1, ..., t_k\}/\mathbf{x}\}$ stands for $\{t_1/x_1, ..., t_k/x_k\}$.
- *t_i stands for the term*

$$\mathbf{equiv}(\alpha\{\xi_i/\mathbf{x}\}, \beta\{\xi_i/\mathbf{x}\})$$

Proof. (Sketch) Similar to the proof of proposition 14.

Definition 17. Let E be an equational axiom system. An n-fold cover set M is *sufficient* to prove that an equation e is an inductive theorem of E w.r.t. $\{x_1, \ldots, x_n\}$ if for every $(\xi_1, \ldots, \xi_n) \in M$

$$E \cup \{e\{\xi'_1/x_1, \ldots, \xi'_n/x_n\} \mid (\xi'_1, \ldots, \xi'_n) \prec (\xi_1, \ldots, \xi_n)\} \vdash e\{\xi_1/x_1, \ldots, \xi_n/x_n\}.$$

where $(\xi_1', \ldots, \xi_n') \prec (\xi_1, \ldots, \xi_n)$ means $\xi_1' \preceq \xi_1, \ldots, \xi_n' \preceq \xi_n$ and for some i $\xi_i' \prec \xi_i$, where $\xi_j' \prec \xi_j$ means that ξ_j' is obtained from a subterm η of ξ_j by replacing some subterms of η by meta-variables occurring in the corresponding subterms.

Proposition 16 specializes into the following proposition as proposition 14 into proposition 15:

Proposition 18. *Let M, E, e, T, x be as the same as proposition 16 where R_E is in the form of a complete TRS and T is strongly normalizing. Then M is sufficient to prove that e is an inductive theorem of E w.r.t. x if and only if T has a normal form like $[\ldots](\text{true})$.*

Proof. (Sketch) Similar to the proof of proposition 15.

Note that even if R_E is in the form of a complete TRS and T is strongly normalizing and does not have the expected normal form $[\ldots](\text{true})$, we cannot say that cover set induction cannot prove that e is an inductive theorem of E. We can only say that the cover set M is not sufficient to prove that e is an inductive theorem of E by cover set induction. The proof depends on the selection of the variables x, the constructors and the cover set M.

Finally we show an example for two-fold cover set induction.

Example 2. Suppose we want to prove that $x + y = y + x$ is an inductive theorem of $\{0 + y = y, s(x) + y = s(x + y)\}$, the same axiom system as example 1. We choose 0 and s as constructors again.

Taking $\{(0, 0), (s(p), 0), (0, s(q)), (s(p), s(q))\}$ as a two-fold cover set, we can code the proof by two-fold cover set induction as the following DTRC term.

```
[ 0,s,+, true,equiv,eq,and; u,v,t:
  equiv(u,v) → [; x,y:
                 0+y → y,
                 s(x)+y → s(x+y) ] (eq(u,v)),
  eq(t,t) → true,
  [; x,y: 0+y → y,
     s(x)+y → s(x+y) ](true)
  → true,
  and( [;:](true),
     [;: p+0 → 0+p ](true),
     [;: 0+q → q+0 ](true),
     [;: p+q → q+p,
       p+s(q) → s(q)+p,
       s(p)+q → q+s(p) ](true))
  → true ]
(and( [;:](equiv(0+0,0+0)),
     [;: p+0 → 0+p](equiv(s(p)+0,0+s(p))),
     [;: 0+q → q+0](equiv(0+s(q),s(q)+0)),
```

```
[;: p+q → q+p,
    p+s(q) → s(q)+p,
    s(p)+q → q+s(p) ](equiv(s(p)+s(q),s(q)+s(p)))))
```

By proposition 18, $x + y = y + x$ is an inductive theorem of $\{0 + y = y,\ s(x) + y = s(x + y)\}$ for this DTRC term is strongly normalizing and has unique normal form like $[\ldots](\textbf{true})$.

Remark. For readability, the examples chosen here are already known as inductive theorems, but processing the procedures will help us to prove an arbitrary equation to be an inductive theorem of an equational axiom system. It seems that it is easy to extend the procedures to handle more complicated inductive theorems, i.e., those inductive theorems needing lemmas for proving, by utilizing the feature of hierarchical declaration. From this point of view, we can prove non trivial inductive theorems.

6 Concluding remarks

We have proposed Dynamic Term Rewriting Calculus for the purpose of formal treatment of meta-computation of TRSs, and shown that DTRC has enough expressive power to automate the inductive equational reasoning. As we have shown in section 5, for an equational axiom system in the form of a complete TRS, we can decide whether the structural induction can prove that an equation is an inductive theorem and whether a cover set is sufficient to prove that an equation is an inductive theorem by executing the corresponding DTRC term when the term is strongly normalizing. It is an interesting future work to apply DTRC to other approaches to inductive equational reasoning such as induction-less induction.

The examples shown in this paper suggest that we can implement in DTRC in a natural way those algorithms which manipulate TRSs as well as those procedures which verify such algorithms. Other such examples can be found in [2] in which we showed that conditional term rewriting systems can be interpreted by DTRC terms.

As illustrated in section 4, we can write hierarchical programs by using DTRC. In this sense, DTRC will be a powerful tool for software modularization. Further, we can write parameterized programs by using DTRC in the sense that a term with no bound function symbols can be considered as a formal parameter and an actual parameter is passed dynamically during the computation. Both formal and actual parameters may be terms containing programs as subterms. We can naturally do reuse of components in such terms. This kind of modularization and parameterization is somehow stronger than the usual, since the parameters are terms instead of function symbols.

One might consider to use DTRC as a programming language since it inherently has the mechanism of modularization and parameterization as stated above. But DTRC is only a basic model and has no features necessary for easy programming such as macro definition. It is a future work to design and implement a programming system based on DTRC in which we can describe and analyze meta-computation on TRSs.

The work of [8] was also motivated by importance of meta-computation. In [8] a meta-environment was proposed, in which a user-defined language definition is edited and a programming environment is generated from the language definition. The meta-environment has several fixed meta-functions for modifying the modules which define a language. Compared to the meta-environment, DTRC is a basic computation model oriented to arbitrary meta-computation rather than fixed meta-computation. It may be possible to use DTRC to give semantics to or implement the meta-environment.

Acknowledgement

The authors would like to thank T. Naoi, M. Sakai and S. Yuen for their helpful discussions. We also thank the referees of the paper for their valuable comments.

References

1. Feng S., Sakabe T. and Inagaki Y. : "Dynamic Term Rewriting Calculus and its Application", IEICE Technical Report, COMP 91-47, pp. 31–40,1991. (in Japanese)
2. Feng S., Sakabe T. and Inagaki Y. : "Interpretation of Conditional Term Rewriting System by DTRC", Proc. of 9th Conference of Japan Society for Software Science and Technology, pp. 313-316, September 1992
3. Feng S., Sakabe T. and Inagaki Y. : "Confluence and Termination of Dynamic Term Rewriting Calculus", to be submitted.
4. Huet G. and Hullot J. M. : "Proofs by Induction in Equational Theories with Constructors", Rapports de Recherch, INRIA,28(1980).
5. Huet G. and Lang B. : "Proving and applying program transformations expressed with second-order logic", Acta Informatica, 11, pp. 31-55, 1978.
6. Huet G. and Oppen D. : "Equations and rewrite rules: A survey", In R. Book,ed.,*Formal Language Theory: Perspectives and Open Problems*, pp. 349–405, Academic Press, New York, 1980.
7. Jouannaud J. P. and Kounalis E. : "Proof by induction in equational theories without constructors", Proc. of 1st Symp. on Logic In Computer Science, pp. 358–366, Boston, USA, 1986
8. Klint P. : "A meta-environment for generating programming meta-environments", In J. A. Bergstra and L. M. G. Feijs, eds., *Algebraic Methods II: Theory, Tools, and Applications*, pp. 105–124. LNCS 490, 1991.
9. Knuth D. E. and Bendix P. : "Simple word problems in universal algebra", In J. Leech, ed., *Computational Problems in Abstract Algebra*, pp. 263–297. Oxford, Pergamon Press, 1970.
10. Kounalis E. and Rusinowitch M., "Mechanizing inductive reasoning", Proc. Eighth National Conference on Artificial Intelligence, AAAI-90, pp. 240–245, July, 1990
11. Nadathur G. and Miller D.: "An overview of λProlog", In K. Bowen and R. Kowalski,ed., *Fifth International Conference and Symposium on Logic Programming*, MIT Press, 1988.

12. Reddy U. S., "Term Rewriting Induction", Proc. 10th International Conference on Automated Deduction, Kaiserslautern, FRG, LNCS 449, pp. 162–177, July 1990.

13. Sakai M., Sakabe T. and Inagaki Y. : "Cover Set Induction for Verifying Algebraic Specifications", The transactions of the institute of electronics, information and communication engineers, Vol. J75-D-I No. 3, pp. 170–179, March 1992. (in Japanese)

14. Zhang H., Kapur K. and Krishnamoorthy M. S. : "A Mechanizable Induction Principle for Equational Specification", Proc. of 9th International Conf. on Automated Deduction at Argonne,Illinois, USA, LNCS 310, pp. 162–181, May 1988.

Distributed Deduction by Clause-Diffusion: the Aquarius Prover *

Maria Paola Bonacina and Jieh Hsiang

Department of Computer Science
SUNY at Stony Brook
Stony Brook, NY 11794-4400, USA
bonacina@loria.fr   hsiang@sbcs.sunysb.edu

Abstract. Aquarius is a distributed theorem prover for first order logic with equality, developed for a network of workstations. Given in input a theorem proving problem and the number n of active nodes, Aquarius creates n deductive processes, one on each workstation, which work cooperatively toward the solution of the problem. Aquarius realizes a specific variant of a general methodology for distributed deduction, which we have called *deduction by Clause-Diffusion* and described in full in [6]. The subdivision of the work among the processes, their activities and their cooperation are defined by the Clause-Diffusion method. Aquarius incorporates the sequential theorem prover Otter, in such a way that Aquarius implements the parallelization, according to the Clause-Diffusion methodology, of all the strategies provided in Otter.
In this paper we give first an outline of the Clause-Diffusion methodology. Next, we consider in more detail the problem of *distributed global contraction*, e.g. normalization with respect to a distributed data base. The Clause-Diffusion methodology comprises a number of schemes for performing distributed global contraction, which avoid the *backward contraction bottleneck* of purely shared memory approaches to parallel deduction. Then, we describe Aquarius, its features and we analyze some of the experiments conducted so far. We conclude with some comparison and discussion.

1 Introduction

In this paper we describe the Clause-Diffusion methodology for distributed theorem proving, its implementation in the theorem prover Aquarius and we analyze the performances of Aquarius on some experiments.

A theorem proving problem consists in deciding, given a set of clauses S and a clause φ, whether φ is a theorem of S. A theorem proving strategy $\mathcal{C}$ is specified by a set of *inference rules* I and a *search plan* Σ. The inference rules

---

* Research supported in part by grant CCR-8901322, funded by the National Science Foundation. The first author is also supported by a fellowship of Università degli Studi di Milano, Italy. First author's current address: INRIA-Lorraine, 615 Rue du Jardin Botanique, B.P. 101, 54602 Villers-les-Nancy, France.

can be further separated into two classes. The *expansion inference rules*, such as resolution and paramodulation, derive new clauses from existing ones and add them to the data base. The *contraction inference rules*, such as simplification and subsumption, delete clauses or replace them by smaller ones. The search plan Σ chooses the inference rule and the premises for each step, so that the repeated application of Σ and I generates a derivation. A derivation is successful if it reaches a solution of the input problem. If the strategy is *complete*, the derivation is guaranteed to succeed whenever the input goal is indeed a theorem. In practice, however, a derivation by a complete strategy may fail to prove a theorem, because it generates so many clauses that it exhausts the available memory before succeeding. In other words, it generates too large a portion of the *search space* of the problem. *Contraction-based strategies* try to reduce the incidence of such failures by applying eagerly powerful contraction rules to keep the data base, and thus the search space, as reduced as possible. These strategies, as implemented for instance in the provers Otter [20], RRL [16] and SBR3 [1], have obtained very encouraging results.

In this paper we present a methodology for parallelizing contraction-based deduction strategies. The main feature of contraction-based strategies is that existing data may be deleted or replaced by others through contraction. For instance, an equation may be reduced to another equation via rewriting. Although such a behaviour is the main reason why contraction-based strategies are effective, it is also the major source of difficulty in parallelization. To illustrate this point, we consider the parallelization of Prolog technology theorem proving (PTTP) methods. In goal-reduction methods such as PTTP, the set of axioms remains static during the course of the derivation. Thus, it is possible to pre-process all the axioms into elaborate data structures before the derivation starts. Such structures are used to exploit parallelism of different granularities. The cost of building them is limited to the pre-processing phase. Contraction-based strategies, on the other hand, are not likely to take advantage of such approaches, because axioms will be added and deleted during the derivation, so that pre-processing is not sufficient.

The basic idea of the Clause-Diffusion methodology, which we present here, is to parallelize a strategy *at the search level*, by partitioning the search space among many concurrent deductive processes, which search in parallel for a solution. As soon as one of them succeeds, the whole distributed derivation succeeds. The deductive processes are asynchronous and work in a largely independent fashion: each process has its own local data base, constructs its own derivation and interacts with the others through *message-passing*.

The Clause-Diffusion approach has a few features which we consider as new:

- It is a general methodology intended for implementing contraction-based strategies in distributed environments.
- The problem of keeping data inter-contracted is dealt with through a notion of *image set* – an approximation of the global data base. This avoids the difficulty of the *backward contraction bottleneck* which often occurs in shared memory implementation of contraction-based strategies [18, 27].

274

– It is a general methodology not confined to a specific architecture, topology or inference system. Depending on the parameters chosen, our method can be easily adopted in different environments.

Aquarius, a prototype built on top of the sequential theorem prover Otter, is completed. Aquarius implements a few of the variants of the Clause-Diffusion methodology for all the theorem proving strategies offered by Otter. Thus, Aquarius inherits most of Otter's valuable features. First, it exploits the high efficiency of basic operations and data structures, for which Otter is well-known. Second, Aquarius maintains the philosophy of Otter of providing the user with a wealth of options to experiment with. In addition to all those of Otter, new parameters related to distributed execution are added, for a total of 121 options. This flexibility allows the user to tailor the prover to different classes of theorems, to use it to "simulate" to some extent other methods, e.g. the team-work method of [3, 12], and to apply it to other problems, such as Knuth-Bendix completion. Third, Aquarius is highly portable, since it has been written in C and PCN [13], under the Unix operating system, for a network of Sun workstations. In such an environment, each deductive process runs on a different node of the network. We have run Aquarius on many problems and we report a selection of results, including both positive and negative ones. For the latter, we analyze the possible causes, especially in terms of performance of communication, duplication of clauses and ways of partitioning the data base. We feel that negative experimental results are important, because they highlight the difficulties which remain to be solved and may contribute to further work.

The paper is organized as follows. First, we describe briefly the Clause-Diffusion methodology and we define the distributed derivations generated by a Clause-Diffusion strategy. While our methodology applies to theorem proving strategies in general, we designed it keeping contraction-based strategies in mind. Thus, we discuss some of the problems related to contraction in a distributed data base and the solutions adopted in Aquarius. The remaining sections are devoted to Aquarius and the experimental results. A full treatment of the issues in parallel theorem proving and a formal presentation of the Clause-Diffusion methodology are beyond the scope of this paper and therefore we refer to [6] for a complete description.

2 The Clause-Diffusion methodology

Given a complete theorem proving strategy $\mathcal{C} =< I; \Sigma >$, we describe how $\mathcal{C}$ is executed according to the Clause-Diffusion methodology. We consider a network of computers or a loosely coupled, asynchronous multiprocessor with distributed memory. The latter may be endowed with a shared memory component. Our methodology does not depend on a specific architecture; it can be realized on different ones. Parameters such as the amount of *memory at each processor*, the availability of *shared memory* and the *topology* of interconnection of the processors or *nodes*, are variable.

The basic idea in our approach is to have a deductive process running at each node and *to partition the search space* among these processes. We use $p_1 \ldots p_n$ to denote ambiguously both the deductive processes and the nodes. The search space is determined by the input clauses and the inference rules. At the clauses level, the input and the generated clauses are distributed among the nodes. For this purpose we need an *allocation algorithm*, which decides where to allocate a clause. Once a clause ψ is assigned to processor p_i, ψ becomes a *resident* of p_i. In this way each node p_i is allotted a subset S^i of the global data base. The union of all the S^i's, which are not necessarily disjoint, forms the current *global data base*. Each processor is responsible for applying the inference rules in I to its residents, according to the search plan Σ. Since the global data base is partitioned among the nodes, no node is guaranteed to find a proof using only its own residents. To assure that a solution will be found when one exists, the nodes need to exchange information, by sending each other their residents in form of messages, called *inference messages*. Each node uses the received inference messages to perform inferences with its own residents. The inference messages issued by a process p_i let the other processes know which clauses belong to p_i, so that they can use them for inferences. In a purely distributed system, inference messages are implemented as messages, which may be routed or broadcast. Depending on the broadcasting algorithm, there may be several inference messages, all carrying the same clause, active at different nodes. In a system with a shared memory component, inference messages may be communicated through the shared memory.

The separation of residents and inference messages is also used to partition the search space at the inference level. Using the paramodulation inference rule as an example of expansion step, we establish that the inference messages are paramodulated *into* the residents, but not vice versa. This restriction has two purposes. First, it distributes the expansion inference steps among the nodes. Second, it prevents a systematic duplication of steps: if this restriction were not in place, then each paramodulation step between two residents ψ_1 of p_1 and ψ_2 of p_2 would be performed twice, once when ψ_1 visits p_2 and once when ψ_2 visits p_1. Other expansion inference rules can be treated in a similar way [6]. While subdividing the expansion steps serves its purpose, it is not productive to subdivide the contraction steps, since the motivation behind contraction is to keep the data base always at the minimal. In a contraction-based strategy, an expansion step should be performed only if all the premises are fully reduced, at least with respect to the local data base. To ensure this, we require that each processor keep both its residents and received inference messages fully contracted.

We call *raw clause* a clause newly generated from an expansion step. Input clauses are also considered as raw clauses. In the presence of contraction rules, a raw clause should not become a resident until it has been fully contracted. Thus, our methods also feature a number of *distributed global contraction schemes* to reduce a raw clause with respect to the global data base. We shall describe these schemes in Section 2.1. After contraction, a raw clause becomes a *new settler*. New settlers are given to the allocation algorithm to be assigned to some node.

Every process executes the allocation algorithm for its new settlers: it may decide either to retain a new settler or to send it to another node. The purpose of the allocation algorithm is to partition the search space and keep the work-load balanced as much as possible.

This is the basic working of the Clause-Diffusion methodology: local contraction and local expansion inferences at the nodes among residents and inference messages, distributed global contraction, allocation of new settlers and mechanisms for passing inference messages. By specifying the inference mechanism I, the search plan Σ to schedule inference steps and communication steps, the allocation algorithm, the distributed contraction scheme and the mechanisms for the communication of messages, one obtains a specific strategy. These elements are summarized in the following notion of *distributed derivation*: every processor p_k, $1 \leq k \leq n$, computes a derivation

$$(S; M; CP; NS)_0^k \vdash_c (S; M; CP; NS)_1^k \vdash_c \ldots (S; M; CP; NS)_i^k \vdash_c \ldots$$

where S_i^k is the set of *residents*, M_i^k is the set of *inference messages*, CP_i^k is the set of *raw clauses* and NS_i^k is the set of *new settlers* at p_k at stage i. A distributed derivation is the collection of the asynchronous derivations computed by the nodes. The *state* of the derivations at processor p_k and stage i is represented by the tuple $(S; M; CP; NS)_i^k$. More components may be added if indicated by a specific strategy. A distributed derivation succeeds as soon as the derivation at one node finds a proof. A step in a distributed derivation can be either an *expansion* step or a *contraction* step or a *communication* step. For instance, sending an inference message for $\psi \in S^k$ from node p_k to an adjacent node p_j can be written as $(S^k \cup \{\psi\}, M^j) \vdash (S^k \cup \{\psi\}, M^j \cup \{\psi\})$. Settling a new settler at node p_k can be written as $(S^k, NS^k \cup \{\psi\}) \vdash (S^k \cup \{\psi\}, NS^k)$. This representation assumes that communication between any two adjacent nodes is instantaneous. It does *not* assume, however, that communication between *any* two nodes is instantaneous. If an inference message sent by p_i reaches p_j through $p_{x_1} \ldots p_{x_m}$, it appears first in M^{x_1}, then in M^{x_2} and so on. The time elapsed in going from the source to the destination is captured in our description, by showing the message stored, at successive stages, in the appropriate component of all the nodes on the path.

2.1 Distributed global contraction

In distributed theorem proving, we term *global contraction* the task of reducing a clause with respect to the global data base, i.e. the union of the sets of residents of the parallel deductive processes. In [6], we have proposed several schemes for *distributed global contraction*. The distributed approach which is common to these schemes provides a solution to an important implementation problem of parallel theorem proving in shared memory, which we term the *backward contraction bottleneck*. In this section, we describe first this problem and then the schemes for distributed global contraction implemented in Aquarius. More details can be found in [6].

What is the backward contraction bottleneck Operationally, contraction steps can be separated into *forward contraction* and *backward contraction*. Informally speaking, forward contraction uses existing data to contract new data (raw clauses and incoming inference messages, for instance), while backward contraction uses new data to contract existing ones (the residents). Designing an effective and efficient method for parallel backward contraction is a much more complicated task than for parallel forward contraction. Indeed, backward contraction has turned out to be a critical problem for shared memory implementations [18, 27] of parallel theorem proving with contraction, while some other implementations simply do not implement backward contraction (e.g. DARES [11] and PARROT [15]). In a related area, parallel implementations of the *Buchberger algorithm* [14, 22, 25] have also suffered from this problem.

Forward contraction amounts to the normalization of a raw clause with respect to the *static* data base of all the clauses existing when the raw clause is generated. Thus, the task can be done once and for all when the raw clause is derived. Backward contraction involves the *normalization of any clause with respect to all the clauses which may be generated afterwards*. The normalization tasks need to be repeated as new clauses are generated. It follows that the data base is *highly dynamic* and there is *no read-only data*, i.e. all the items in the data base need be accessible not only for reading but also for writing. In turn, this implies that the clauses cannot be pre-processed into fast, specialized data structures, such as those used in approaches to parallel rewriting in equational programs, e.g. [17].

Furthermore, in contraction-based strategies, raw clauses are not used for expansion steps. Therefore, forward contraction does not enter in conflict with expansion. But backward contraction does, because it affects clauses that are already being used as parents of expansion steps. Finally, a clause which is reduced by a backward contraction step, should be tested for further contraction with respect to all the other clauses. Thus, a single backward-contraction step may induce many. In shared memory implementations such as [18, 27], this avalanche growth of contraction steps causes a write-bottleneck, the *backward contraction bottleneck*, since all the backward contraction processes ask write-access to the shared memory, where residents reside. Not all of them may be served and an otherwise unnecessary sequentialization is imposed. The clauses which are supposed to be subject to backward contraction may not be made available for other tasks, e.g. expansion steps, so that these are delayed as well.

Distributed global contraction schemes In [6], we have given two classes of schemes for distributed global contraction: *global contraction by travelling* and *global contraction at the source*. In the first, we assume that no node has access to the global data base $\bigcup_{i=1}^{p} S^i$ and thus global contraction employs messages. In global contraction at the source, we assume that every node has access to an approximation of the global data base, so that a raw clause can be contracted at the node where it was born (its "source"). By an approximation of the global data base, we mean a set of copies of the residents in the systems, which may be used as simplifiers. We call such a set an *image set*. An image set is an approximation, because it is not guaranteed to be identical to the global data base $\bigcup_{i=1}^{p} S^i$ at any stage of the execution.

In *global contraction at the source by localized image sets*, we assume that the local memory of each node p_i is large enough to hold a *localized image set* SH^i of $\bigcup_{i=1}^{p} S^i$. Each node uses its localized image set as set of simplifiers to perform global contraction of residents, raw clauses and incoming messages. The localized image sets can be built by utilizing the inference messages: *whenever a node p_i receives an inference message, it stores the clause carried by the message in SH^i*. The identities $SH^j = \bigcup_{i=1}^{n} S^i$ for all j, $1 \leq j \leq n$, do not hold in general, because the sets of residents S^i's keep evolving. Thus a localized image set is just an approximation of the global data base. However, each of the SH^i's is logically equivalent to the global data base $\bigcup_{i=1}^{p} S^i$, if all the persistent residents, i.e. those not deleted by contraction, are broadcast as inference messages. In *global contraction at the source by global image set in shared memory*, a single, global image set is held in shared memory. The choice of the appropriate global contraction scheme is related to the available resources: global contraction by travelling requires very fast communication, while global contraction at the source requires either sufficiently large local memories or a shared memory component. In this paper we consider global contraction at the source by localized image sets, because it is the scheme implemented in Aquarius, while we refer to [6] for the others.

Our global contraction schemes do not suffer from the backward contraction bottleneck, because *the clauses being rewritten by contraction are held in the local memories of the nodes*. Therefore, concurrent contractions are done independently in the local memories at the nodes, with no need to wait to get write-access to a shared memory. An additional advantage of image sets is that such large sets of simplifiers can be implemented as *discrimination nets* [10, 23] for the purpose of fast simplification.

Maintenance of the image sets A fundamental issue in global contraction at the source is whether contraction of the simplifiers in the SH^i's should be allowed. The question is whether the advantage of maintaining the SH^i's fully reduced is worth the cost of updating them. In [6], we have proposed several different approaches. One possibility is to have each p_i performing contraction on SH^i just like on S^i ("maintenance by direct contraction"). Each node contracts its own raw clauses, but all nodes execute independently contraction of all residents: if ψ is a resident at p_i which is also stored in the SH^j components at the other nodes, contraction of ψ is performed at all nodes which have a copy of ψ. If contraction inferences are sufficiently fast, this may be a reasonable choice.

At the other extreme, one may forbid contraction on the SH^i's and allow only insertion of new elements ("no contraction" policy). If $\psi \in SH^j - S^j$ is reducible, it may be reduced at the node p_i, such that $\psi \in S^i$, and a reduced form of ψ will be added to SH^j eventually. If SH^j is used only as a data base of simplifiers, the presence of both ψ and a reduced form ψ' should not represent serious redundancy. In fact, especially if the SH^j's are implemented as discrimination nets, frequent updates of the elements in the net may not be cost-effective. However, if no element is ever deleted from the SH^j's, their sizes may grow up to compromise their performances. Furthermore, if the elements in SH^j are used for expansion steps, redundant clauses in SH^j would induce the generation of more redundant clauses. In order to avoid such phenomena, we may design mechanisms to update the SH^i's without resorting to the direct application of contraction inferences.

We associate to every resident of a node a unique *identifier*: for every node p_i and for every resident ψ of p_i, ψ receives an identifier a, so that a is the unique identifier of ψ at p_i and $< p_i, a >$ is the unique *global identifier* of ψ. We also establish that a resident ψ at p_i has another attribute, the *birth-time*, i.e. the time at p_i's clock when ψ was recorded as a resident of p_i. Overall the format of a resident is $< \psi, a, x >\in S^i$, where a is the identifier and x is the birth-time. The global identifiers of the residents can be used to index the clauses in the image sets. For instance, an image set may be implemented as a *hash table*, with the global identifier as key. We require that inference messages carry a clause together with its global identifier and birth-time. An inference message for a resident $< \psi, a, x >\in S^i$ has the form $< \psi, p_i, a, x >$. These additional fields allow a node to recognize that an inference message is carrying a reduced form of a previously received clause. If $< \psi, a, x >$ is reduced to $< \psi', a, y >$, where $y > x$, at p_i, a new inference message $< \psi', p_i, a, y >$ will be broadcast eventually. Whenever a node p_j receives an inference message, e.g. $< \psi', p_i, a, y >$, it checks whether an element ψ with the same global identifier $< p_i, a >$ is stored in SH^j. If this is the case, node p_j compares ψ and ψ' according to the ordering on clauses and saves the smallest in SH^j. If the two clauses are not comparable, the one with most recent birth-time is saved. We call this solution "update by inference messages".

In case of update by inference messages, the situation is less favorable if $< \psi, a, x >\in S^i$ is deleted, rather than replaced, by a contraction step. In such case, no more messages with identifier $< p_i, a >$ will be issued and therefore, localized image sets may never be updated. However, inference messages may still help: whenever an inference message $< \psi, p_i, a, x >$ is deleted at a node p_k, it is possible to check whether SH^k contains any clause with identifier $< p_i, a >$ and delete it. This is not sufficient in general to update all the localized image sets, because clause ψ may not be deleted at p_k. Then, if performance is hindered by not updating the localized data bases with respect to deletions, one may consider broadcasting a special *deletion message* with identifier $< p_i, a >$ to inform all the nodes that the resident at $< p_i, a >$ has been deleted. It is also possible to integrate different policies: for instance, the strategies implemented in Aquarius apply first update by inference messages and then direct contraction. Thus, deletion messages are not necessary. Also, fewer direct contraction steps will be performed if direct contraction is preceded by update by inference messages.

2.2 Clause-Diffusion strategies

In the above we briefly presented the Clause-Diffusion methodology by describing its objectives, essential operations and various unique features. We give a summary of operations performed by a strategy designed according to our methodology:

- local expansion inferences between *residents* and between residents and *inference messages* (resulting in the generation of *raw clauses*),
- local contraction of residents and inference messages,
- global forward contraction of raw clauses,
- global backward contraction of residents,
- allocation of *new settlers*,
- communication of messages.

For most of the operations we outlined a number of possibilities. A specific clause-diffusion theorem proving strategy can be formed by making specific choices from the various options described. In other words, a *clause-diffusion strategy* is specified by choosing

- a set of inference rules,
- a search plan that specifies the order of performing expansion, contraction and communication steps at each process,
- the algorithm to allocate new settlers,
- the scheme for global contraction,
- the mechanism for message-passing.

In [6, 7], we proved that the Clause-Diffusion methodology is correct: if $C =< I; \Sigma >$ is a complete sequential strategy, its parallelization by Clause-Diffusion yields complete distributed strategies. Since in Clause-Diffusion all the concurrent processes have the given inference system I, parallelization does not affect the completeness of the inference system. Therefore, our correctness result consisted in proving that parallelization by Clause-Diffusion preserves the completeness of the search plan, i.e. its *fairness*. In [6, 7], we gave a set of formal properties that the algorithms and policies handling messages in a distributed strategy need to satisfy. We proved that these properties imply fairness and we showed that the specific policies of the Clause-Diffusion method satisfy those properties.

In the following we describe a specific class of Clause-Diffusion strategies that we have implemented.

3 The Aquarius theorem prover

Aquarius implements a version of the Clause-Diffusion methodology with global contraction at the source by localized image sets. Each of the concurrent deduction processes executes a modified version, called *Penguin* [6], of the code of the theorem prover *Otter* (version 2.2) [20]. Otter is a resolution-based theorem prover for first logic with equality. Its basic mechanism works as follows: select a *given_clause* from the *Sos* (Set of Support), execute all the expansion inferences with the *given_clause*, pre-process (forward contraction) and post-process (backward contraction) all the generated raw clauses and then iterate. Different strategies may be obtained by setting several options. Penguin is structured into a *communication layer* and a *deduction layer*. The communication layer, written in PCN [13], implements the message-passing part. The deduction layer, written in C, incorporates the code of Otter, so that each Penguin process executes the basic cycle on the *given_clause*. In addition, the deduction layer implements the features required by the Clause-Diffusion methodology, e.g. the partitioning of the expansion steps based on the ownership of clauses, the distributed allocation of clauses and so on.

The Aquarius program is invoked with two main parameters, the name of the input problem and the number of requested Penguin processes. Each Penguin process reads its own input file and creates its own output and error files. The format of the files is the same as in Otter. The user may set a very high number of options: Aquarius has 121 options, 99 flags (boolean-valued options) and 22 parameters (integer-valued options), including all those of Otter. These options determine the components of the executed strategy: the inference mechanism, e.g. by selecting which inference rules are active, the search plan at each node, e.g. by sorting the lists of clauses and messages, the allocation of clauses as residents and the interleaving of inference steps and communication steps at each node. Since each Penguin process reads its own input file, the user may set different options patterns, and thus different strategies, at different Penguins. This flexibility allows the user to set Aquarius to reproduce interesting features of other methods. For instance, by having different strategies at different nodes, Aquarius may "simulate" to some extent the *team-work method* of [3, 12], albeit without the "referee processes" and the periodical reconstruction of a common data base that are characterizing parts of the team-work method. The KNUTH-BENDIX option (inherited from Otter), allows to perform Knuth-Bendix completion, so that Aquarius executes Knuth-Bendix completion in parallel. The STAND-ALONE option induces each Penguin to work by itself as a sequential prover, with no message-passing. One purpose of this option is to try in parallel different strategies on a given problem. Another application is to give to each Penguin a different input and have the nodes working in parallel on different problems. For instance, one may want to give to each Penguin a different lemma from a large problem and have the lemmas proved independently. While it provides the user with many input options, Aquarius is not interactive: like for Otter, the emphasis is on obtaining fully automated proofs, with no human intervention during the run. In the following, we analyze some experiments with Aquarius, while we refer to [6] for a complete description of the prover.

3.1 Experiments with Aquarius

In the following table, we give the performances of Aquarius on some problems. Aquarius-*n* is Aquarius with *n* nodes, where each node is a Sun Sparcstation.

So far we have been able to experiment with up to 3 of them. They communicate over the departmental Ethernet at Stony Brook. The sparcstations used for our experiments were not isolated from the rest of the network and were simultaneously used by other users. Therefore the reported run times (in seconds) represent the performances under realistic working conditions. For Aquarius-1 the run-time is that of the best run found. For $n > 1$, the run-time of Aquarius-n is the run time of the first Penguin to succeed, which includes both inference time and communication time. However, it includes neither the initialization time spent to set up the Penguin processes at the nodes nor the time spent to close all the PCN processes upon termination. Thus, the turn-around time observed by a user is usually longer than the run time. The other Penguins run till either they receive a halting message or also find a proof, whichever comes first. Among the listed problems, two are propositional (*pigeon* and *salt*), four are purely equational (*luka5*, *robbins2*, *s7* (a problem in algebraic logic) and *w-sk*), two are in first order logic with equality (*ec* and *subgroup*) and the remaining ones are in first order logic.

Problem	Aquarius-1	Aquarius-2	Aquarius-3
andrews [8, 21]	18.00	25.40	24.39
apabhp [18]	11.86	18.11	14.18
bledsoe [4, 18]	12.29	21.53	23.00
cd12 [18]	104.18	50.98	47.56
cd13 [18]	98.79	45.32	51.07
cd90 [18]	3.10	0.63	11.87
cn [18]	5.04	8.63	14.50
ec	3.03	1.96	1.77
imp1 [18]	6.63	2.64	3.54
imp2 [18]	7.25	3.31	7.43
imp3 [18]	32.05	17.92	38.89
luka5 [5]	844.20	299.24	1079.45
pigeon (ph4) [21]	8.21	7.66	8.14
robbins2 [18]	21.62	22.91	24.12
s7 [2]	630.62	208.37	192.54
salt	3.89	4.45	5.49
sam's lemma	6.35	5.40	3.90
subgroup [26]	15.55	9.36	17.40
w-sk [19]	3.50	3.52	3.34

3.2 Analysis of the experiments

The significance of these experiments is limited by having only up to 3 nodes. Also, the problems which can be solved sequentially in a few seconds are probably too easy for the parallelization to pay off. Furthermore, this problems set may not be the most ideal to test Clause-Diffusion. Most of the above problems are taken from the input sets for Otter and ROO and therefore problems in first order logic prevail over problems with equality. On the other hand, problems with equality are those where the impact of backward contraction is most dramatic. Aquarius-1 is generally slower than Otter, which indicates that the overhead induced merely by having linked the PCN part with the C part is not irrelevant. As can be seen, the experimental results are quite unstable. There may be many factors for such mixed results. One reason is of course the prototype nature of Aquarius, which was developed in a short 5 months period. In the following, we try to analyze the performances of Aquarius in terms of *communication*, *duplication* and *distribution of clauses*.

Observations of communication problems in Aquarius Communication in Aquarius is very slow. An immediate evidence of this is the following. In the current version, only one Penguin, i.e. Penguin0, finds the input clauses in its input file and broadcasts them to the other Penguins. In many cases, Penguin0 is the first one to succeed. Also, it happens that Penguin1 and Penguin2 have shorter run time than Penguin0, since the start of the derivations by Penguin1 and Penguin2 is delayed by the necessity of waiting for the input clauses. This observation suggests that a simple improvement would be to have each Penguin reading the input clauses from its input file. Another evidence that communication is hindering the performances is the following. Let ψ be a clause which can be derived independently at two nodes, e.g. Penguin0 and Penguin1. In most runs, it happens that Penguin0 generates and broadcasts ψ, but Penguin1 derives it on its own, *before* receiving the inference message from Penguin0. The intuitive idea of inference messages in the Clause-Diffusion methodology is that in general the clause carried by the message is "new" for the receiver. Therefore, when the above phenomenon happens in Aquarius the purpose of the inference messages is sort of defeated.

Communication among Penguin processes is handled by PCN. PCN [9] is a logic-programming-like language built on top of a sequential language such as C to serve as the communication layer. The performance of Aquarius is affected by PCN in at least two ways:

1. The current implementation of PCN gives priority to the execution of C code over the execution of PCN code.
2. The communication done through PCN and Unix is hampered by too many levels of software, causing too much copying for each message.

The effect of the first problem is that no PCN message-passing will take place until the C code completes. The producers of messages, i.e. the deduction layers of the Penguins, are written in C, while the consumers, i.e. the communication layers, are written in PCN. It follows that a consumer may not be scheduled from the active queue to get its pending messages while the C code is being executed at the node. Therefore communication, which is already likely to be the potential bottleneck in a distributed implementation, is at a strong disadvantage with respect to inference. The producers generate messages at a much faster pace than the consumers may consume them. Indeed, we observed executions, where the inference part of the computation halts upon finding a proof and then several pending messages are delivered all together.

We countered this problem by reducing the size of the C processes, i.e., the deduction layer of each Penguin. However, this does not seem to have been sufficient. An alternative approach is to synchronize the communication and deduction layers within each Penguin. Currently, they are largely asynchronous. A possible synchronization is to let the deduction layer proceed only when all the pending messages have been received by the communication layer.

Duplication After having experienced the problem with communication, we resorted to try to reduce the amount of communication by empowering the single nodes. Because communication is so slow, it is better that all nodes are able to work as independently as possible. Some of the reported experiments have been done by setting the flags in such a way that each node owns most of the input clauses. In other experiments, the flags for the allocation of clauses have been set in such a way that each node retains most of its raw clauses as residents. None of the reported results, however, refer to executions under a combination of flags equivalent to the $STAND\_ALONE$ mode. In other words, in all the listed experiments, there is some partitioning of the search space.

While reducing communication, these settings of flags, together with the use of localized image sets, induce a strong increase in duplication. It appears from the trace files of the experiments, that often most of the clauses needed in the proof are generated independently at all nodes. For instance, in one run of the problem $cd90$, the clause $P(e(e(x,y),e(x,y)))$ appeared in the trace of the execution at Penguin2 as follows: first, it is generated and sent as new settler to Penguin0; second, it is generated again and kept as resident; third, it is received as inference message from Penguin0; fourth, it is generated one more time and sent as new settler to Penguin1; fifth, it is received as new settler. Finally, $P(e(e(x,y),e(x,y)))$ is subsumed by $P(e(x,x))$. This amount of duplication may explain the lack of speed-up in many experiments. The Clause-Diffusion methodology and Aquarius are sufficiently flexible to provide combinations of different degrees of communication and duplication. However, the current version of Aquarius realizes a highly duplication-oriented version of the Clause-Diffusion methodology, which was not intended to be the main one, since it reduces the significance of partitioning the search space. The basic idea in the Clause-Diffusion methodology is to partition the search space. Indeed, the cases where Aquarius-2 speeds-up significantly over Aquarius-1 are exactly those where partitioning the search space helps. More precisely, in most of the positive results, one Penguin finds a shorter proof than the one found by the sequential prover, because it does not retain some clauses. An example is $cd90$, where Aquarius-2 has super-linear speed-up over Aquarius-1. The latter finds an 8-steps proof, which uses first $P(e(e(x,y),e(x,y)))$ and then $P(e(x,x))$. Aquarius-2 finds a 5-steps proof, which uses $P(e(e(x,y),e(x,y)))$, but does not even generate $P(e(x,x))$.

Distribution of clauses The third issue, i.e. the distribution of clauses, is more of a conceptual nature. The criteria for distributed allocation of clauses implemented in Aquarius try to balance the work-load by balancing the number of residents at the nodes. They keep into account neither the contents of a message, i.e. the clause, nor the history of the derivation, in order to decide its destination. The design of more informed allocation policies, e.g. policies which use informations about the clause being allocated and the history of the derivation, may be an important progress. As an example, one may think of heuristics of the form: if more than n clauses with property Q have been allocated to node p_i, then the next clause with property Q will also be allocated to node p_i. Such criteria, however, will be more expensive to compute and it may not be simple to devise them. More generally, the question is how to find better ways to partition the search space of a theorem proving problem.

4 Discussion

In the first part of the paper, we outlined our methodology for distributed deduction by Clause-Diffusion. This approach realizes a sort of coarse-grain parallelism, that we have termed *parallelism at the search level* [6]. Our methodology does not exclude the application of techniques for fine-grain parallelism, such as those employed for parallel rewriting languages, e.g. [17]. While the Clause-Diffusion methodology applies to theorem proving strategies in general, we have devoted special attention to contraction-based strategies. We formulated the problem of global contraction with respect to a distributed data base, clarifying the differences between forward global contraction and backward global contraction. We indicated in the *bottleneck of backward contraction* a critical problem in the shared memory approaches to parallel automated deduction. This source of inefficiency had not been identified before. In [6], we proposed as solutions several schemes for distributed global contraction. Here we have focused on *global contraction at the source by localized image sets*, since it is the scheme implemented in Aquarius.

In the second part of the paper, we described Aquarius and analyzed some experiments. Other parallel theorem provers have obtained better experimental results than Aquarius. For instance, ROO [18] shows linear speed-up on most non-equational problems, while its performances on equational problems suffer from the backward contraction bottleneck. ROO uses *parallelism at the clause level*, since each concurrent process consists in selecting and processing a *given_clause*. A common data base of clauses is kept in shared memory and thus the search space is not partitioned. Such a purely shared approach to parallel theorem proving, with parallelism at the term/clause level, does not modify the search space (and does not intend to). Thus, the parallel prover works on a search space which is basically the same as in the sequential case and it is likely to find a similar proof. The parallel prover speeds-up over the sequential one by generating faster the same proof and the results are rather regular.

Our philosophy is very different, because by partitioning the search space, we aim at parallelism at the search level. Then the concurrent processes deal with search spaces that may be radically different from that of the sequential prover. For instance, in Aquarius, it is sufficient that a Penguin does not retain a certain clause and sends it to settle at another node to change dramatically the search space for that Penguin. By considering a different portion of the search space, a shorter proof may be found. In such cases, the distributed theorem prover speeds-up considerably. However, if the search space turns out to be partitioned in a way that does not reveal a shorter proof, the distributed prover is at a strong disadvantage, as it may be trying to generate the sequential proof from a fragmented search space. The irregular results are the consequence of this kind of phenomena.

In summary, at the operational level, the main cause for the mixed results of Aquarius is the inefficiency of communication. At least part of the problem seems to be related to the choice of the PCN language, which perhaps was not designed for the parallelization of a large, computation-bound C program, such as Otter. The problem with communication may represent evidence in favor of a less distributed version of the Clause-Diffusion methodology. Because of the use of localized image sets, Aquarius implements a *distributed duplication-oriented approach*. If a shared memory component is available, one may choose global contraction at the source by image set in shared memory (see Section 2.1 and [6]) and obtain a *mixed shared-distributed approach*. This approach reduces the amount of both communication, because exchange of messages may be replaced in part by access to the shared memory, and duplication, because just one image set is maintained. On the other hand, if a single image set in shared memory is used, the search spaces considered by the different concurrent processes may turn out to be less differentiated than in the more distributed approach of Aquarius. Thus, the results might be more regular, but also, in a sense, less challenging than in Aquarius. The latter probes a radically new approach to parallelization, whose success will require a better understanding of the parallelization of search.

References

1. S.Anantharaman, J.Hsiang, Automated Proofs of the Moufang Identities in Alternative Rings, *JAR*, Vol. 6, No. 1, 76–109, 1990.
2. A.Wasilewska, Personal communication, March 1993.
3. J.Avenhaus and J.Denzinger, Distributing Equational Theorem Proving, in C.Kirchner (ed.), *Proc. of the 5th RTA Conf.*, Springer Verlag, LNCS, to appear.
4. W.Bledsoe, Challenge problems in elementary calculus, *JAR*, Vol. 6, No. 3, 341–359, 1990.
5. M.P.Bonacina, Problems in Lukasiewicz logic, *Newsletter of the AAR*, No. 18, 5-12, Jun. 1991.
6. M.P.Bonacina, Distributed Automated Deduction, Ph.D. Thesis, Dept. of Computer Science, SUNY at Stony Brook, Dec. 1992.
7. M.P.Bonacina and J.Hsiang, On fairness in distributed deduction, in P.Enjalbert, A.Finkel and K.W.Wagner (eds.), *Proc. of the 10th STACS*, Springer Verlag, LNCS 665, 141-152, 1993.

8. D.Champeaux, Sub-problem finder and instance checker: Two cooperating pre-processors for theorem provers, in *Proc. of the 6th IJCAI*, 191–196, 1979.

9. K.M.Chandy, S.Taylor, An Introduction to Parallel Programming, Jones and Bartlett, 1991.

10. J.D.Christian, High-Performance Permutative Completion, Ph.D. Thesis, Univ. of Texas at Austin and MCC Tech. Rep. ACT-AI-303-89, Aug. 1989.

11. S.E.Conry, D.J.MacIntosh and R.A.Meyer, DARES: A Distributed Automated REasoning System, in *Proc. of the 11th AAAI Conf.*, 78–85, 1990.

12. J.Denzinger, Distributed knowledge-based deduction using the team-work method, Tech. Rep. SR-91-12, Univ. of Kaiserslautern, 1991.

13. I.Foster, S.Tuecke, Parallel Programming with PCN, Tech. Rep. ANL-91/32, Argonne Nat. Lab., Dec. 1991.

14. D.J.Hawley, A Buchberger Algorithm for Distributed Memory Multi-Processors, in *Proc. of the International Conference of the Austrian Center for Parallel Computation*, Oct. 1991, Springer Verlag, LNCS, to appear.

15. A.Jindal, R.Overbeek and W.Kabat, Exploitation of parallel processing for implementing high-performance deduction systems, *JAR*, Vol. 8, 23–38, 1992.

16. D.Kapur. H.Zhang, RRL: a Rewrite Rule Laboratory, in E.Lusk, R.Overbeek (eds.), *Proc. of CADE-9*, LNCS 310, 768–770, 1988.

17. C.Kirchner, P.Viry, Implementing Parallel Rewriting, in B.Fronhöfer and G.Wrightson (eds.), *Parallelization in Inference Systems*, Springer Verlag, LNAI 590, 123–138, 1992.

18. E.L.Lusk, W.W.McCune, Experiments with ROO: a Parallel Automated Deduction System, in B.Fronhöfer and G.Wrightson (eds.), *Parallelization in Inference Systems*, Springer Verlag, LNAI 590, 139–162, 1992.

19. W.W.McCune, L.Wos, Some Fixed Point Problems in Combinatory Logic, *Newsletter of the AAR*, No. 10, 7–8, Apr. 1988.

20. W.W.McCune, OTTER 2.0 Users Guide, Tech. Rep. ANL-90/9, Argonne Nat. Lab., Mar. 1990.

21. F.J.Pelletier, Seventy-five problems for testing automatic theorem provers, *JAR*, Vol. 2, 191-216, 1986.

22. K.Siegl, Gröbner Bases Computation in STRAND: A Case Study for Concurrent Symbolic Computation in Logic Programming Languages, M.S. Thesis and Tech. Rep. 90-54.0, RISC-LINZ, Nov. 1990.

23. M.E.Stickel, The Path-Indexing Method for Indexing Terms, Tech. Note 473, SRI Int., Oct. 1989.

24. S.Tuecke, Personal communications, May 1992 and Dec. 1992.

25. J.-P.Vidal, The Computation of Gröbner Bases on A Shared Memory Multiprocessor, in A.Miola (ed.), *Proc. of DISCO90*, Springer Verlag, LNCS 429, 81–90, Apr. 90 and Tech. Rep. CMU-CS-90-163, Carnegie Mellon Univ., Aug. 1990.

26. L.Wos, *Automated Reasoning: 33 Basic Research Problems*, Prentice Hall, 1988.

27. K.A.Yelick and S.J.Garland, A Parallel Completion Procedure for Term Rewriting Systems, in D.Kapur (ed.), *Proc. of the 11th CADE*, Springer Verlag, LNAI 607, 109–123, 1992.

The Design of the SACLIB/PACLIB Kernels*

Hoon Hong and Andreas Neubacher and Wolfgang Schreiner

{hhong,aneubach,schreine}@risc.uni-linz.ac.at
Research Institute for Symbolic Computation (RISC-Linz)
Johannes Kepler University, Linz, Austria

Abstract. This paper describes the design of the kernels of two variants of the SAC-2 computer algebra library: SACLIB and PACLIB. SACLIB is a C version of SAC-2, supporting automatic garbage collection and embeddability. PACLIB is a parallel version of SACLIB, supporting lightweight concurrency, non-determinism, and parallel garbage collection.

1 Introduction

In this paper, we report an on-going work on developing two variants of the SAC-2 computer algebra library [3]: SACLIB and PACLIB, where SACLIB is a C version of SAC-2, and PACLIB is a parallel version of SACLIB. In particular we concentrate on the design and implementation of the kernels of these two systems. We decided to report both systems in the same paper since they are closed related.

The SAC-2 computer algebra library is a collection of carefully designed algorithms for various operations over integers, rationals, finite fields, algebraic numbers, and polynomials with such coefficients. The algorithms in the library are coded in the programming language ALDES [11]. In order to produce an executable program using the library, one usually translates the ALDES code into FORTRAN and uses a FORTRAN compiler.

The first part of this paper describes the work on developing a C version of SAC-2, namely SACLIB [1]. Other people have worked on translating SAC-2 into languages such as LISP and MODULA-2 [10, 6]. We chose the C language for our work since it is most wide-spread, portable, and efficient. Portability was an important issue since we plan to distribute our work in the public domain so that any researchers in computer algebra or engineers can freely obtain a copy and use it. While designing the SACLIB kernel, we desired to meet various requirements. Among them, the most important ones were automatic garbage collection and embeddability. These requirements were met mainly by adopting a conservative garbage collection scheme where the system stack is used for identifying the potential roots of active list cells.

The second part of this paper describes the work on developing a parallel version of SACLIB, namely PACLIB [5]. The kernel of PACLIB was designed in

---

* Supported by the Austrian Science Foundation (FWF) grant S5302-PHY "Parallel Symbolic Computation".

order to fulfill the following requirements: upward compatibility to the SACLIB kernel, high-level parallel programming model, light-weight concurrency, non-determinism and speculative parallelism, communication by heap references, and parallel garbage collection. For this purpose, we adapted the μSYSTEM package [2] that supports light-weight concurrency on shared-memory multiprocessors and workstations running under UNIX. The current kernel has been implemented on a 20 processor SEQUENT SYMMETRY computer.

2   The SACLIB Kernel

This section gives an overview of the kernel of SACLIB [1], the C version of the computer algebra library SAC-2 [3, 11].

2.1   Design Goals

The main points taken into consideration when SACLIB was designed were the following:

1. **Compatibility to** SAC-2: As we intended to build upon the work done for SAC-2 by using its wide range of algorithms, the kernel should be such that it would support the SAC-2 algorithms without making changes to the existing code (apart from syntactic translations) necessary.
2. **Efficiency:** The system (and especially the low-level kernel functions) should be computationally as efficient as possible.
3. **Portability:** The system should be easily portably among UNIX systems, thus making it available to a large number of users.
4. **Programming environment:** The system should be easy to use both for the algorithm designer and the applications programmer. This demand encompasses a simple interface to the system's functions and the availability of program development tools.
5. **Embeddability:** It should be possible to link SACLIB to application programs and to call the functions directly, thereby avoiding time-consuming communication protocols.
6. **Hidden garbage collection:** Developers should not be required to do special coding for bookkeeping tasks such as garbage collection.

2.2   Overall Design

Aiming at the goals described above resulted in the following major design decisions:

1. **Programming Language:** By using C as the development language of SACLIB, several goals were already met to a high degree: C allows the production of highly efficient code and applications based on the standard C functions are highly portable. Furthermore, C is the basic language of the UNIX operating system, thus the powerful UNIX source code debugging and maintenance tools can be used.

2. **Conservative Garbage Collection:** Since SACLIB may be embedded in all kinds of applications, we must not make any assumptions about the context in which SACLIB functions are called, in particular not about the layout of program data. However, we have to identify all potential roots of active list cells for garbage collection. Since we do not want to burden the program (respectively the programmer) with extra code for the bookkeeping of heap references, we use a *conservative* garbage collection scheme where it is safe to mistake a memory item for a heap reference.

In the following section, we describe how the programmer can use the system.

2.3 Programming Interface

To the C programmer, SACLIB is simply another library of functions which are linked to a program and can be called (after proper initialization) from anywhere in the code.

Figure 1 shows the basic layout of a program using SACLIB functions. Note

```
#include "saclib.h"

int sacMain(int argc, char *argv[])
{
        Word    I,F;
Step1:  /* Input. */
        SWRITE("Please enter an integer: "); I = IREAD();
Step2:  /* Compute factorial. */
        F = IFACTL(I);
Step3:  /* Output. */
        IWRITE(F); SWRITE("is the factorial of "); IWRITE(I);
        return(0);
}
```

Fig. 1. A sample program.

that the standard **main** routine used in C is replaced by **sacMain**, which causes all the necessary initializations to be done before the main program is executed. In this way, coding overhead for using SACLIB functions is minimized.

Furthermore, note that the programmer need not insert any statements regarding garbage collection, and neither are they added by some preprocessor. A minimal amount of bookkeeping is necessary for the handling of global variables. An example is given in Figure 2. See also the following section for details.

As an alternative to using **sacMain** it is also possible to call the initialization function explicitly at some later point in the program, as shown in Figure 3. Here

```
#include "saclib.h"

Word    F = NIL;

int sacMain(int argc, char *argv[])
{
        Word I;
Step0:  /* Declare global variables. */
        GCGLOBAL(&F);
Step1:  /* Input. */
        ...

}
```

Fig. 2. Fragment of a sample program using global variables.

some_subroutine is a module of a larger application program. Calls to SACLIB's algebraic algorithms are embedded in this subroutine, so that the resources (esp. memory) used by SACLIB are only allocated when needed. Thereby, the use of SACLIB is completely encapsulated and has no impact on the design of the main program.

In SAC-2 and its programming language ALDES, the kernel's list processing and garbage collection facilities are accessed by the three primitive functions COMP, FIRST, and RED. Implementing C versions of these ALDES functions allowed us to convert all other algorithms of the SAC-2 library by a simple syntactic translation from ALDES to C. Below we describe the semantics of these SAC-2 / SACLIB list processing primitives.

The basic data type in SACLIB is called Word. This is similar to the type int from C, with some restrictions:

− *Atoms* (also called BETA-*digits* or simply *digits*) are all integer constants or variables of type Word of value greater than -BETA and less than +BETA, where BETA is an integer constant.
− *Lists* are either empty, which is indicated by the constant value NIL, or they are constructed by the function COMP(x,L), which puts the atom or list x at the head of the (possibly empty) list L and returns the resulting list.
− The two primitive list operations which are dual to COMP are FIRST(L), which returns the first element of a (non-empty) list L, and RED(L), which returns the reductum of a (non-empty) list L, i.e. the list L without its first element.

This scheme originally employed by SAC-2 allows the use of the standard C operators +, -, *, /, %, etc. for atoms provided that the results stay in the allowed range. Furthermore, the distinction between lists and atoms can be made based on the value of the variable, so it is not necessary to allocate and access

```
#include "saclib.h"

... some_subroutine(...)
{
        Word dummy;
        ...             /* Other variable declarations. */

Step1:  /* Initialize SACLIB. */
        BEGINSACLIB(&dummy);

Step2:  /* Use SACLIB. */
        ...

Step3:  /* Remove SACLIB. */
        ENDSACLIB(SAC_FREEMEM);
        return(...);
}
```

Fig. 3. Embedding SACLIB in a subroutine.

additional memory for storing the type of an object. Its influence on garbage collection is made clear in the following sections.

2.4 Heap Management

In order to keep the SAC-2 semantics of list processing, its way of implementing lists was also used in SACLIB:

When SACLIB is initialized, the array SPACE containing $NU + 1$ Words (with NU being a global variable defined in the kernel) is allocated. This array is used as the memory space for list processing.

Lists are built from *cells*, which are pairs of consecutive Words the first of which is at an odd position in the SPACE array. *List handles* ("pointers" to lists) are defined to be BETA plus the index of the first cell of the list in the SPACE array. As mentioned above, the handle of the empty list is NIL (which is a constant equal to BETA). Figure 4 shows the structure of the SPACE array.

The first Word of each cell is used for storing the handle of the next cell in the list (i.e. the value returned by RED), while the second Word contains the data of the list element represented by the cell (i.e. the value returned by FIRST). Figure 5 gives a graphical representation of the cell structure for a sample list, with arrows representing list handles.

As already mentioned in the previous section, atoms are required to be integers a with $-BETA < a < BETA$. This allows the garbage collector and other functions operating on lists to decide whether a variable of type Word contains

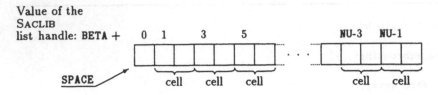

Fig. 4. The SPACE array.

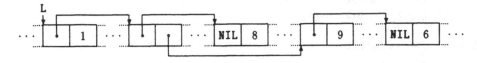

Fig. 5. The cell structure of the list $L = (1, (9, 6), 8)$.

an atom or a list handle. Note that list handles can only be odd values between BETA and BETA + NU.

The Words of a cell addressed by a list handle L are SPACE[L − BETA] and SPACE[L − BETA + 1]. The use of the C programming language allows us to simplify these computations, by setting the C pointers SPACEB and SPACEB1 to the memory addresses of SPACE[−BETA] and SPACE[−BETA + 1], respectively. This is used by the functions FIRST(L), which returns SPACEB1[L], and RED(L), which returns SPACEB[L], thus avoiding time-consuming address computations.

Most of the well-known garbage collection schemes require adding bookkeeping statements to the program code, e.g. for counting the number of references to a list cell or for declaring variables referencing list cells to the garbage collector. While such statements can be automatically inserted and need not increase the amount of work done by the programmer, they do increase the running time of the program, because they are executed even if there is a sufficient amount of memory available so that garbage collection is not necessary.

In SACLIB, time is spent for garbage collection only when its conservative garbage collector is invoked by COMP in the case of the kernel's internal list of available (i.e. unreferenced) cells being empty. A mark-and-sweep method is employed for identifying unused cells:

1. **Mark:** The processor registers, the system stack, and the global variables declared by a call to GCGLOBAL are searched for list handles. If a list handle is found, the cells of the list are traversed and marked. If the second Word of a cell contains a list handle, the corresponding cells are marked recursively.
2. **Sweep:** In the sweep step, the list of available cells is built: Unmarked cells in the SPACE array are linked to the list. If a cell is marked, it is only unmarked and not changed in any other way.

By this scheme it is guaranteed that all cells which are referenced (directly or indirectly) by any of the program's variables are not linked to the list of available cells. The problem here is that the stack in general does not only contain variables from SACLIB code, but also C pointers, floating point numbers, arbitrary integer values, etc. It might happen that some of these fall into the range of a legal list handle when interpreted as a Word. This may result in some cells, which should go into the list of available cells, being left unchanged. While this does not influence the correctness of the program, it does incur a trade off of memory needs against execution speed.

Extensive testing and heavy use of the system have shown that in general this does not impose a problem, i.e. the number of cells being kept allocated due to wrongly interpreted references is not significant compared to the amount of memory used.

3 The PACLIB Kernel

In this section, we describe the design of the PACLIB kernel which is a parallel variant of the SACLIB kernel for shared memory multiprocessors [14, 15].

3.1 Design Goals

In the design phase of the PACLIB kernel, we raised several demands the parallel computer algebra kernel should fulfill:

1. **Upward compatibility to the SACLIB kernel:** The PACLIB kernel should provide the same interface to the heap and the same efficiency of heap access as the SACLIB kernel. Each function of the SACLIB library should therefore run without change and without loss of speed when linked with the PACLIB kernel.

2. **High-level parallel programming model:** The system should be based on a high-level parallel programming model. It should be possible to spawn each function of the SACLIB library as a concurrent task and to retrieve the result of this function without any change of the interface. Tasks should be first-class objects.

3. **Light-weight concurrency:** Since many parallel algebraic algorithms are rather fine-grained, the runtime overhead for the creation of a new task should be very small and it should be possible to spawn thousands of tasks.

4. **Non-determinism and speculative parallelism:** Since the runtime of algebraic algorithms can often not be predicted, there should be a synchronization construct waiting for a whole set of tasks and returning the first available result. If the results of the other tasks are not needed any more, it should be possible to abort these tasks.

5. **Communication by heap references:** On a shared memory machine, task arguments and task results should be heap references only. Neither the creation of a new task nor its termination should require to copy whole SACLIB structures (which would increase the runtime and might duplicate originally shared substructures).

6. **Parallel garbage collection:** There should be a global garbage collector that on demand reclaims all of the no more referenced heap cells. The garbage collector should be parallelized in order to profit from the multi-processor architecture.

In the following subsections, we will describe how and to which extent we met these demands. A comprehensive description of the kernel design can be found in [12].

3.2 Overall Design

The PACLIB kernel consists of the following main components (depicted in Figure 6):

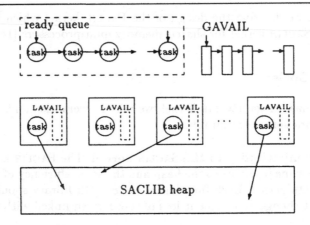

Fig. 6. PACLIB Kernel Design

1. **Ready Queue:** This queue contains all tasks that are *active* but not *running*. The scheduler selects tasks for execution from the head of this queue; tasks that are preemptively descheduled are inserted at the tail.
2. **Shared Heap:** The SACLIB heap serves as the only communication medium between tasks. There is no communication bottleneck since all PACLIB tasks may read all heap cells simultaneously.
3. **Virtual Processors:** Each virtual processor is a UNIX process that owns a local list LAVAIL of free heap cells. Each task executed on this processor allocates new heap cells from this list.
4. **Global Available List:** GAVAIL is a global list of free lists in which the unused heap cells are linked together. Processors whose local lists run out of free cells receive their new LAVAIL from GAVAIL.

The next section describes the programming interface to the system in more detail.

3.3 Programming Interface

```
Word PIPROD(Word a, Word b) {          /* parallel integer product */
   Word l, n, i, j, b0, b1;
   Word t, tlist;
   l = LENGTH(b);                      /* length of b */
   n = l / N + (l % N != 0);           /* number of tasks */
   tlist = NIL;                        /* list of tasks */
   for (i = 0; i <= n; i ++) {         /* start tasks */
      ADVANCE(N, b, &b0, &b1);         /* split off subinteger b0 */
      j = i*N;                         /* shifting factor */
      t = pacStart(IPRODS, 3, a, b0, j); /* start task for (a*b0)<<j */
      tlist = COMP(t, tlist);          /* put task into list */
      b = b1;                          /* rest of b (without b0) */
   }
   for (i = 0; i < n-1; i++) {         /* collect results */
      b0 = pacWaitListRm(&tlist);      /* 1st result, remove task */
      b1 = pacWaitListRm(&tlist);      /* 2nd result, remove task */
      t = pacStart(ISUM, 2, b0, b1);   /* start task for b0+b1 */
      tlist = COMP(t, tlist);          /* put new task into list */
   }
   return(pacWaitOne(FIRST(tlist)));   /* result of last task */
}
```

Fig. 7. Parallel Integer Multiplication

Most algebraic algorithms are based on purely mathematical functions that are entirely defined by their argument/result behavior. The basic PACLIB model of parallelism reflects this view. A PACLIB task receives certain SACLIB objects as input arguments and returns a SACLIB object as its result. This result may be retrieved later by a handle to this task.

$$task = \texttt{pacStart}(fun, \, args)$$

creates a new task that executes *fun(args)* and can be referenced via the handle *task*. *task* is a first-order SACLIB object i.e. it can passed to other functions or tasks or stored in a SACLIB list. The result of this task (which is the value returned by *fun*) may be retrieved *an arbitrary number of times* by any task that knows *task*.

$$value = \texttt{pacWait}(tptr, \, tasks)$$

returns the result *value* of one of the *tasks* and writes the handle of the delivering task into the location *tptr*. If all denoted tasks are still active, pacWait blocks until one task terminates and then delivers the result of this task. There exist several variations of this function: `pacWaitListRm` takes as its single argument a SACLIB list of task handles and removes from the list the task whose result is delivered; `pacWaitOne` waits for the result of a single task.

Figure 7 illustrates the PACLIB model of parallelism by a parallel version of the school algorithm for the multiplication of two integers a and b. PIPROD splits b into subintegers b_0 of N digits and multiplies a with each b_0 in parallel. Each task executes the function IPRODS that multiplies a with b_0 and shifts the result by j digits. PIPROD iteratively waits for the results delivered by any two tasks and starts a new task that adds the results. Finally, there remains a single task left that returns the total result.

> `pacStop(tasks)`

terminates the execution of all denoted *tasks* and of all subtasks created by these tasks. If some task has already delivered its result, it is not affected any more. `pacStop` can be considered as an annotation that does not influence the correctness of a program (modulo termination) but only its operational behavior.

There are several parallel algebraic algorithms (e.g. for *critical pair completion*) for which the purely functional model is not adequate. These algorithms can be better described in terms of tasks each of which iteratively receives a sequence of input objects, processes each object according to its internal state (which is changed by the computation) and forwards a sequence of results to other tasks. Therefore PACLIB supports an additional form of task communication by *streams* i.e. automatically synchronized lists that may be used for this purpose. For a comprehensive description of the PACLIB programming model, see [5].

3.4 Task Management

The task management of the PACLIB kernel is built on top of the μSYSTEM [2]. This free software package consists of a library of C functions supporting lightweight concurrency on shared memory multiprocessors and UNIX workstations. The μSYSTEM kernel distributes *tasks* (light-weight processes) among *virtual processors* which are (heavy-weight) UNIX processes scheduled by the operating system. On a multi-processor, virtual processors are therefore executed by multiple hardware processors truly in parallel. The μSYSTEM task management is quite efficient and allows the utilization of rather *fine-grained parallelism*.

We applied and considerably extended the features of this package in order to implement the PACLIB parallel programming model which is higher-level and more general than that provided by the μSYSTEM. There were two major problems that had to be solved:

1. How to represent the task handles returned by `pacStart` and
2. How to implement the non-deterministic functionality of `pacWait`.

The implementation of pacWait required the development of a fairly complicated algorithm whose description is outside the scope of this overview paper. The details of this algorithm can be found in [12], its correctness has been formally verified in [13].

For implementing task handles, there basically exist two possibilities: The first one is to view such a handle as a *task descriptor* i.e. as a reference to the workspace of the task. Before a task terminates, it stores its result in this workspace for the later retrieval by other tasks. Then either the workspace is not reclaimed as long as there are references to the task or the workspace is reclaimed when the result is received by another task. However, the first approach would waste too much space while in the second approach (implemented in the μSYSTEM) tasks are not first-order objects any more.

Therefore we decided to split the description of a task into two separate entities: a *task descriptor* and a *result descriptor*. *task* is then a reference to the result descriptor which is a SACLIB list with the following elements:

sem	task	wait	value

- **sem** is a pointer to a semaphore that provides mutual exclusion on the descriptor.
- **task** is a pointer to the workspace of the task that computes the result. If the task has already terminated, task is the constant AVAILABLE.
- **wait** contains, if task $\neq$ AVAILABLE, the head of a queue of tasks blocked on an attempt to receive the result.
- **value** contains, if task $\neq$ AVAILABLE, the tail of the wait queue and the result of the task, otherwise.

On a call of pacStart, a result descriptor and a task descriptor are allocated and mutually linked. This new task executes a function pacShell that calls the denoted function with the corresponding arguments. When the function returns, pacShell stores the result value in the result descriptor and awakes all tasks blocked on this descriptor. Finally, pacShell deallocates its task descriptor and the task terminates.

The large task descriptor therefore occupies space only as long as the task is actually alive. After the termination of the task, only the small result descriptor is left by which other tasks can receive the result. Since this descriptor is a SACLIB object, it is subject to garbage collection and is (together with the result) reclaimed when no task has the *task* handle any more. Hence, task results do not occupy memory longer than necessary.

3.5 Heap Management

The PACLIB kernel provides an operation COMP with the same interface and efficiency as the SACLIB kernel. However, the organization of free lists in the PACLIB kernel is much more sophisticated: There is a free list LAVAIL attached to each (virtual) processor and there is a global list of free lists GAVAIL. When a task executes COMP, the new cell is allocated from the LAVAIL of that processor that currently executes this task. If LAVAIL runs out of free cells, the processor picks the first list in GAVAIL as its new LAVAIL. If GAVAIL is also empty, a global garbage collection is triggered. This two-level memory management scheme of PACLIB is a compromise between two extremes:

1. **Global Available List Only:** In order to keep GAVAIL consistent, it must not be simultaneously updated by different processors. Without local available lists, each call of COMP would require a semaphore operation that would considerably increase the runtime overhead of the function. Moreover, due to the serial access to GAVAIL, this shared resource would represent a significant bottleneck for parallelization.

2. **Local Available Lists Only:** Without a global available list, there arises the danger of wasting heap space. Some processor might consume cells from its local available list LAVAIL faster than the other processors. It would then trigger garbage collection even if there were still plenty of free cells available (since these cells were attached to the local available lists of other processors).

In order to trigger garbage collection, the processor puts the currently executed task back into the ready queue and interrupts by a UNIX signal the execution of the other (virtual) processors. After all processors have acknowledged the interruption, garbage collection takes place in two phases:

1. **Mark:** All cells that can be referenced by any task are marked. Since all user tasks are descheduled, all heap references are on the stacks of these tasks. Each virtual processor picks a task descriptor, scans the associated stack and marks all heap cells that can be reached.

2. **Sweep:** All unmarked cells are reclaimed to construct a new list GAVAIL. Each processor scans a different portion of the heap, unmarks the marked cells and reclaims the unmarked cell. When a new free list of the desired length is built, the list is linked into GAVAIL.

The work of the sweep phase is well balanced among all processors since they operate on different heap sections of the same size. The balance of load in the mark phase depends on the number of available tasks and the number of cells that can be reached by each task. Several experiments [14] show that the efficiency of the garbage collector is actually very high (a speedup of more than 14 could be achieved with 18 processors compared to the sequential SACLIB collector).

3.6 Performance Monitoring

For visualizing the dynamic aspects of a PACLIB program, the kernel optionally generates profiling information during a program run. This information consists of trace records that are written on each context switch of a task into a buffer which is constantly flushed into a file. We have developed two programs,

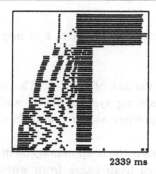 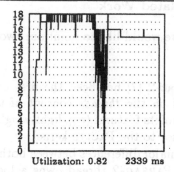

2339 ms   Utilization: 0.82   2339 ms

Fig. 8. Performance Monitoring

pacgraph and pacutil, that generate from this information LaTeX pictures that visualize the behavior of the program and the utilization of the processors during the program run, respectively.

Figure 8 shows the output of these tools for some PACLIB program with 18 processors used. The left picture consists of a set of horizontal lines each of which represents one task and shows in the horizontal direction the times during which the task was scheduled for execution. The right picture shows how many processors were saturated with work during each moment of the program run. The output of these visualization tools has been proved to be a valuable help for the tuning of PACLIB programs.

3.7 Future Work

Currently, we are working on the integration of the following features into PACLIB:

1. **Task Priorities:** The user will be given the possibility of controlling the task scheduling mechanism be the introduction of different task priorities.
2. **Virtual Tasks:** pacStart will create "virtual" tasks that only turn into "real" tasks when processors are not saturated with work. The efficiency of task creation should increase and the space consumption of tasks should significantly drop.
3. **Visualization Environment:** An X11-based visualization environment is developed that will allow the programmer to learn in an interactive way many additional aspects of the dynamic behavior of a PACLIB program (e.g. memory consumption).
4. **Functional Language Compiler:** A compiler is developed that in a semi-automatic fashion generates parallel PACLIB code from high-level functional specifications.

A variety of parallel algebraic algorithms has already been implemented using the PACLIB kernel [16, 4].

3.8 Related Work

Finally, we would like to point out two projects that preceded and inspired our work:

- **Parsac-2** [7] has been the first parallel variant of SAC-2. The system is based on the *threads* concept of the operating system MACH and has by some of its features (e.g. light-weight concurrency and *virtual tasks* [9]) stood sponsor to the PACLIB kernel.

  The major differences between both systems is the heap management scheme: each PARSAC-2 thread owns a local set of heap pages from where it can allocate new cells and which may be extended from a global page set. Since free lists are attached to threads, the COMP operation needs one indirection to determine the head of the current free list (in the PACLIB kernel the head has the same location in the virtual address space of each processor).

  PARSAC-2 does not use a global garbage collection mechanism but it has introduced the concept of *preventive garbage collection* [8]: a terminating thread copies a SACLIB result structure into the page set of the parent thread (in contrast to PACLIB tasks that just return heap references) and frees its own page set.

  This heap management scheme is the reason for another major difference to the PACLIB kernel: PARSAC-2 threads are not first-class objects, since the result of a thread may be retrieved only once (by the parent task). Threads may be aborted but there is no PARSAC-2 equivalent to the PACLIB support of non-determinism.

- **Network-Sac** [17] is a parallel version of SAC-2 for computer networks and the coarse-grain parallelization of algebraic programs. Each computer runs a single SAC-2 process (the *algorithm server*) that receives from a scheduler messages describing tasks to be executed.

  Task arguments and results are converted into a character representation, transferred over the network and then transformed back into the internal representation. Since each algorithm server uses a separate heap with only local references, no global garbage collector is required.

  The concept of *multibags* is applied to express non-deterministic computations; futures (i.e. task references) are put into a multibag, any access to the multibag then returns the first available result. However, tasks are not really first-class objects (futures must not be passed as arguments to other tasks) and there is no possibility to abort tasks.

References

1. B. Buchberger, G. Collins, M. Encarnation, H. Hong, J. Johnson, W. Krandick, R. Loos, A. Mandache, A. Neubacher, and H. Vielhaber. A SACLIB Primer. Technical Report 92-34, RISC-Linz, Johannes Kepler University, Linz, Austria, 1992.

2. P. A. Buhr and R. A. Stroobosscher. The μSystem: Providing Light-weight Concurrency on Shared-Memory Multiprocessor Computers Running UNIX. *Software — Practice and Experience*, 20(9):929–964, September 1990.

3. G. E. Collins and R. Loos. ALDES/SAC-2 now available. *ACM SIGSAM Bull.*, 1982.

4. H. Hong. Parallel Computation of Modular Multivariant Polynomial Resultants on Shared Memory Machine. In *Submitted to the 2nd International Conference of the Austrian Center for Parallel Computation*, Gmunden, Austria, October 4–6, 1993.

5. H. Hong, W. Schreiner, A. Neubacher, K. Siegl, H.-W. Loidl, T. Jebelean, and P. Zettler. PACLIB User Manual. Technical Report 92-32, RISC-Linz, Johannes Kepler University, Linz, Austria, May 1992. Also: Technical Report 92-9, ACPC Technical Report Series, Austrian Center for Parallel Computation, July 1992.

6. H. Kredel. From SAC-2 to Modula-2. In *Proceedings of ISSAC*, pages 447–455. Springer, Berlin, 1988. Volume 358 of Lecture Notes in Computer Science.

7. W. Küchlin. The S-Threads Environment for Parallel Symbolic Computation. In *Computer Algebra and Parallelism, Second International Workshop*, pages 1–18, Ithaca, USA, May, 1990. Springer, Berlin. Volume 584 of Lecture Notes in Computer Science.

8. W. Küchlin and N. Nevin. On Multi-Threaded List-Processing and Garbage Collection. In *3rd IEEE Symposium on Parallel and Distributed Processing*, Dallas, TX, December, 1991. IEEE Press.

9. W. Küchlin and J. Ward. Experiments with Virtual C Threads. In *4th IEEE Symposium on Parallel and Distributed Processing*, Arlington, TX, December, 1992. IEEE Press.

10. L. Langemyr. Converting SAC-2 Code to Lisp. In *Proceedings of EUROCAL 87*, pages 50–51. Springer-Verlag, 1989.

11. R. G. K. Loos. The Algorithm Description Language ALDES (Report). *ACM SIGSAM Bull.*, 10(1):15–39, 1976.

12. W. Schreiner. The Design of the PACLIB Kernel. Technical Report 92-33, RISC-Linz, Johannes Kepler University, Linz, Austria, 1992.

13. W. Schreiner. The Correctness of the PACLIB Kernel — A Case Study in Parallel Program Verification by Temporal Logic. Technical report, RISC-Linz, Johannes Kepler University, Linz, Austria, 1993. To appear.

14. W. Schreiner and H. Hong. A New Library for Parallel Algebraic Computation. In *Sixth SIAM Conference on Parallel Processing for Scientific Computing*, Norfolk, Virginia, March 22-24, 1993. Also: Technical Report 92-73, RISC-Linz, Johannes Kepler University, Linz, Austria, 1992.

15. W. Schreiner and H. Hong. PACLIB — A Library for Parallel Algebraic Computation on Shared Memory Computers. In *Parallel Systems Fair at the Seventh International Parallel Processing Symposium*, Newport Beach, CA, April 13–16, 1993.

16. W. Schreiner and V. Stahl. The Exact Solution of Linear Equation Systems on a Shared Memory Multiprocessor. In *Submitted to the PARLE 93*, Munich, Germany, June 14–18, 1993.

17. S. Seitz. Algebraic Computation on a Local Net. In *Computer Algebra and Parallelism, Second International Workshop*, pages 19–32, Ithaca, USA, May, 1990. Springer, Berlin. Volume 584 of Lecture Notes in Computer Science.

The Weyl Computer Algebra Substrate

Richard Zippel*

Department of Computer Science
Cornell University, Ithaca, NY 14853

Abstract. Weyl is a new type of computer algebra substrate that extends an existing, object oriented programming language with symbolic computing mechanisms. Rather than layering a new language on top of an existing one, Weyl behaves like a powerful subroutine library, but takes heavy advantage of the ability to overload primitive arithmetic operations in the base language. In addition to the usual objects manipulated in computer algebra systems (polynomial, rational functions, matrices, etc.), domains (*e.g.*, $\mathbf{Z}$, $\mathbf{Q}[x, y, z]$) are also first class objects in Weyl.

1 Introduction

In the last twenty years the algorithms and techniques for manipulating symbolic mathematical quantities have improved dramatically. These techniques have been made available through a number of interactive algebraic manipulation systems. Among the most widely distributed systems are Reduce [4], Macsyma [11], Maple [6] and Mathematica [12]. These systems contain comparable mathematical functionality and possess relatively similar designs. They are typical of most computer algebra systems currently being developed. Nonetheless, we feel that these systems are deficient in three respects:

- They are monolithic structures intended for interactive use as self-contained computational systems.
- These systems are not designed in a "functorial" fashion, so their existing algorithms and mathematical data structures cannot be easily extended to new algebraic structures.
- The systems cannot represent or manipulate mathematical "domains" like groups, rings and function spaces.

Weyl is a new computer algebra substrate designed to address these deficiencies. The rest of this section discusses these points in a bit more detail.

Macsyma, Maple and Mathematica are intended to be complete systems within which one does mathematical computations. In addition to the basic symbolic algorithms, each of these systems contains a complete, and often sophisticated, user interface and a programming language in which the user describes their computations and extends the system.

* This research was supported in part by the Advanced Research Projects Agency of the Department of Defense under ONR Contract N00014–92–J–1989, by ONR Contract N00014–92–J–1839, NSF Contract IRI–9006137 and in part by the U.S. Army Research Office through the Mathematical Science Institute of Cornell University.

In contrast, stand-alone linear algebra systems like Matlab [8] are used for research, but large scale applications such as those used in computation fluid dynamics are built using libraries such as LINPACK [3] and EISPACK [9], rather than on top of Matlab.

We argue that, in addition to the previously mentioned interactive packages, we need symbolic computing libraries that can be incorporated in other software and linked with existing numerical and graphical libraries. The results of symbolic computations should be as easy to manipulate as the matrices of linear algebra, which are used so often in scientific computing. This type of system would make symbolic computation useful to an even larger scientific computing community.

The second limitation of current symbolic mathematics systems is that they deal with a fixed set of algebraic types, which the user is not expected to extend significantly. For instance, most systems contain an implementation of multivariate polynomials. In some problems, it is convenient to deal with polynomials whose coefficients are rational functions (quotients of polynomials). Adding this new type usually requires a new implementation of each of the arithmetic operations for polynomials with rational function coefficients, even though the algorithms are no different than those that are already implemented. Experimenting with mathematical objects that are not already provided, such as polynomials with power series coefficients or Poisson series, often requires a great deal of effort, even if the system's source code is available.

Alternatively, the basic algorithms can be implemented in a "functorial" fashion. Rather than implementing polynomials with integer coefficients, the functorial approach implements polynomials over an arbitrary ring where the particular arithmetic operations used for computing with coefficients are extracted from the coefficient ring in an object oriented fashion. This implementation can then be used for polynomials over the integers, over algebraic function fields, over rings of matrices, etc.

The third limitation of current systems is their lack of "domains." Consider the ring of polynomials in x, y and z with integer coefficients: $Z[x, y, z]$. While most algebraic systems can represent and manipulate the elements of this ring, they cannot represent the ring itself. Objects like $Z[x, y, z]$ are called "domains." While many properties can be attached to the elements of domains, others are best attached to the domains themselves. For example the fact that the integers modulo 6 is not an integral domain is best represented as a property of $Z/6Z$. In addition, when reasoning about mathematical algorithms, the deductions refer to the domains more than the domain elements.

Axiom [5] is a notable exception to that of the computer algebra systems mentioned above. Its algebra code was developed in a "functorial" fashion that makes the whole system quite extensible. In addition, Axiom provides a mechanism for representing mathematical domains. The Axiom developers achieved this by developing their own language for describing algebraic algorithms and then compiling it into a lower level language. In effect, they created their own object oriented programming language.

In contrast, Weyl uses the object oriented mechanisms of Common Lisp [10] directly. The support it requires is but a few pages of standard class definitions. Weyl users write programs in standard Common Lisp and continue to use the user interface tools, libraries, debuggers and software development tools that they are familiar with. However, their environment is augmented by new types representing algebraic objects and new control structures to ease their use.

One of the most important concepts in Weyl is that of a *domain*. Most of the design decisions in Weyl derive from the premise that domains should be first class citizens in the programming environment. Section 2 gives a few examples that illustrate the need for domains. Section 3 explains the type system that is used in Weyl, the role of domains in the type system and how coercion is implemented. How domains are created is discussed inSection 4. Section 5 illustrates how numbers are implemented in Weyl. A similar approach is used for dealing with polynomials. In Appendix A we give an example of how Weyl is used to solve basic computer algebra problems.

2   The Domain Concept

Consider a routine called `solve` that returns one of the zeroes of a polynomial in one variable. Given $x^2 - 5$ it should return $\sqrt{5}$. But how should the symbol $\sqrt{5}$ be interpreted, *i.e.*, how does one know that $(\sqrt{5})^2 = 5$, and does $\sqrt{5} = 2.2361$ or -2.2361? Furthermore, after `solve` has been asked to solve $x^2 - 5$, what is returned for $x^2 - 20$? Either $\sqrt{20}$ or $2\sqrt{5}$ could be correct and have different meanings. Algebraically, the latter is probably better than the former.

These issues are addressed by *domains*. Domains are collections of objects that possess an underlying mathematical organization. For example, the set of rational integers $\mathbf{Z}$ is a domain, as is the set of rational numbers $\mathbf{Q}$. The polynomials $x^2 - 5$, $x^2 - 20$ and $x^2 - x - 1$ could all be elements of the domain $\mathbf{Q}[x]$, the set of polynomials in x with rational number coefficients. Other common domains are the real numbers $\mathbf{R}$ and the complex numbers, $\mathbf{C}$. When necessary we indicate the domain of an element as a subscript, *e.g.*, $(x + 1)_{\mathbf{Q}[x]}$ and $(x^2 - 1)_{\mathbf{R}(x,y,z)}$.

The first invocation of `solve` is given $x^2 - 5$, which is an element of $\mathbf{Q}[x]$. When `solve` returns $\sqrt{5}$ it must also create a domain that contains $\sqrt{5}$. The simplest such domain is $\mathbf{Q}[\alpha]/(\alpha^2 - 5)$, the ring of polynomials in α with coefficients in $\mathbf{Q}$ reduced modulo the relation $\alpha^2 - 5 = 0$. Notice that α can correspond to either of the zeroes of $x^2 - 5$. To be precise `solve` must also set up an embedding of $\mathbf{Q}[\alpha]/(\alpha^2 - 5)$ into $\mathbf{C}$—that is, a map that sends each element of $\mathbf{Q}[\alpha]/(\alpha^2 - 5)$ to a complex number. Since there is a natural embedding of $\mathbf{Q}$ into $\mathbf{C}$ it is only necessary to indicate the image of α in $\mathbf{C}$.

This final algebraic structure, along with the embedding into $\mathbf{C}$, is abbreviated as $\mathbf{Q}[\sqrt{5}]$. Solve returns $(\sqrt{5})_{\mathbf{Q}[\sqrt{5}]}$, thus conveying all the simplification rules that $\sqrt{5}$ must obey as well as the embedding of $\sqrt{5}$ in $\mathbf{C}$. Notice that this information is associated with the domains, rather than attached directly to $\sqrt{5}$.

The second invocation of `solve` is interesting. Is `solve` invoked on $(x^2 - 20)_{\mathbf{Q}[x]}$ or $(x^2 - 20)_{\mathbf{Q}[\sqrt{5}][x]}$? It is not possible to determine which is the case by just examining the polynomial. Some reference to the enclosing domain is needed. If `solve` is called on $(x^2 - 20)_{\mathbf{Q}[x]}$ then $(\sqrt{20})_{\mathbf{Q}[\sqrt{20}]}$ is returned, just as in the previous discussion.

When solve is called on $(x^2-20)_{Q[\sqrt{5}][x]}$, the coefficient field needs to be extended and $(2\sqrt{5})_{Q[\sqrt{5}]}$ is returned. This eliminates the potential need of determining the relationship between $\sqrt{5}$ and $\sqrt{20}$.

One might not expect a need for two isomorphic domains in a single computation. For instance, why would one want two copies of Z during a computation? One answer is to allow more complete type checking in the programs and is discussed a bit later. Another motivation is provided by algebraic extensions. The three zeroes of $x^3 - 2$ are $\sqrt[3]{2}$, $\zeta_3 \sqrt[3]{2}$ and $\zeta_3^2 \sqrt[3]{2}$, where

$$\zeta_3 = \frac{1 + \sqrt{-3}}{2}.$$

Each generates an algebraic extension of Q of degree 3. Algebraically each extension is of the form $Q[\alpha]/(\alpha^3 - 2)$, i.e.,

$$Q[\sqrt[3]{2}] = Q[\alpha]/(\alpha^3 - 2),$$

$$Q[\zeta_3 \sqrt[3]{2}] = Q[\beta]/(\beta^3 - 2),$$

$$Q[\zeta_3^2 \sqrt[3]{2}] = Q[\gamma]/(\gamma^3 - 2).$$

That is, the polynomials that represent elements in the three different fields are manipulated in precisely the same fashion. The only difference is their embedding in C, i.e.,

$$\alpha \longrightarrow 1.259921,$$

$$\beta \longrightarrow -0.629960 + 1.091124i,$$

$$\gamma \longrightarrow -0.629960 - 1.091124i.$$

Finally, having multiple isomorphic domains allows us to type check "functorial" code for missing or incorrect conversions while still using simple examples. For instance, consider the representation of polynomials over the integers, $Z[x]$. Internally, such polynomials may be represented as lists of exponent-coefficient pairs:

$$2x^3 + 5 \approx ((3, 2), (0, 5)).$$

The coefficients are elements of Z, as are the exponents. For most operations, the exponents never come in contact with the coefficients. For instance, when multiplying polynomials, the coefficients are multiplied with each other; and the exponents are compared and added to each other; but there are no operations that mix exponents and coefficients.

Many implementations use the same representation for both coefficients and exponents. Instead, we suggest that there should be a domain of integers for the exponents in addition to the domain associated with the coefficients. More generally, $k[x]$ would have two associated domains, k for the coefficients and a copy of Z for the exponents.

The value of this approach is apparent when examining polynomial differentiation, where exponents must be multiplied by coefficients. By having two isomorphic, but different, copies of Z we are forced to coerce an exponent into the domain of the coefficients even though it may not initially appear to be necessary for polynomials in $Z[x]$, thus catching the need for coercion immediately.

3 Weyl's Type Structure

The objects manipulated by Weyl include *domains*, *domain elements* and *morphisms* between domains. Domain elements are the objects with which algebraic manipulations systems normally deal: integers, polynomials, algebraic numbers, etc. Examples of domains are the rational integers ($\mathbf{Z}$), polynomial rings ($\mathbf{Q}[x, y]$, $\mathbf{F}_p[t]$) and vector spaces ($\mathbf{R}^3$, $P^3(\mathbf{Z})$). The elements of these domains are domain elements.[2] Each domain element is the element of a single domain, and this domain can be determined from the element. Morphisms are maps between domains and are used in the coercion mechanism described later.

The "type" of a domain element consists of two components, the domain that contains the object and the type of Lisp structure used to represent the object. For instance, the polynomial $x + 2$ might be implemented as a list, but it is an element of the domain $\mathbf{Z}[x]$. We say that the *structural type* of $x + 1$ is list and that its *domain type* is $\mathbf{Z}[x]$. Elements of $\mathbf{Z}[x]$ could also be represented as vectors or in other ways, and lists could be used to represent objects in other domains besides $\mathbf{Z}[x]$. Thus the structural type is truly orthogonal to the domain type of an object. Additional examples are given in Section 5.

Both domains and domain elements are implemented as instances of CLOS (Common Lisp Object System) classes. These classes are part of a conventional, multiple inheritance class hierarchy. The class hierarchy used for domains is completely separate from and orthogonal to the class hierarchy for domain elements. The domain class hierarchy includes classes corresponding to groups, rings, fields and many other familiar types of algebraic domains.

This approach allows us to provide information about domains that is often difficult to indicate if we could only specify information about the type of the domain's elements. Whether a domain is an integral domain or just a ring does not change how the domain's elements are represented or the operations that can be performed on them. What it may affect is the algorithms that are used to implement the operations.

Domains also provide information about the permissible operations involving their elements. For instance, a Group must have an operation * that can be applied to pairs of its elements and which yields an element of the group. This parallels the mathematical definition of a group where a group is a pair $(G, \times)$ consisting of a set of elements (G) and a binary operation ($\times$) that maps pairs of elements of G into G. Contrast this with the definition of a type as a set of objects that obey a predicate.

Finally, domains are objects that can participate in computation. There are (constructive) functors that take an integral domain and produce its quotient field, take a ring and produce its spectrum and take a field and produce a vector space. The characteristic of a ring R is determined by applying the characteristic function to R. In the implementation of certain algorithms, the presence of domains seems to be a big improvement over the organization of systems like Macsyma. At the moment, this type of information is maintained in an *ad hoc* fashion using the type hierarchy

---

[2] To be strictly correct, we are referring of the Lisp objects that represent the different domains and domain elements. Thus the Lisp object that represents $\mathbf{Z}$ or $\mathbf{Q}[x, y]$ are domains, and the Lisp object that represents $x^2 + 2y + 1$ is a domain element.

of CLOS and properties attached to the domains themselves. In the future, we plan to use a full theorem prover along the lines of Nuprl [1] or Ontic [7].

3.1 Domain Hierarchy

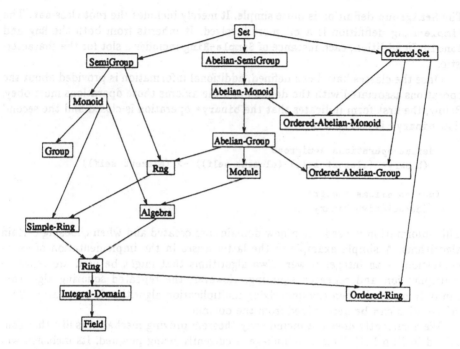

Fig. 1. Section of Weyl's Domain Hierarchy

The domain hierarchy of Weyl closely follows the conventions of modern algebra and is modeled on the category hierarchy used in Axiom [5]. The root of the hierarchy is the class Set. A section of the hierarchy is shown in Figure 1. As in Axiom, distinct classes of domains are used in the abelian (commutative) and non-abelian cases. This allows us to be a bit more explicit about the operations they use. That is, the binary operation of a Group is *, while the operation of an Abelian-Group is +.

Actually, the *binary* operations are binary+ and binary*. However, to match Common Lisp's usage, where + and * are *n*-ary functions, + and * are macros that expand into repeated applications of binary+ and binary* respectively. For example,

$$(+ \ a \ b \ c) = (binary+ \ a \ (binary+ \ b \ c)).$$

Similar transformations are done for *, -, / and other binary functions.

Domains are defined using three forms. The CLOS defclass form is used to define the class used for the domain, *e.g.*, the following code is used to define a Semigroup and a Simple-Ring.

```
(defclass Semigroup (set)
  ())

(defclass Simple-Ring (rng monoid)
  ((characteristic :initarg :characteristic
                   :reader ring-characteristic)))
```

The Semigroup definition is quite simple. It merely includes the root class set. The Simple-Ring definition is a bit more involved. It inherits from both the Rng and Monoid. In addition, each instance of Simple-Ring includes a slot for the characteristic.

Once the classes have been defined, additional information is provided about the operations associated with the domain and the axioms those operations must obey. Below, the first form indicates that the binary* operation is closed and the second that binary* is associative.

```
(define-operations semigroup
  (binary* (element self) (element self)) -> (element self))

(define-axioms semigroup
  (associative binary*))
```

This information is used when new domains are created and when choosing certain algorithms. A simple example of the latter arises in the implementation of exponentiation to an integer power. Two algorithms that might be used are repeated multiplication and repeated squaring. However, the repeated squaring algorithm can only be used when the underlying multiplication algorithm is associative. This information can be determined from the domain.

Weyl currently does not include any theorem proving mechanisms like that contained in Nuprl [1]. This type linkage is currently being pursued. Its inclusion will make the reasoning processes of the previous paragraph much more powerful.

None of the domains in Figure 1 can actually be instantiated. They are "abstract domains" that are intended to be included in instantiable domains. For instance, the domain used for polynomial rings inherits from the Ring domain and the domain used for the rational integers includes the Unique-Factorization-Domain class.

3.2 Morphisms

Morphisms are maps between domains and are essential to the coercion mechanisms used in Weyl. Bare Lisp functions could have been used to map elements of one domain into another. However, we have decided to encapsulate the mapping functions in an object so that additional information can be attached to the map itself. This also allows us to take advantage of the polymorphism and overloading mechanisms provided by CLOS.

Morphisms have two user visible components, the domain of the morphism and the range of the morphism. The printed representation of a morphism from A to B is A->B. Like other objects in Weyl, morphisms have a structural type and domain type. The domain type of a homomorphism A->B is $\text{Hom}(A, B)$, the set of morphisms of A to B. The structural type depends on how it is implemented. At this time there are

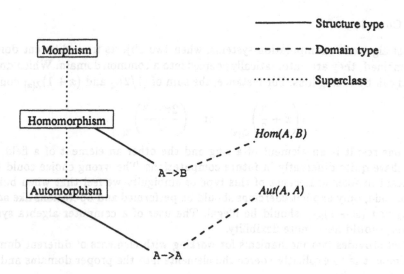

Fig. 2. Piece of Morphism Domain Hierarchy

only three choices, morphism, homomorphism and automorphism. This is illustrated in Figure 2.

A few abstract classes of morphisms have been provided to indicate that more is known about the map than it is a simple morphism. In particular, a morphism can be a homomorphism, isomorphism or automorphism. Finer structure, such as if the morphism is injective, bijective, continuous, exact, etc. awaits further work.

Some Morphisms are created automatically when new domains are created. For instance, when the polynomial ring $k[x, y, z]$ is created from k, a homomorphism is crated that embeds elements of k in $k[x,y,z]$, *i.e.* k->k[x, y, z]. Similarly, when F_{163} (the finite field of 163 elements) is created the homomorphism Z->GF(163) is also created.

The function apply-morphism applies a morphism to an element of its domain and yields an element of its range. The main operation on morphisms is compose, which constructs a new morphism that maps elements of the domain of its first argument into the range of its second. Thus,

```
> (compose Z->GF(163) GF(163)->GF(163)[x, y, z])
Z->GF(163)[x, y, z]
```

In the future, we plan to develop image, kernel and co-kernel routines, which will create new algebraic domains. Needless to say, computing the image and kernel of an arbitrarily defined morphism can be difficult or impossible. However, when the morphism is defined by reducing by an ideal, by localization or some combination of these two operations image, kernel and co-kernel computations can often be made effective.

3.3 Coercions

In most algebraic manipulation systems, when two objects from different domains are combined, they are automatically coerced into a common domain. Which domain is used can be ambiguous. For instance, the sum of $(1/2)_Q$ and $(x+1)_{Z[x]}$ could be either

$$\left(x + \frac{3}{2}\right)_{Q[x]} \quad \text{or} \quad \left(\frac{2x+3}{2}\right)_{Z(x)}.$$

Since one result is an element of a ring and the other an element of a field, they will behave quite differently in future computations. The wrong choice could cause significant problems. Because of this type of ambiguity we feel that when building system code, only explicit coercions should be performed and operations like adding $(1/2)_Q$ and $(x+1)_{Z[x]}$ should be illegal. The user of a computer algebra system, however, should have more flexibility.

Weyl provides two mechanisms for working with elements of different domains. The simplest is to explicitly coerce the elements into the proper domains and then combine them, *i.e.*,

$$(\text{+ (coerce } (1/2)_Q \text{ } Q[x]) \text{ (coerce } (x+1)_{Z[x]} \text{ } Q[x]))$$

The coerce function searches the set of homomorphisms that have been defined to determine a set of homomorphisms whose composition goes from the domain of the first argument to the target domain. If only one such set is found, then the composition is applied. If more than one such set is found, then an error is signaled because of the ambiguity.

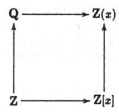

Fig. 3. Sample Computational Context

The second approach is used by applications developers. When a binary operation is applied to two elements of different domains, and a switch is set, Weyl attempts to find a minimal common domain into which both elements can be coerced. If such a domain can be uniquely determined, the operation will be performed there.

The set of domains and homomorphisms that are in force at any point in time is called a *computational context*. Recalling the previous problem of adding $(1/2)_Q$ and $(x+1)_{Z[x]}$, if the user has established the computational context shown in Figure 3 then the value of the sum will be well defined. Application specific systems that use the Weyl substrate use different computational contexts for different problems.

4 Domain Creation

Domains are created in Weyl using functions called *functors*. Examples of these functors are get-polynomial-ring and get-quotient-field, which generates a domain that is the quotient field of its argument. Functors are coded as regular CLOS methods. For instance, the following forms are used to define get-quotient-field.

```
(defmethod get-quotient-field ((ring Field))
  ring)

(defmethod get-quotient-field ((ring Integral-Domain))
  (let ((qf (make-instance 'quotient-field :ring ring)))
    (with-slots (zero one) qf
      (setq zero (make-quotient-element qf (zero ring) (one ring)))
      (setq one (make-quotient-element qf (one ring) (one ring))))
    qf))
```

The first form takes care of the trivial case of the quotient field of a field. The second form creates quotient fields from integral domains. There is no get-quotient-field constructor for more general rings, since the resulting structure is not necessarily a field. In addition, we have decided to cache the values of zero and one in the quotient field for efficiency reasons.

If there is a gcd routine for the ring, then the following, more efficient routine is used. The elements of efficient-quotient-field use the gcd routine to reduce the numerator and denominator of quotient elements to lowest terms.

```
(defmethod get-quotient-field ((ring Gcd-Domain))
  (let ((qf (make-instance 'efficient-quotient-field :ring ring)))
    (with-slots (zero one) qf
      (setq zero (make-quotient-element qf (zero ring) (one ring)))
      (setq one (make-quotient-element qf (one ring) (one ring))))
    qf))
```

Algebraic extensions provide additional complications. Consider the case of extending $\mathbf{Q}[\sqrt{2}]$ by a zero of $f(x) = x^4 - 10x^2 + 1$. Although $f(x)$ is irreducible over $\mathbf{Q}$, over $\mathbf{Q}[\sqrt{2}]$ it factors into

$$f(x) = (x^2 - 2\sqrt{2}x - 1)(x^2 + 2\sqrt{2}x - 1).$$

(The zeroes of $f(x)$ are $\pm\sqrt{2} \pm \sqrt{3}$.) There are several potential problems here. The algebraic extension constructor could return one of several different results. In the following table we have indicated the zero of $f(x)$ that would be returned, the image of $\sqrt{2}$ in the resulting field and field that would contain both.

zero	$\sqrt{2}$	field
α	$\sqrt{2}$	$A = \mathbf{Q}[\sqrt{2}][\alpha]/(\alpha^4 - 10\alpha^2 + 1),$
β	$\sqrt{2}$	$B = \mathbf{Q}[\sqrt{2}][\beta]/(\beta^2 - 2\sqrt{2}\beta - 1),$
γ	$\frac{\gamma^3 - 9\gamma}{2}$	$C = \mathbf{Q}[\gamma]/(\gamma^4 - 10\gamma^2 + 1),$
β, β'	$\sqrt{2}, \sqrt{2}$	$B' = \{\mathbf{Q}[\sqrt{2}][\beta]/(\beta^2 - 2\sqrt{2}\beta - 1),\ \mathbf{Q}[\sqrt{2}][\beta']/(\beta'^2 + 2\sqrt{2}\beta' - 1)\}.$

The first answer has the problem that A is not a field since it contains zero divisors. In the second answer, B is a field but we have arbitrarily chosen to include a root of one of the two factors of $f(x)$. Furthermore, B is not generated by a primitive element over $\mathbf{Q}$. This can complicate some algorithms. C is generated by primitive element over $\mathbf{Q}$, and has the proper degree over $\mathbf{Q}$. However, it also has selected a root from a particular quadratic factor of $f(x)$. Which zero is chosen depends upon the embedding of γ and $\mathbf{Q}[\sqrt{2}]$ in C.

In the final answer, two roots (and two fields) are returned corresponding to the two factors of $f(x)$. This is essentially the approach taken by Duval [2]. Notice that to provide this answer it is necessary to create two distinct, but isomorphic fields.

5 Number Types

The different numerical types provided by Weyl illustrate some of the issues that arise in symbolic computations as we have organized it. The domains we will consider are the rational integers, $\mathbf{Z}$, the rational numbers, $\mathbf{Q}$, the real numbers, $\mathbf{R}$, and the complex numbers $\mathbf{C}$. The domain $\mathbf{Z}$ is a Ring, (*i.e.*, commutative ring with unit), while $\mathbf{Q}$, $\mathbf{R}$ and $\mathbf{C}$ are Fields.

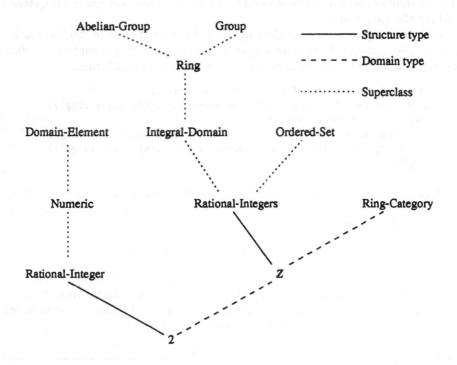

Fig. 4. Structure/Domain hierarchy for $2_\mathbf{Z}$

Each element of the rational integers, $\mathbf{Z}$, is represented by an instance of the

CLOS class Rational-Integer:

```
(defclass rational-integer (numeric)
    ((value :initarg :value
            :reader integer-value)))
```

The class Numeric is the base class for all domain elements that are numbers. One slot is provided to hold the Lisp number that is used to represent the rational integer. (In a language that does not have built-in variable precision integers, this structure would be more complex.)

A similar structure type is used to represent floating point numbers in R. However, since Z is a sub-ring of R we have chosen to represent exact integers in R using rational-integer. Thus there can exist two instances of rational-integer that come from completely different domains. Notice that once again we have a domain whose elements can have more than structure type.

Figure 4 illustrates the structural and domain type hierarchies that lie over 2_Z. The structure types are indicated by solid lines and the domain types by dashed lines. Thus the integer 2_Z has the structure of a Rational-Integer, while Z has the structure of Rational-Integers. The domain of 2_Z is Z, while the domain of Z is the category of rings.[3]

Lying above Rational-Integer and Rational-Integers is the CLOS class hierarchy that is used to implement them. The most interesting is that lying over Rational-Integers. Z is an ordered integral domain. In Common Lisp the predicate (typep *obj type*) can be used at run-time to determine if *obj* is of type *type*. Because Rational-Integers inherits from the domains shown, we have

```
(typep Z 'Abelian-Group) ⟵ T
(typep Z 'Ordered-Set)   ⟵ T
(typep Z 'Field)         ⟵ NIL (false)
```

In this fashion, we can use the type inheritance structure of CLOS to provide some very rudimentary reasoning mechanisms about domains. When a real theorem prover is added to Weyl this will not be necessary and there may be further changes.

Methods are provided for the basic operations with rational integers. For instance, a simple version of the binary addition operator might be defined as

```
(defmethod binary+ ((x Rational-Integer) (y Rational-Integer))
  (make-element (domain-of x)
                (lisp:+ (integer-value x) (integer-value y))))
```

This method returns an element of the domain of x, which need not be isomorphic to Z. The function lisp:+ refers to the built in Lisp routine for adding variable precision integers.

However, since x and y are not necessarily elements of the same domain, this approach is not strictly correct. Instead, we have chosen to use the following approach

```
(defmethod binary+ :around ((x Rational-Integer) (y Rational-Integer))
  (let ((domain (domain-of x)))
```

---

[3] The category of rings is not currently implemented. Numerous issues will arise when this is done that we have not yet resolved.

```
(cond ((eql domain (domain-of y))
       (make-element domain
                     (lisp:+ (integer-value x) (integer-value y))))
      (t (call-next-method x y)))))
```

This :around method ensures that x and y are elements of the same domain. If they are not, then this routine "passes the buck" to whoever else might have an idea on how to add two rational-integers.

The buck stops with the following default routine:

```
1  (defmethod binary+ ((x domain-element) (y domain-element))
2    (when (null *coerce-where-possible*)
3      (error "No way to add ~S and ~S" x y))
4    (let ((domain-x (domain-of x))
5          (domain-y (domain-of y))
6          common-domains)
7      (when (eql domain-x domain-y)
8        (error "No way to add ~S and ~S" x y))
9      (setq common-domains (find-common-domains domain-x domain-y))
10     (cond ((null common-domains)
11            (error "No way to add ~S and ~S" x y))
12           ((null (rest common-domains))
13            (binary+ (coerce x (first common-domains))
14                     (coerce y (first common-domains))))
15           (t (error "Ambiguous coercion for addition (~S, ~S)"
16                     x y)))))
```

Notice that the specializers are domain-element, so this method is the default method for addition of any two elements of Weyl domains.

The switch *coerce-where-possible* enables automatic coercion between domains, as described in Section 3.3. In most system code this switch is turned off, and an error is signaled, indicating that one cannot add two integers from different domains (lines 2 and 3).

The routine find-common-domains tries to determine the smallest domains into which both x and y can be coerced. If there is just one such domain then both x and y are coerced into the common domain and the addition is tried again (lines 9–14). If there is more than one common domain, an error is signaled (lines 15–16).

Finally, if this routine is invoked when two elements are already in the same domain, but there is no binary+ method, then an error is signaled (lines 7 and 8). These lines eliminate the possibility of infinite recursion. Similar routines are present for most binary operations.

5.1 Structural Type and Domain Type Orthogonality

Now consider the elements of Q. Rational numbers like 1/3 and 3/7 are represented using the following CLOS class:

```
(defclass rational-number (numeric)
  ((numerator :accessor qo-numerator
              :initarg :numerator)
```

```
(denominator :accessor qo-denominator
             :initarg :denominator)))
```

However, we have chosen to represent the integers in **Q** using the `Rational-Integer` class. Thus elements of **Q** can have one of two different representations: as `Rational-Integers` or as `Rational-Numbers`. Thus we could have two elements, 2_Z and 2_Q, that use the same representation but that are elements of different domains.

However, we now have an interesting problem. If we try to divide 2_Q by 3_Q we should get $2/3_Q$. But, under most circumstances, the quotient of 2_Z by 3_Z should signal an error. In this case, the class based dispatching used by CLOS fails us, since it only pays attention to the structural types of 2_Z and 3_Z. Both of these quantities have the same structural types as 2_Q and 3_Q, for which division is proper.

Weyl deals with this problem by including additional checks in the quotient routine:

```
(defmethod binary/ :around ((x rational-integer) (y rational-integer))
  (let ((domain (domain-of x)))
    (cond ((not (eql domain (domain-of y)))
           (call-next-method x y))
          ((1? y) x)
          ((typep domain 'field)
           (make-element domain
                         (lisp:/ (integer-value x) (integer-value y))))
          (t (call-next-method x y)))))
```

It would be far better if the domain type could be used in the generic function dispatch.

6 Conclusions

Weyl was designed to be used as a component of larger software systems. Its initial implementation in Common Lisp demonstrates that when properly used, existing objected oriented programming technology is sufficient to write fully functorial symbolic computing code with relatively little overhead, either to the programmer or to the user. Only 5 pages of additional "system code" is required to support Weyl. Furthermore, the current performance has proven to be adequate for many applications, even though we have been relying on CLOS for all of the required dispatching.

Though the type system used by Weyl is supported at run time, there is no reason why it could not be supported by the compiler. This would provide all the advantages of strong typing, and might improve performance somewhat. We have opted for the run time version used here for convenience in development.

Much of this work is the direct result of discussions with members of the IBM symbolic computing group, particularly Barry Trager and Dick Jenks, whose assistance I gratefully acknowledge.

References

1. R. L. CONSTABLE, S. F. ALLEN, H. M. BROMLEY, W. R. CLEAVELAND, J. F. CREMER, R. W. HARPER, D. J. HOWE, T. B. KNOBLOCK, N. P. MENDLER, P. PANANGADEN, J. T. SASAKI, AND S. F. SMITH, *Implementing Mathematics with the Nuprl Proof Development System*, Prentice-Hall, Englewood Cliffs, NJ, 1986.

2. C. DICRESCENZO AND D. DUVAL, *Algebraic extensions and algebraic closure in scratchpad ii*, in ISSAC '88, P. Gianni, ed., vol. 358 of Lecture Notes in Computer Science, Berlin-Heidelberg-New York, 1988, Springer-Verlag, p. ??

3. J. DONGARRA, J. R. BUNCH, C. B. MOLER, AND G. W. STEWART, *LINPACK User's Guide*, SIAM Publications, Philadelphia, PA, 1978.

4. A. C. HEARN, *Reduce 3 user's manual*, tech. rep., The RAND Corp., Santa Monica, CA, 1986.

5. R. D. JENKS AND R. S. SUTOR, *AXIOM: The Scientific Computation System*, Springer-Verlag, New York and NAG, Ltd. Oxford, 1992.

6. MAPLE GROUP, *Maple*, Waterloo, Canada, 1987.

7. D. A. MCALLESTER, *Ontic: A Knowledge Representation System for Mathematics*, MIT Press, Cambridge, MA, 1987.

8. C. B. MOLER, *Matlab user's guide*, Tech. Report CS81-1, Dept. of Computer Science, University of New Mexico, Albuquerque, NM, 1980.

9. B. T. SMITH, J. M. BOYLE, Y. IKEBE, V. C. KLEMA, AND C. B. MOLER, *Matrix Eigensystem Routines: EISPACK Guide*, Springer-Verlag, New York, NY, second ed., 1976.

10. G. L. STEELE JR., *Common Lisp, The Language*, Digital Press, Burlington, MA, second ed., 1990.

11. SYMBOLICS, INC., *MACSYMA Reference Manual*, Burlington, MA, 14th ed., 1989.

12. S. WOLFRAM, *Mathematica: A System for Doing Mathematics by Computer*, Addison-Wesley, Redwood City, CA, 1988.

A   A Session with Weyl

This section gives a short example of how Weyl can be used to perform a simple algebraic calculation, calculating the coefficients of the "F and G series." These coefficients are defined by the following recurrences:

$$f_0 = 0$$

$$f_n = -\mu g_{n-1} - \sigma(\mu + 2\epsilon)\frac{\partial f_{n-1}}{\partial \epsilon} + (\epsilon - 2\sigma^2)\frac{\partial f_{n-1}}{\partial \sigma} - 3\mu\sigma\frac{\partial f_{n-1}}{\partial \mu}$$

$$g_0 = 1$$

$$g_n = f_{n-1} - \sigma(\mu + 2\epsilon)\frac{\partial g_{n-1}}{\partial \epsilon} + (\epsilon - 2\sigma^2)\frac{\partial g_{n-1}}{\partial \sigma} - 3\mu\sigma\frac{\partial g_{n-1}}{\partial \mu}$$

Each of the f_i and g_i are polynomials in μ, ϵ and σ with integer coefficients and can be viewed as elements of the domain $Z[\mu, \epsilon, \sigma]$.

Our first task is to generate the *domain* $Z[\mu, \epsilon, \sigma]$. In this problem, not much is required of the domain, it is only used to generate the desired polynomials. This done by first getting a copy of the rational integers, Z, using the function get-rational-integers. The polynomial ring $Z[\mu, \epsilon, \sigma]$ is then created using get-polynomial-ring:

```
> (setf R (get-polynomial-ring (get-rational-integers) '(m e s)))
Z[m, e, s]
```

where we have used the symbols m, e and s to represent the greek letters μ, ϵ and σ. The line proceeded by the ">" is what is typed into Lisp by the user. The following line is Weyl's response.

At this point we can create polynomials that are elements of the domain R. To do this it is convenient to bind the Lisp variable mu to μ, eps to ϵ and sigma to σ. This is done by coercing the symbols m, e and s into polynomials in the domain R.

```
> (setf mu (coerce 'm R))
m
> (setf eps (coerce 'e R))
e
> (setf sigma (coerce 's R))
s
```

The coefficients of the first terms of the both recurrences are the same, so rather than computing them afresh each time, they are stored in global variables

```
> (setq x1 (- mu))
-1 m
> (setq x2 (* (- sigma) (+ mu (* 2 eps))))
-1 s m + -2 s e
> (setq x3 (+ eps (* -2 (expt sigma 2))))
e + -2 s^2
> (setq x4 (* -3 mu sigma))
-3 s m
```

Notice that the usual Lisp functions for performing arithmetic operations are used, and normal Lisp variables can have polynomials as values. This is one of the convenient features of Weyl. We have simply extended the basic Lisp functions to deal with the algebraic objects of Weyl.[4] Also note that the values are printed in a linear infix notation so that can be easily interpreted, although the output form could be improved upon.

We can now write the recursion formulae.

```
(defun f (n)
  (if (= n 0) (coerce 0 R)
      (+ (* x1 (g (1- n)))
         (* x2 (partial-deriv (f (1- n)) eps))
         (* x3 (partial-deriv (f (1- n)) sigma))
         (* x4 (partial-deriv (f (1- n)) mu)))))
```

Notice that the function partial-deriv is used to compute the partial derivative of a polynomial. The recursion formula for g_n is similar.

Finally, we can compute the coefficients by invoking the f function.

```
> (f 3)
m^2 + (3 e + -15 s^2) m
> (f 4)
-15 s m^2 + (-45 s e + 105 s^3) m
```

This simple example illustrates that writing programs that use the Weyl substrate is not much different from writing ordinary Lisp programs. We have extended some of the data types and we have introduced the concept of domains, but the same control structures, abstractions and programming tools will continue to work.

---

[4] Actually, the symbols "+", "-" and "*" are in a different package so that a clean Common Lisp world can be preserved.

On the Uniform Representation of Mathematical Data Structures *

Carla Limongelli and Marco Temperini

Dipartimento Informatica e Sistemistica - Università "La Sapienza"
Via Salaria 113, I-00198 Roma, Italy
e-mail: {limongel,marte}@disco1.ing.uniroma1.it

Abstract. Topics about the integration of the numeric and symbolic computation paradigms are discussed. Mainly an approach through a uniform representation of numbers and symbols is presented, that allows for the application of algebraic algorithms to numeric problems. The p-adic construction is the basis of the unifying representation environment. An integrated version of the Hensel algorithm is presented, which is able to perform symbolic and numeric computations over instances of ground (concrete) and parametric structures, and symbolic computations over instances of abstract structures. Examples are provided to show how the approach outlined and the proposed implementation can treat both cases of symbolic and numeric computations. In the numeric case it is shown that the proposed extension of the Hensel Algorithm can allow for the exact manipulation of numbers. Moreover, such an extension avoids the use of simplification algorithms, since the computed results are already in simplified form.

1 Introduction

Systems for symbolic mathematics are based on the availability of powerful methods and techniques, which have been developed for numeric computation, symbolic and algebraic computation and automated deduction. But those different computing paradigms really work independently in such systems. Each of them represents an individual computing environment, while they are not integrated to support a uniform environment for computation. The problem of the integration of numeric and symbolic computation is still open [2, 8].

In this paper, a possible solution to this problem is considered, taking into account the well-known need for abstraction in the definition of mathematical data structures. Mainly we focus on the possibility of using a uniform representation of those structures and of extending the use of algebraic algorithms to numerical settings. Such a representation is based on Truncated Power Series. It allows, through the p-adic arithmetic tool, for the integration of the numeric and symbolic capabilities of algorithms defined at high level of abstraction. In

* This work has been partially supported by Progetto Finalizzato "Sistemi Informatici e Calcolo Parallelo" of CNR under grant n. 92.01604.69.

this work we describe how a uniform representation for functions and numbers, based on the p-adic construction, can be used in order to obtain a homogeneus approach to symbolic and numeric computation.

The well known Hensel algorithm is extended to cover both symbolic and numeric operations: polynomial factorization, finding of the n-th root of an analytical function, root finding of polynomial equations, and p-adic expansion of an algebraic number are examples of operations that can be performed by this extension. The extended algorithm works on a common p-adic representation of algebraic numbers and functions. In the numeric case the computations are performed by the p-adic arithmetic. Actually, numeric computations by the extended Hensel algorithm provide the exact representation of the resulting algebraic numbers, by means of their p-adic expansion.

2 A proposal of integration

The design of new generation symbolic computation systems [9, 13] has benefited from the use of abstraction [7] and, later, of object-oriented programming [1]. The use of abstraction and of hierarchical definition of mathematical data structures can improve the expressive power of symbolic computation systems and their characteristics of correctness. The abstraction supported by object-oriented techniques can be seen as primary feature of a system able to support the integration of numeric and symbolic computation.

In [10] the approach through abstraction to the classification of algebraic structures has been described; an appropriate inheritance mechanism has been defined for their implementation by an object-oriented paradigm. Abstract structures (like ring or field algebraic structures), Parametric structures (like matrix of ring elements, or polynomials over ring coefficients) and Ground structures (like integer or real) can be distinguished by the different completions of their algebraic definition. Each parametric or ground structure can be derived from the appropriate abstract structure through inheritance of features. The mechanism of strict inheritance plus redefinition also allows one structure to be derived from another.

Actually, a uniform computing environment is obtained once each algebraic structure is implemented by its appropriate class template (namely abstract, parametric or ground classes). The behavioural compatibility of data structures is fixed by their inheritance relationships. The algorithms which have been included in a (class) structure S, designed to be functional attributes for the instances of S, will be available on every other structure connected to S by inheritance. Hence, some kind of "implicit coercion" has the effect of polymorphism induced by inheritance.

But this can still be unsatisfactory. Indeed, it is possible that ground classes of the same abstract type may have incompatible representation. This would happen when two classes have no direct inheritance connection (one is not derived from the other), even if they are derived from the same abstract class. For example, numeric computations involving both real and complex numbers are not directly possible, although, from an algebraic point of view, they are instances of structures which are compatible. At present such operations can be performed only by means of "explicit coercion", that is applying special algorithms (functions) purposely devised for such borderline cases. The design of those functions implies that some peculiarities of the operands in a numeric operation must be predicted and explicitly treated, before their eventual runtime occurrence. Normally such operations are defined outside of the designed hierarchy of data types. Because of this, mathematical structures may be removed from their natural algebraic setting, making them prone to erroneous use [10]. Explicit coercions can be avoided once a uniform representation of data is provided to support numeric and algebraic computations.

The main drawback in using classical domains of numeric computing, such as floating point, is the lack of algebraic features. Once the structures used for numeric computations have a common representation and possess algebraic properties, an appropriate check is guaranteed on the operations in which they are involved.

The first requirement for our aim of integration is that each structure involved in numeric operations is an instance of a defined algebraic structure (with all its properties and characteristics). The second requirement is the uniform representation of both numbers and symbols in the computed expressions. Both requirements must be satisfied to reach an effective integration of numeric and symbolic computation.

The first requirement can be met through the use of abstraction and inheritance. The second requirement can be met by using truncated power series expansion to uniformly represent numbers and functions. Focusing on polynomial representation, it is well known that $D[[x]]$ is isomorphic to $D[x]$, for a domain D. Since any fixed analytical function can be represented by truncated power series succession, we can choose in $D[[x]]$ such approximated representation.

3   The p-adic representation

The approach to numeric computation through truncated p-adic arithmetic [6] has been widely analysed [5, 4]. As a matter of fact, the truncated p-adic representation can be viewed as the truncation of the p-adic series representing the rational number. The degree r, chosen for the approximation, depends on the characteristics of the problem. The same approach can be used to treat problems in abstract structures. In particular it is possible to exploit the uniformity between the truncated p-adic expansion of a rational number (Hensel code) and the p-adic polynomial expansion by power series. The theory and construction of Hensel Code for arithmetic of a rational function over a finite field is analogous to the truncated p-adic arithmetic. So a rational polynomial function can be represented in the form of Hensel code like the Example 1 shows.

Example 1. Given the univariate polynomial $P(x) = x^3 + 3x^2 - 2x + 1$ we fix the base x and the approximation $r = 4$, the related Hensel code is: $H(x, 4, p(x)) = (1 - 2\ 3\ 1,\ 0)$

Given the bivariate polynomial $P_1(x,y) = yx^2 - 2yx + 2y + y^2x - y^2 - 3$ we fix the base $(x-1)$ and the approximation $r = 4$, the related Hensel Code is: $H(x-1, 4, P_1(x,y)) = (y^2 \; y \; y - 3, \; 0)$ □

The means to operate on such uniformly represented structures are provided by a variety of algebraic techniques, based on the application of the Extended Euclidean Algorithm (EEA henceforth), as shown in [8]. The possibility of a unifying strategy used to treat problems by approximated p-adic construction methods and truncated p-adic arithmetic is confirmed, since both are based on the same algebraic properties of the EEA. As it is well known, the approximated p-adic construction method is founded on the following computational steps:

- starting from an appropriate initial approximation;
- computes the first order Taylor series expansion;
- solve the obtained equation;
- find an update of the solution.

Following Newton's idea [11], the Hensel method gives an iterative mechanism to solve equations of the following type: $\Phi(G, H) = 0$ where is

$$\Phi : \mathbf{D}[x] \times \mathbf{D}[x] \to \mathbf{D}[x]$$

with $\mathbf{D}$ an abstract Euclidean Domain. Starting from the initial approximations, $G_1(x)$ and $H_1(x)$ such that $\Phi(G_1, H_1) \equiv 0 \bmod I$, at each step k the approximate solution G_k, H_k can be obtained, such that $\Phi(G_k, H_k) \equiv 0 \bmod I^k$, where I is an ideal in $\mathbf{D}[x]$. Then the bivariate Taylor series expansion of Φ at the point (G_k, H_k), is computed as:

$$\Phi(G, H) = \Phi(G_k, H_k) + (G - G_k)\frac{\partial \Phi(G_k, H_k)}{\partial G} + (H - H_k)\frac{\partial \Phi(G_k, H_k)}{\partial H}.$$

Dividing by p^k and assuming

$$C = \Phi(G_k, H_k), \quad A_k = -\frac{\partial \Phi(G_k, H_k)}{\partial G}, \quad B_k = -\frac{\partial \Phi(G_k, H_k)}{\partial H},$$

$$\frac{G - G_k}{p^k} = \Delta G_k, \quad \frac{H - H_k}{p^k} = \Delta H_k,$$

we solve the following equation

$$A_k \Delta G_k + B_k \Delta H_k \equiv \frac{C}{p^k} \bmod p \tag{1}$$

in order to find the $(k+1)$-th updates

$$G_{k+1} = G_k + p^k B_k, \quad H_{k+1} = H_k + p^k A_k.$$

Here G_i (respectively H_i) stands for the sum of the first i terms in the p-adic series expansion of G (respectively H). This derivation is precisely established by the inductive construction based on the following lemma:

Lemma 1 (Hensel lemma). *Given a prime number $p \in \mathbf{Z}$, and given $F(x) \in \mathbf{Z}[[x]]$, if $G_1(x)$ and $H_1(x)$ are two polynomials that are relatively prime in $\mathbf{Z}_p[x]$ such that $F(x) \equiv G_1(x) \cdot H_1(x)$ mod p then, for every integer $k \equiv 1$ two polynomials, $G_k(x), H_k(x) \in \mathbf{Z}_{p^k}[x]$, do exist such that $F(x) \equiv G_k(x) \cdot H_k(x)$ mod p^k where $G_k \equiv G_1(x)$ mod p and $H_k \equiv H_1(x)$ mod p.*

The proof of this lemma is found in [14]. This lemma ensures that, starting from an appropriate initial approximation, we can obtain higher order approximations.

The generality of this method makes it an apt tool to deal with approximated representations in a uniform way and in the same algebraic context. The Hensel algorithm can perform different computational methods according to different actual specializations of the parameters G and H (i.e. different forms of the equation and of the initial approximations). The following scheme provides some examples:

(i) Factorization: $\qquad\qquad\qquad \Phi(G, H) = F - GH = 0;$

(ii) n-th root of a function G: $\qquad \Phi(G) = F - G^n = 0;$

(iii) Legendre polynomial determination: $\Phi(G, t) = (1 - 2x + t^2)G^2 - 1 = 0;$

(iv) Newtons method: $\qquad\qquad\qquad \Phi(x) = F = 0.$

A more detailed description of the Hensel algorithm is reported in [10]. At each step we rebuild the p-adic coefficients related to series expansion of the solution. We must note that the hypothesis of the relative primality of G_1 and H_1 in the previous lemma is necessary only to solve the above diophantine equation (1). When F has a specialized form, like in (ii), (iii) and (iv), the equation to be solved is just a modular univariate equation. In these cases the hypothesis of relative primality is no longer needed. A unifying view of different data structures and algebraic methods comes out by their treatment through a generalized p-adic representation. The Figure 1 shows a set of application areas of the Hensel algorithm, distinguishing the numeric from the symbolic cases.

The p-adic representation allows for a unified form to treat numbers and functions by truncated power series, and it constitutes the mathematical background for the definition of basic abstract data structures for our unique computing environment [12].

4   Hierarchies of structures and methods

Once the treatment of numeric data structures is founded on the p-adic representation and arithmetic, it is possible to devise a hierarchy of data structures, which represent the computational domain of symbolic (and numeric) computations. An approach, based on abstraction, to the classification of algebraic structures has been described in [10]. Abstract structures (like ring or field algebraic structures), Parametric structures (like matrix of ring elements, or polynomials

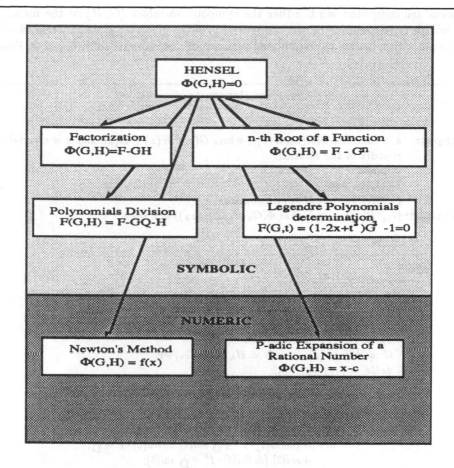

Fig. 1. Possible specializations of the Hensel method.

over ring coefficients) and Ground structures (like integer or real) can be distinguished by the different completion of their algebraic definition. Among all these structures, the mechanism of strict inheritance plus redefinition allows to derive each one by another. Following this approach, we can locate in the appropriate planes the algebraic methods and structures, which have been cited previously as main devices in the unified view of symbolic and numeric computations. The first structure is the truncated power series. It is straightforwardly a parametric structure whose coefficients are defined over an Euclidean domain. In Fig. 2 a specification of the Hensel algorithm is given. Notice that $+_D$, $=_D$, 0_D, represent, respectively, the addition operation, the equality relation and the zero element in the domain D. The function call $EVAL(x = xk, H = Hk, G = Gk, F)$, evaluates Φ in the point (G_k, H_k, x_k). The function $SOLVE$ computes the expansion in Taylor series of a function F, and the related zeros. In particular these zeros are computed w.r.t. either the symbolic variables (G, H) or the numeric variable (x) depending on the configuration of the given input parameters. Hence the algorithm is able to compute either symbolic or numeric solutions of Φ. From

---

EXTENDED HENSEL ALGORITHM

Input: a: a function $\Phi(G(x), H(x))$, where $G(x),\ H(x) \in \mathbf{D}[x]$, and x is a variable;
            r: order of the solution;
            n: possible exponent of G;
            $I \in \mathbf{D}[x]$: base;
            G_0, H_O, x_0: initial approximations;
Output: $G_k,\ H_k,\ x_k$, such that $\Phi(G_k(x_k), H_k(x_k)) \equiv 0\ mod\ I^k$;

Begin
    $k := 1$;
    $\Phi := F -_{\mathbf{D}} G^n \cdot H$;
    $sol := [x_0, G_0, H_0]$;
    $a[1] := sol$;
    $x_k := x_0$; $H_k := H_0$; $G_k := G_0$;
    $C := EVAL(x = x_k,\ H = H_k,\ G = G_k, \Phi)$;
    $delta := [{}^0{\mathbf{D}}, {}^0{\mathbf{D}}, {}^0{\mathbf{D}}]$;
    While$(k \le r) \wedge (C \neq_{\mathbf{D}} {}^0{\mathbf{D}})$
        do
            $SOLVE(F, G, x, x_0, G_0, H_0, \Phi, C, I)$;
            $a[k + 1] := delta$;
            $sol := [delta[1] \cdot I^k +_{\mathbf{D}} sol[1], delta[2] \cdot I^k +_{\mathbf{D}};$
                $+sol[2], [delta[3] \cdot I^k +_{\mathbf{D}} sol[3]$;
            $x_k := sol[1]$; $G_k := sol[2]$; $H_k := sol[3]$;
            $k := k + 1$;
            $C := EVAL(x = x_k,\ H = H_k,\ G = G_k, \Phi)$;
        od
End;

Fig. 2. Specification of the extended Hensel algorithm.

---

the previous description we can see that the Hensel method takes polynomials as input and returns either a truncated power series or a Hensel code. It applies the EEA (in $SOLVE$, in order to solve either the diophantine equation, or the modular equation). It should be located where all the properties of an Euclidean Domain are available (see Fig. 3). Hence the Euclidean Domain is the structure of highest level of abstraction, where the Hensel method can be defined. Let us note that, in this case, the algorithm is not a characterizing attribute of the abstract structure of Euclidean Domain; in fact it is added to the definition, just enriching it.

The p-adic arithmetic is represented by the Hensel-Code which is conveniently located in the plane of ground structures. Hensel codes are particular instances of truncated power series whose coefficients are defined over $\mathbf{Z}_p$ (p a prime number).

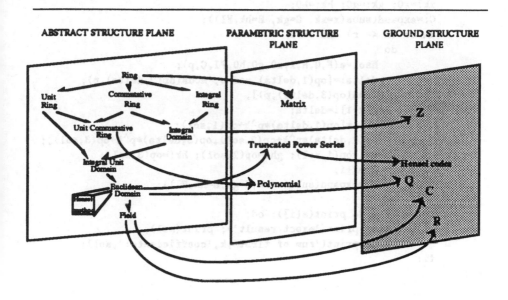

ABSTRACT STRUCTURE PLANE

PARAMETRIC STRUCTURE PLANE

GROUND STRUCTURE PLANE

Fig. 3. Specification levels.

5   Use of an integrated algorithm in the numeric case

In this section we propose a first implementation of the "integrated" algorithm of Fig. 2. The programming language included in the system MAPLE [3] has been used. The procedure **hgen** implements the plain Hensel algorithm. In it the procedure **hsolve**, which we do not show here, performs the selection between numeric and symbolic case. Actually, as in every other existing system for symbolic computation, the specification of data structure is not allowed as described in Sect. 3. So we had to design special structures in order to simulate organization of abstract parametric and ground planes. In the rest of this section we will provide examples. Firstly we show how the specialization of the Hensel algorithm in the numeric case can provide error-free results.

```
hgen := proc(F,G,H,x,x0,g0,h0,n,p,r);
        a:=array(1..r+1);
        k:=1; FI:=F-G^ n*H;
        sol:= [x0,g0,h0]; a1:=sol;
        xk:=x0; gk:=g0; hk:=h0;
        C:=expand(subs(x=xk, G=gk, H=hk,FI));
        while (k <= r) and C<>0
             do
                hsolve(F,G,H,x,x0,g0,h0,FI,C,p);
                delta:=[op(1,delta) mod p, mods(op(2,delta),p);
                mods(op(3,delta),p)];
                a[k+1]:=delta;
                sol:=[op(1,delta)*p^k+op(1,sol);
                op(2,delta)*p^k+op(2,sol),op(3,delta)*p^k+op(3,sol)];
                xk:=op(1,sol); gh:=op(2,sol); hk:=op(3,sol);
                k:=k+1;
                C:=expand(subs(x=xk,G=gk,H=hk,FI));
             od
        for i to k do print(a[i]); od;
        if C=0 then print('exact result'); print(sol);
                else print('sum of first',k,'coefficients:',sol);
        fi;
end;
```

Fig. 4. Implementation of extended Hensel algorithm.

Example 2. Finding the root of $x^2 - 2$, means to obtain the exact representation of the number $\sqrt{2}$. Let be $F(x) = x^2 - 2$.

Input:
$$F(x); \ G = 0; \ H = 0; \ p = 7; \ G_0 = 0; \ H_0 = 0; \ x_1 = 3; \ r = 3;$$
$$(F(3) \equiv 0 \ mod \ 7)$$
$$\Delta x_0 = 3; \ x_1 = 3;$$

We want to compute its roots in $\mathbb{Z}_7[x]$. Following the computational steps described in the extended Hensel algorithm we obtain:

1st iteration
$$F(x) = F(x_1) + (x - x_1) \cdot F'(x_1)$$
$$\frac{F(3)}{7} + \frac{x-3}{7} \cdot F'(3) \equiv 0 \ mod \ 7$$
$$1 + \Delta x_1 \cdot 6 \equiv 0 \ mod \ 7 \ \Delta x_1 = 1, \ x_2 = 3 + 1 \cdot 7^1 = 10$$

2nd iteration
$$\frac{F(10)}{7^2} + \frac{x-10}{7^2} \cdot F'(10) \equiv 0 \ mod \ 7$$
$$2 + \Delta x_2 \cdot 6 \equiv 0 \ mod \ 7 \ \Delta x_2 = 2, \ x_3 = 10 + 2 \cdot 7^2 = 108$$

3rd iteration
$$\frac{F(108)}{7^3} + \frac{x-108}{7^3} \cdot F'(108) \equiv 0 \bmod 7$$
$$6 + \Delta x_2 \cdot 6 \equiv 0 \bmod 7 \quad \Delta x_3 = 6, \quad x_3 = 108 + 6 \cdot 7^3 \in \mathbf{Z}_{7^4}[x].$$

At the end of the iterations the algorithm gave :

$$x_4 = \sum_{i=0}^{3} \Delta x_i \cdot p^i = 3 + 1 \cdot 7 + 2 \cdot 7^2 + 6 \cdot 7^3$$

and the coefficients of this p-adic expansion (3 1 2 6) represent the mantissa of the Hensel code related to the irrational number $\sqrt{2}$ in the p-adic arithmetic. In fact, in this arithmetic, (3 1 2 6, 0) × (3 1 2 6, 0) = (2 0 0 0, 0), which is the representation of the number 2 in the integer arithmetic. □

Example 3. Following a procedure like in the previous example the computation of $F(x) = x^2 - 5$ is briefly shown:
Input:

$$F(x); \ G = 0; \ H = 0; \ p = 11; \ G_0 = 0; \ H_0 = 0; \ x_1 = 7; \ r = 3;$$

Output:

$$x_3 = \Delta x_0 \cdot p^0 + \Delta x_1 \cdot p^1 + \Delta x_2 \cdot p^2 + \Delta x_3 \cdot p^3 = 7 \cdot 11^0 + 6 \cdot 11^1 + 0 \cdot 11^2 + 6 \cdot 11^3 = 8059.$$

The coefficients of this 11-adic expansion, (7 6 0 6), represent the number $\sqrt{5}$ in the p-adic arithmetic. It is simple to verify that, in this arithmetic, is (7 6 0 6, 0) × (7 6 0 6, 0) = (5 0 0 0, 0), which is the representation of the integer number 5. □

The importance of dealing with exact results is well known: an example is the following case, involving the analytical expression of the n-th number:

$$F_n = \frac{1}{\sqrt{5}} \cdot \left[\left(\frac{1 + \sqrt{5}}{2} \right)^n - \left(\frac{1 - \sqrt{5}}{2} \right)^n \right]$$

By using a p-adic representation, we are guaranteed to obtain the exact result, say for F_3, with $p = 11$, $r = 4$ and $\sqrt{5}$ represented by (7 6 0 6, 0).

$$F_3 = \frac{(1\,0\,0\,0,\ 0)}{(7\,6\,0\,6,\ 0)} \cdot$$

$$\left[\left(\frac{(1\,0\,0\,0,\ 0) + (7\,6\,0\,6,\ 0)}{(2\,0\,0\,0,\ 0)} \right)^3 - \left(\frac{(1\,0\,0\,0,\ 0) - (7\,6\,0\,6,\ 0)}{(2\,0\,0\,0,\ 0)} \right)^3 \right],$$

$$F_3 = \frac{(1\,0\,0\,0,\ 0)}{(7\,6\,0\,6,\ 0)} \cdot \left[\left(\frac{(8\,6\,0\,6,\ 0)}{(2\,0\,0\,0,\ 0)} \right)^3 + \left(\frac{(6\,6\,0\,6,\ 0)}{(2\,0\,0\,0,\ 0)} \right)^3 \right],$$

$$F_3 = \frac{(1\,0\,0\,0,\ 0)}{(7\,6\,0\,6,\ 0)} \cdot \left[(4\,3\,0\,3,\ 0)^3 + (3\,3\,0\,3,\ 0)^3 \right]$$

where $(4\,3\,0\,3,\,0)^3 = (9\,6\,0\,6,\,0)$ and $(3\,3\,0\,3,\,0)^3 = (5\,6\,0\,6,\,0)$

$$F_3 = \frac{(1\,0\,0\,0,\,0)}{(7\,6\,0\,6,\,0)}\cdot[(9\,6\,0\,6,\,0) + (5\,6\,0\,6,\,0)\,], \quad F_3 = \frac{(3\,2\,1\,1,\,0)}{(7\,6\,0\,6,\,0)} = (2\,0\,0\,0,\,0)$$

It is worthwhile to stress that the previous computation cannot be performed by the existing systems, without applications of some simplification algorithm. In fact this is not needed once the proposed p-adic approach is followed.

Other examples can be shown here, about the application of the "integrated" Hensel algorithm to numeric computation problems.

Example 4. In this example we find, starting from an appropriate initial approximation, a real root of the following polynomial: $f(x) = 6x^2 - 11x + 3$, the zeros of which are 3/2 and 1/3. Starting from $x_0 = 2$, $p = 5$, and choosing three iterations for the algorithm, at each iteration we can observe the following output:

$$\Delta x_0 = 2, \quad \Delta x_1 = 3, \quad \Delta x_2 = 1, \quad \Delta x_3 = 3,$$

which are the first four coefficients of the 5-adic expansion of the rational number 1/3. The other solution, 3/2, can be found starting from $x_0 = 4$. □

6 Remarks and conclusions

By means of a uniform representation of mathematical data structures, it is possible to import into a numeric setting the precision guaranteed by the algebraic one. On this basis the integration of numeric and symbolic computation is defined also from a computational point of view.

This integration is founded on the possibility of establishing a precise isomorphism guaranteeing the validity of the approximation method for objects that can be uniformly represented through truncated power series (i.e. numbers and polynomials).

The examples of Sect. 4 show the effectiveness of extending the use of algebraic algorithms to numeric algorithms. Moreover complex numbers can be represented and treated, as the following example shows.

Example 5. Applying the "integrated" algorithm to the polynomial $x_2 + 1$ with $x_0 = 2$, $p = 5$, and $r = 3$, we obtain the following experimental results:
hgen(x^2+1,G,H,x,2,0,0,1,5,3);

$$[2, 0, 0]$$
$$[1, 0, 0]$$
$$[2, 0, 0]$$
$$[1, 0, 0]$$

summation of the first, 4, coefficients with base, 5, [182, 0, 0]
Let us note that $(2\,1\,2\,1,\,0) \times (2\,1\,2\,1,\,0) = (4\,4\,4\,4,\,0)$ which represents the rational number -1. □

Finally let us discuss some remarks about the choice of initial approximation in the algorithm that we have presented. It must be noted that some problems do still exist. Presently the initial assignments G_0, H_0 or x_0 are chosen by following a heuristic approach. For instance, in each one of the "numeric" examples which have been presented, we have chosen always the smallest base such that an initial approximation does exist.

The analysis of the computational complexity of the algorithm will complete the study of this algorithm, as soon as the authors will finish the implementation of the very first step of this algorithm, that is the choice of the suitable base p.

References

1. S.K. Abdali, G.W. Cherry, and N. Soiffer. An object-oriented approach to algebra system design. In B.W. Char, editor, *ACM SYMSAC '86*, 1986.
2. B. Caviness. Computer algebra: Past and future. *Journal of Symbolic computation*, 2, 1986.
3. B.W. Char, G.J. Fee, K.O. Geddes, G.H. Gonnet, B.W. Monagan, and S.M. Watt. A tutorial introduction to maple. *Journal of Symbolic Computation*, 2, n.2, 1986.
4. A. Colagrossi, C. Limongelli, and A. Miola. Scientific computation by error-free arithmetics. *Journal of information Science and Technology*, July-October 1993. to appear.
5. R.T. Gregory and E.V. Krishnamurthy. *Methods and Applications of Error-Free Computation*. Springer Verlag, 1984.
6. K. Hensel. *Theorie der Algebraischen Zehlen*. Teubner, Leipzig-Stuttgart, 1908.
7. R. Jenks and B. Trager. A language for computational algebra. Technical report, T.J.Watson Research Center, Yorktown, 1987.
8. C. Limongelli and A. Miola. Abstract specification of numeric and algebraic computation methods. *Journal of information Science and Technology*, July-October 1993. to appear.
9. C. Limongelli, A. Miola, and M. Temperini. Design and implementation of symbolic computation systems. In P.W. P.W. Gaffney and E. N. Houstis, editors, *IFIP TC2/WG 2.5 Working Conference on Programming Environments For High Level Scientific Problem Solving*. North-Holland, September 1991.
10. C. Limongelli and M. Temperini. Abstract specification of structures and methods in symbolic mathematical computation. *Theoretical Computer Science*, 104:89–107, October 1992. Elsevier Science Publisher.
11. J. D. Lipson. Newton s method: a great algebraic algorithm. In *1976 ACM Symposium on Symbolic and Algebraic Computation*, 1976.
12. G. Mascari and A. Miola. On the integration on numeric and algebraic computations. In *Proc. AAECC-5*, volume 307 of *LNCS*, Karlsrhue, Germany, 1986. Springer Verlag.
13. A. Miola. Tasso: a system for mathematical problem solving. In B. Sushila E. Balagurusami, editor, *Computer Systems and Applications, ICC Series*. TaTa Mc Graw-Hill, 1990.
14. D. Y. Y. Yun. *The Hensel Lemma in Algebraic manipulation*. PhD thesis, Massachusetts Institute of Technology, November 1974.

Compact Delivery Support for REDUCE

A. C. Norman

University of Cambridge Computer Laboratory
New Museums Site, Cambridge, England
and Codemist Ltd

Abstract. The CSL Lisp system is one designed primarily for delivering
Lisp applications to users. It thus emphasises robustness, portability and
small size rather than support for an integrated programming environment.
Both portability and compactness are served by making CSL compile the
bulk of applications code into a compact byte-code instruction set which is
then emulated. The speed penalties inherent in this are offset by providing
instrumentation that makes it easy to identify code hot-spots, and a second
compiler that translates critical parts of the original Lisp into C for incorporation in the CSL kernel. For use with REDUCE it is found that compiling
about 5% of the source code into C led to overall performance competetive
with other Lisp implementations.

1   Introduction

Over twenty years ago, Hearn documented a clean and simple subset-style dialect
of Lisp [9] for use as the output from the parser for REDUCE's implementation
language, RLISP [8]. A significant revision of the definition of this dialect was made
in [12], but the object remained: this was not a leading edge Lisp for direct use by
programmers but a conservative design, intended to be reasonable and practical for
implementation on a wide range of computers and with major use via a front-end
parser that could provide whatever syntactic niceties seemed useful. A number of
independent implementations of this dialect emerged. Among these were Portable
Standard Lisp (PSL) [6], produced in Utah by people associated with REDUCE
itself, and Cambridge Lisp [2][3] in England. In those days such Lisp systems could
attract use (and hence justify the effort of supporting them) over and above their use
for REDUCE. More recently Common Lisp [15] has displaced other dialects of Lisp
for most purposes. A quite modest amount of code makes it easy to build REDUCE
on top of any Common Lisp (in effect by building a model of Standard Lisp within
Common Lisp), and for development purposes on reasonably configured workstations
that is entirely satisfactory. For delivering computer algebra to the masses it seems
less good, for a number of reasons:

1. Since REDUCE uses Lisp just as a portable machine-code the large range of extra
   features (e.g. all the keyword-driven options in the sequence functions, Common
   Lisp's iteration macros, the CLOS object system, ...) will not be used. At the
   very least this mis-match is an inelegance, and for many implementations of
   Common Lisp it represents a significant wasted cost in store and possibly speed
   of the final system.

2. For most potential users of algebra systems the elaborate and high powered development environments packaged with modern Lisp systems are not needed — and indeed in the case of REDUCE the fact that the main REDUCE sources are coded in RLISP [11] not Lisp will make Lisp-level source management and debugging tools even less helpful.

3. Major expected uses of Common Lisp are in general knowledge engineering applications. The aspects of Lisp most critical to these are not always the same as those crucial to computer algebra — for instance the performance of the REDUCE code to find roots of polynomials or to calculate Groebner bases is to a large extent dominated by the speed of the bignum arithmetic code in the underlying Lisp, and at least some Common Lisp implementations go rather slowly in this area.

4. Despite the rapid growth in the size and power of computers, school and college use of computer algebra is still constrained by the amount of memory (in some cases file-server bandwidth, or swap-space needs) available, and the amount of code that it would be nice to have loaded seems to grow almost as fast as the amount of space to put it in. Gross inefficiency in space usage can be entirely sensible in a well-resourced development environment but is not appropriate when systems are delivered to end-users.

5. Many Common Lisp systems are quite expensive to purchase.

The system discussed here addresses all these issues by providing a Lisp environment just good enough for the delivery of REDUCE. Instead of effort being put into compatibility with the (quite elaborate) Common Lisp specification and into development and debugging tools emphasis is on very easy portability, compactness and performance tuned to give good speed for those operations most important to computer algebra. The extent to which this has been successful can be judged by comparisons reported below where the new system (CSL) is compared with other Lisps that support REDUCE, but the main observation is that CSL provides a very flexible arrangement for the system builder to make speed/space trade-offs, which for general use lead to a set of REDUCE binaries dramatically smaller than those seen with most other Lisps and with only modest degradation of speed even when compared with the best alternatives. As well as its use with REDUCE, CSL is being used experimentally with a simulation system coded in RLISP88 [10], where again it can complement the other Lisp systems that are (and will continue to be) used.

2 Basic Strategy and Storage Organisation

CSL aims for easy portability to lower-end machines of the 1990s by being coded in ANSI C, making non-guaranteed assumptions that it is on a machine where pointers are 4 bytes wide and there is an integral datatype of the same width available. It further assumes that, when a pointer is cast to an integer, addresses appear as if consecutive integers address consecutive bytes of memory; thus words in memory have addresses with the lower two bits zero. By keeping all CSL objects double-word aligned this leaves three low-order bits in each word for object tagging. By demanding that less than half of the 32-bit address space gets used for the Lisp heap the top bit of every Lisp pointer can be assumed to have the same value (be it 0 or 1) so that this bit is used as a mark bit during garbage collection. Similar schemes have, of course, been used elsewhere.

When work on CSL started it was envisaged that the kernel would be compiled using a C compiler [14] that could be extended to support a few non-standard features that would help performance. For instance, the global dedication of a CPU register to refer to an extra stack was investigated. It was furthermore expected that various low-level code fragments would exist as hand-crafted machine code. The experience of [2] made it seem probable that control over the underlying C compiler would greatly ease system building. In the event it was found that a design using normal unadorned C, without any hand-written assembly code at all, was feasible.

One of the first design decisions made for CSL was the global layout of memory. The heap is divided into three regions: the first contains CONS cells, each of which is a doubleword; the second contains all segments of compiled binary code (of which more later); while the last holds vectors, all of which are prefixed by a header word containing both type information and a length code. The vectors represent strings, identifiers, big integers, boxed floating point values and a variety of classes of user-accessible vectors. Both tagging and storage layouts were designed to provide for the full Common Lisp range of data types (eg. 4 types of floating point number: short, single, double and long) but for the purposes of supporting REDUCE the code to activate some of the more elaborate cases is disabled. Garbage collection is a separable module of code, so that different strategies can be implemented on different machines. For machines with small real memory or swap discs a pointer-reversing mark phase (which is quite delicate to implement in the face of a rich selection of vector datatypes) is followed by a compacting phase. When garbage collection is triggered all valid pointers must be stored either on a special Lisp stack, or in locations that can be recognised as list-bases. Because the Lisp value 'nil' is so important, a pointer to it will often be maintained in a (register) variable, so CSL exploits this by using nil as a base for its main vector of list-bases. On modern RISC architectures the effect of this is to improve performance over a system where each key variable was a separate external variable, since it guarantees that just one base register is needed to provide addressability to all key workspace.

In order to support the 80286 a scheme that allocated heap space in 64 Kbyte segments was implemented. The only significant problem was that it limited the size of the largest possible Lisp vector to one that occupies a little under 64 Kbytes, but otherwise it had no significant effect on the rest of the code and made it much easier to shrink or enlarge the heaps — it is now used as standard on all versions of the system.

When CSL finds it has plenty of free memory it uses a stop-and-copy garbage collector; when the heap loading is higher (about 25% of the total heap size occupied at the end of a garbage collection) it switches to a mark-slide compacting scheme. This hybrid strategy has been explored by Terashima[16] for a Japanese Lisp implementation and it allows CSL to function reasonably without change on both small memory-limited machines and on large workstations. In each case memory is compacted at each garbage collection, improving subsequent behaviour on virtual memory systems.

The facility for making and reloading checkpoint files is closely related to storage allocation and garbage collection code. After regular garbage collection has compacted the heaps, all pointers can be converted to a form that makes them identify the 64 Kbyte segment they point into and their offset within that segment. The code that writes the heap image to disc is arranged so that the data dumped is independent of the byte ordering used by the computer (eg. DEC or Intel vs Motorola or IBM 370), and it applies file compression techniques as it writes the image file. The compression means that a set of REDUCE executable binaries and the associated checkpoint file can fit reasonably well onto one or two floppy discs: this is most helpful when one wants to deliver the system to users.

3   The Interpreter

As with the main decisions about data representation, the interpreter in CSL was built to make extension to Common Lisp compatibility as painless as possible, if in the future that should be desired. Thus the initial version used deep binding for variables, with lexical closures fully supported. It provided for optional and keyword arguments, catch, throw, unwind-protect and all the rest. Symbols are implemented with a collection of flag bits that (among other things) indicate if the variable has been 'proclaimed special', and use two words to hold a function definition. The first of these is always an entry-point into real executable code, while the other holds some Lisp datastructure. Compiled code uses the second word to keep an environment, either in the sense of a captured set of bindings or just a vector of literal values the code wants to refer to. Interpreted or traced code can be implemented by putting a call to the interpreter into the function pointer, and a Lisp representation of the function in the environment cell. The effect is that the interface between interpreted and compiled code is seamless at the cost of just making all function calls indirect through symbol heads. As the installation of REDUCE on top of CSL proceeded measurements showed that variable look-up in the interpreter represented a hot-spot. A shallow binding scheme was implemented as an alternative to the more general model, and support for the unused argument decoding options, Lisp-level declarations and so on was suppressed.

Non-local exits (including error recovery) are coped with in CSL by adhering to a convention that a function that needs to trigger unwinding sets a global flag to indicate this, plus extra flags and variables to indicate the class of unwinding involved. After every single function call CSL then has to check this flag and either dispatch into an exception handler or proceed in the normal way. This is of course both ugly and a slight expense, but it does avoid the need for CSL to have any detailed knowledge of the layout used by C for its stack, and it avoids a host of unpleasant problems that seem able to arise if the C setjmp/longjump facility is used instead.

For use with REDUCE CSL is a shallow-bound Lisp. Optional compilation in the kernel allows for the construction of a variant that uses deep binding, with full ability to capture lexical scopes including labels and block exits, and support for local 'special' declarations. Such capabilities are not called for in Standard Lisp as needed for REDUCE, and implementing them slows down the interpreter; fortunately the code that deals with them was fairly well localised and easy to discard in favour of the simpler Standard Lisp model.

The parts of Lisp that are absolutely most performance critical for REDUCE are hand-coded as part of the CSL kernel or interpreter. Special treatment was given to speeding up property list access by keeping a bitmap with each symbol to show what flags and properties it has, and the code to dispatch generic arithmetic operations between floating point and small and big integer routines was tuned to give priority to the most frequently occuring cases.

Code that has to use the normal Lisp interpreter will of course be fairly bulky and even with a carefully written interpreter it will run rather slowly, which leads on to the next section ...

4 Bytecodes

It is by now a commonplace observation that a large proportion of any significant program does not get executed very often. It is furthermore well-established that very compact codings for programs can be designed. CSL exploits these ideas by compiling the bulk of REDUCE into a byte-code format that owes its design in part to a series of interpreted codes for BCPL designed by Martin Richards and refined by Bennett [1], and partly to the real instruction set of the Inmos transputer.

Every function, while running, has available to it a vector holding the literal values to which it can refer, and a set of three local registers, A, B and C. These three registers behave as a mini-stack, in that loading A automatically pushes the previous value of A into B and the previous value of B into C. All operations, including function calls, with up to three operands can take their operands from and leave results on the mini-stack. The fine details of the CSL byte-coded instruction set were refined from an initial attempt by building and running REDUCE with the entire system compiled into bytecodes — it was fairly easy to collect both static and dynamic counts relating to bytecode combinations used and only slightly harder to profile the system and identify which byte opcodes corresponded to the greatest proportion of real time spent in the interpreter. The refined design that resulted uses about a quarter of the available bytecodes for conditional branching (including tests to see if an object is an atom or if objects satisfy the Lisp 'eq' predicate), a further quarter for function calls (specifying the number of arguments passed so that caller-callee compatibility can be checked), a further quarter for opcodes that access the first few items in the literal vector and the stack and perform general data transfer operations, with the remainder of the codes being allocated one at a time to the most heavily used Lisp operations, starting with car, cdr and cons.

The compiler that turns Lisp into a stream of bytecodes uses the sort of technology that one would expect to find in a compiler for a (nearly) registerless machine — it turns the source code into a collection of basic blocks and analyses control-flow between these to get rid of redundant chains of branches, to identify tail-calls and so on. The assembly phase of compilation optimises branch instructions to use short forms (as provided by the byte codes) where possible. The speed of a system that uses no more than this technology is, as one would expect, not wonderful but equally it is not a calamity. The compactness of functions that have been compiled into bytecodes is of course excellent.

5   Compilation into C

Other systems that I am aware of that use bytecode (or threaded code) interpreters accept the modest speed loss that is thereby introduced as the cost to be paid for a large space saving or for the excellent portability achieved. CSL provides a further level of flexibility whereby the person installing a package can trade off space for speed. A second compiler can convert Lisp code into C that can be compiled and linked in with the CSL kernel. Since CSL is intended as a Lisp delivery vehicle, not as a base for general-purpose development, this compilation process can afford to be static — ie. the mechanically generated C code is linked in to form a customised version of the main CSL executable binary. No provision is made for dynamic compilation into C and subsequent loading and linking in of object files, which makes everything rather more straightforward than the situation in, say, KCL [17] where the main Lisp compiler works via C. The Lisp to C compiler is similar in intent to that described by Fitch [4], but was in fact independently written. It arranges to ensure that all pointers are safely stored on the Lisp stack whenever anything that could cause garbage collection is to be done. To keep stack use to a minimum it uses data-flow analysis to compute the live ranges of all variables and a graph colouring scheme to associate values with both C temporary variables and Lisp stack locations. A policy of CSL is that errors involving taking car or cdr of atoms should be detected and reported promptly even in compiled code — available expression analysis in a Lisp to C compiler can frequently establish that the value of some variable is not atomic (for instance if car or cdr of it has recently been taken successfully) and avoid the need for a further test. The generated C code has to do a significant amount of book-keeping to keep to the system-wide conventions used to handle unwinding after errors and during throws: overall the generated code is fairly bulky and unreadable, but it runs at a satisfactory speed. From Fitch's previous experiments it is clear that such a system can run as fast as more conventional Lisp systems, but that if all the Lisp of a package such as REDUCE is converted into C the resulting bulk is substantial and the loss of flexibility is painful. As will be seen later on, by measuring REDUCE execution to identify the most critical 5% of the code and compiling that into C, leaving the rest as bytecodes, a good compromise can be reached.

6   Arithmetic

For the support of computer algebra the quality of the implementation of arithmetic in the underlying Lisp can be critical. It turns out that around half of the lines of C that make up the CSL kernel are devoted to this — with the usual range of special support for small integers (28-bit signed values in this case), big integers (held in vectors, 31-bits used per word, and a 2's complement representation for negative values), floating point and modular numbers. The CSL code also makes provision for the Common Lisp rational and complex data types, and for up to 4 widths of floating point values, but these are not enabled when REDUCE is to be supported. The arithmetic can be completely coded in C, but provision is made for the use of assembly code or otherwise optimised versions of functions that multiply pairs of 32 bit values to get a 64 bit result and that divide a 64 bit value by a 32 bit one to get quotient and remainder. At present no implementation of CSL takes advantage of this possibility! Big-integer printing uses special code for dividing by 10^9 (in effect by multiplying by a fixed point version of its reciprocal) to help its binary-to-decimal conversion. Multiplication switches from classical long multiplication to a Karatsuba method for numbers with more than about 12 words of data (ie. somewhat over 100 decimals), while GCD uses a variant on Lehmer's method that manages to get more reduction per step than the version documented by Knuth. While developing CSL it became clear that care to make common cases of small arithmetic fast and to tune up the code for handling big integers was important: as mentioned before Groebner base calculations often have their performance determined by the speed of operations on big numbers.

7 Graphics and the User Interface

Increasingly it is necessary for applications to exist in windowed environments — X-windows on Unix workstations and various more or less individualistic systems on other machines. CSL tries to support these at two levels: the code involved will of necessity be strongly system dependent, and so the initial port of CSL to any architecture will generally not include it.

The model CSL adopts for low-level interaction with a window manager is one where the window system must be polled regularly, and keyboard input and mouse-generated events can only be accepted when the window system is explicitly probed. The main difficulty this gives for a Lisp system is that of ensuring that polling proceeds even while potentially lengthy calculations are under way. This is achieved by making the operating system provide CSL with a regular stream of asynchronous clock-tick events (this seems reasonably easy in many environments). The handler for a tick can not interact in any very sophisticated way with the active Lisp calculation - for instance some Lisp object may be only partially instantiated or pointers may exist solely in C variables and not be saved properly on the Lisp stack. So the tick handler just sets a few (global) flag variables. To be specific, it sets the variables used to record the limit of the region allocated for the Lisp stack and the value ends of the allocatable heap segments. The result is that as soon as the executing Lisp code checks for available stack or heap space it finds there is none, and enters the garbage collector. Such checks naturally appear very well scattered through the Lisp kernel, and a very few extra ones (such as on backwards branches to be taken in compiled code) ensure that CSL will synchronise with a clock tick quite rapidly. Since the condition when a tick is accepted is identical to that which would arise if storage or stack was exhausted in the normal way, all pointers and datastructures are liable to be in tidy states. The entry-code to the garbage collector then has a quick path that notes when it has been called as a result of a tick — it can reset the limit registers, poll the window manager to bring the screen up to date and accept pending requests (eg. for the current computation to be interrupted) and then return to the main computation. Slight extra care is needed to deal with ticks that arrive while garbage collection is active, but the problems are not extreme. An incidental benefit of having a regular stream of clock ticks available is that they can be exploited to allow collection of additional profile data showing which Lisp functions would benefit most from being expanded from bytecodes into directly executable C.

For graphics environments that provide their own asynchronous call-backs on window or mouse updates the tick mechanism still provides a way to get in step with CSL so that any Lisp code needed to satisfy the update request can be executed.

The clock-tick scheme also allows CSL to run properly even under the sorts of window managers that provide non-preemptive 'cooperative' multi-tasking. At present the option is implemented under the RISC-OS system on the Acorn Archimedes range of computers, but other versions are not expected to be especially troublesome to complete. Input windows supporting cut-and-paste operations, scrollback and all the other niceties currently expected then become just(!) so much programming effort.

The fact that CSL is coded in C and thus has very direct and convenient access to the C graphics libraries typically found on current workstations clearly makes this task much easier than it might otherwise have been. Users with graphics needs in typical Unix environments can either link C code to drive X-windows into the CSL kernel, or provide a pipe-style interface between CSL and a stand-alone graphics server. This latter approach has been demonstrated for REDUCE using the GNU-PLOT graphing server.

8 Performance

CSL, like most modern software, claims to be 'highly portable'. In justification of this claim it can be reported that the system (without the clock-tick processing and graphics part) has been ported by simple recompilation of the C code to about a dozen machine architectures, with a range from 80286 (with extended memory) through VAX and 68000 to a good range of RISC processors. Even though most of these ports have been quick checks, and not followed up by full testing of complete REDUCE functionality it seems reasonable to expect that problems will be limited to refinements in operating system interfaces and to the consequences of bugs in the C compilers that are at present available (in this respect access to a C compiler of which I was part author and which I could mend if necessary has been a great help on those machines for which the Norcroft compiler is available). For delivering REDUCE to end-users the fact that checkpoint and pre-compiled 'fasl' files for CSL do not need to be changed from one system to another (even across machines with different orderings of the bytes within a word) greatly speeds up installing code.

The performance of the CSL-based version of REDUCE can be measured in a great many different ways. Both sizes and speeds will depend on the underlying hardware and the characteristics of the C compiler used, and different test cases can lead to different balances between use of hard-compiled and bytecoded functions. It also seems that system costs (eg. disc and screen access times) can make a substantial contribution to the total time reported for many tests. Given all those disclaimers there are a number of figures that I can quote, which overall seem to tell a consistent story.

A first measure is a raw one of speed running the main REDUCE test file. Fitch [5] reports that CSL with all of REDUCE compiled into bytecodes ran at almost exactly the same speed as a KCL-based system, but CSL was happy in around 2 Mbytes while the KCL version ran in about 5. When the top 5% of functions in REDUCE were converted to C and linked in to provide an extended CSL this roughly doubled its speed, making it twice as fast as KCL. Under MSDOS/386 CSL can be compared with a PSL based system: with output from the test directed to file CSL was 25% slower than PSL, while on other tests and with output to screen the ratio could get up to 2:1. On an inadequately configured machine where the PSL version got involved in heavy page disc traffic in support of virtual memory the performance advantage could tip the other way. Overall it seems reasonable to claim that a CSL-based REDUCE is somewhat slower than one built on PSL, but rather faster than one that uses KCL. I would expect various other commercial Common Lisp implementations to approach, but not exceed, the speed of PSL for these purposes. Even though most of REDUCE remains bytecoded the speed of CSL seems tolerable.

The size of the initial executable image involved is another very system dependent issue, and there can be confusions about how much memory is given over to datastructures rather than code. On some machines code that is present but never executed will not count as a serious run-time overhead — but somebody will still need to have written and debugged it, and will have to remain ready to maintain it so there are at least indirect costs associated with even unused code. On a Decstation the following sizes were reported by the Unix 'size' command for the executables for various Lisps installed for general use in Cambridge:

- KCL: 1.1 Mbytes of code plus 1.1 Mbytes initialised data
- Allegro 3.1.12: 4.3 Mbytes code plus 2 Mbytes initialised data
- CSL with no REDUCE: 430 Kbytes total
- CSL with REDUCE hot-spot code: 700 Kbytes total

In general a certain amount of heap-space will have to be added to these figures, and the density of representation of lists will not differ much between the various Lisps. Also the compiled code for REDUCE will have to be added. To estimate sizes for that I look at the size of the fasl (Lisp binary loadable) file for a fairly typical REDUCE module, the factorizer [13]. For PSL on the 80386 this object file (which will mostly consist of executable code to be loaded) is 85 Kbytes. For CSL the equivalent file contains mostly bytecodes and the same version can be used whatever computer REDUCE is running on. It is 22 Kbytes long. Similar ratios, of the order of 4:1, are shown for the other REDUCE modules. Looking at the number of bytes compiled for individual functions rather than just taking an overall measure of object code sizes give similar results.

A final, and perhaps instructive, measure of size is to look at the bulk involved in the source code for CSL: at the time of writing it is 47000 lines, amounting to 1.5 Mbytes. This is a little larger than the source code for Cambridge Lisp [2] (30000 lines, 1 Mbyte), possibly in part because CSL has two separate compilers not just one. A remarkably large amount of this code is dedicated to arithmetic, especially to respectable implementation of calculation involving big integers. For comparison, the whole of the rest of REDUCE, including the associated library of user-contributed code, is just over 100000 lines (3.5 Mbytes), showing that even with a pretty-well cut-down Lisp such as CSL the bulk of code involved in the Lisp is a significant proportion of that in the algebra system. This fact came as something of a surprise to me, but perhaps explains some of the feelings I had previously that the larger scale modern Lisp implementations had grown to a scale where they overshadowed all but the very largest packages, and where one could well imagine that the end-user would have to pay more in either time and effort or in cash for the Lisp underpinning their algebra than for the algebra system itself. Perhaps CSL goes some way towards putting the algebra system back on top.

9 References

1. Bennett, J.P. A Methodology for Automated Design of Computer Instruction Sets: PhD Thesis, University of Cambridge, 1988
2. Fitch, J.P. and Norman, A.C. Implementing LISP in a High Level Language: Software Practice and Experience, 7, 713-725, 1977
3. Fitch, J.P. and Norman, A.C. Cambridge Lisp M68000 manual: University of Bath and Metacomco plc, 1984
4. Fitch, J.P. A Delivery System for REDUCE: Proceedings of ISSAC90, Tokyo, pp 76-81, ACM and Addison Wesley, 1991
5. Fitch, J.P. private communication, 1991
6. Griss, M.L., Benson, E. and Maguire, G.Q. PSL: A Portable LISP System: Proceedings 1982 ACM Symposium on LISP and Functional Programming, 1982
7. Hearn, A.C. REDUCE User's Manual, Version 3.4: RAND Publication CP 78, July 1991
8. Hearn, A.C. The REDUCE Program for Computer Algebra: Proceedings of the Third International Colloquium on Advanced Computing Methods in Theoretical Physics, 1973
9. Hearn, A.C. Standard LISP: SIGPLAN Notices, ACM, 4-9, September 1969, Reprinted in SIGSAM Bulletin, ACM, 13, 28-49, 1969
10. Marti, J. RISE: The RAND Integrated Simulation Environment: in Distributed Simulation: Unger B. and Jefferson, D. (eds), Simulation Councils, Inc., San Diego, California, 1988.
11. Marti, J. RLISP'88: An Evolutionary Approach to Program Design and Re-use: World Scientific Publishing Company, 1993
12. Marti, J. et al Standard Lisp Report: University of Utah, 1978
13. Norman, A.C. and Moore, P.M.A. Implementing a polynomial factorization package: Proceedings SYMSAC-81, Snowbird, Utah. ACM, 1981
14. Norman, A.C. and Mycroft, A. The Norcroft C compiler: versions as used with the Acorn Archimedes, Inmos Transputer, SPARC, mips, IBM mainframe and other processors
15. Steele, G.L. Common LISP, the language: (2nd edition) Digital Press, 1990
16. Terashima, M., Hybrid Garbage Collection: draft report, collection of comparative benchmarks: private communication.
17. Yuasa, T. and Hagiya, M. Kyoto Common Lisp Report: Teikoku Insatsu, Kyoto, 1985

IZIC: a Portable Language-Driven Tool for Mathematical Surfaces Visualization

Robert Fournier   Norbert Kajler   Bernard Mourrain

SEMIR   SAFIR *

2004 routes des Lucioles
06565 Sophia-Antipolis Cedex, France

e-mail:{*fournier,kajler,mourrain*} @*sophia.inria.fr*

Abstract

This paper presents IZIC, a stand-alone high-quality 3D graphic tool driven by a command language. IZIC is an interactive version of ZICLIB, a 3D graphic library allowing efficient curve and surface manipulations using a virtual graphic device. Capabilities of ZICLIB include management of pseudo or true colors, illumination model, shading, transparency, etc. As an interactive tool, IZIC is run as a Unix server which can be driven from one or more Computer Algebra Systems, including Maple, Mathematica, and Ulysse, or through an integrated user interface such as CAS/PI. Connecting IZIC with a different system is a very simple task which can be achieved at run-time and require no compilation. Also important is the possibility to drive IZIC both through its freely-reconfigurable menus-buttons user interface, and through its command language, allowing for instance the animation of surfaces in a very flexible way.

Keywords: Computer Algebra, meshed surfaces, graphic tool, rendering, TCL.

1   Introduction

Most of the surface plotting tools used with Computer Algebra Systems are those provided with Axiom, Maple, Mathematica, etc. Being part of commercial products these plotters are neitheir available separately nor callable from other systems. Also, they do not include important features such as shadowing rendering, transparency, or surfaces composition. In fact, few attempts were made in the Computer Algebra community to develop powerful portable plotting tools. One of the most notable exception appears to be SIG [Wan90], a curve and surface plotting engine under X11 developped in 1990 by Wang at Kent State University. As a surface plotter, SIG is fairly competitive with respect to other plotting engines available at that time, allowing for instance hidden-line removal on surface mesh displays. As a portable tool, SIG include two parts: a system-independent visualization engine called Xgraph and a system-dependent set of

---

*SAFIR is a common project to INRIA, Université de Nice-Sophia Antipolis, and CNRS.

functions, called mgraph, which performs computations of data needed by Xgraph from a given Computer Algebra System. The communication between Xgraph and mgraph uses a file to transfer data and a minimal protocol to initiate displays. Clearly, Wang advocates in [Wan90] that more work is needed to improve communications between these two components.

On the other hand, many plotting engines were developed outside the Computer Algebra community. Most of these tools are either not powerful enough to allow quality display of mathematical surfaces, or necessitate the use of optimized hardware for 3-D manipulations. This includes respectively freely available tools like Gnuplot or commercial applications such as Explorer.

In this paper, we will present IZIC, a graphic server, that can be driven from *outside* and is based on the C-library ZICLIB. *Commands scripts* can be sent to this tool from one or more Computer Algebra Systems, or any other scientific tools, in order to display the geometrical objects and act on their representations. Conversely to SIG, IZIC can be dynamically connected to one or more Computer Algebra Systems and offer the power of a real command language to drive the surface display and/or customize its menu-button user interface.

The first part of this paper is related to the library and algorithms used in order to obtain nice and efficient rendering of surfaces. The second part deals with the incorporating of this code in an interpreter which can be driven from *outside*, including an easily configurable menu-button user interface. The third part presents the way geometrical objects are manipulated formally in a Computer Algebra System and how to link this one with our graphic server.

2 ZICLIB

ZICLIB [Fou92] is a software graphic library written in C by RObert Fournier at INRIA Sophia-Antipolis. By using only memory allocation, ZICLIB can run on any Unix system and requires no specific hardware. All existing applications based on ZICLIB, such as ZICVIS, ZVIS, and IZIC, also use standard Xlib drawing routines, a common simple language defining input data format, and various X11 toolkits, respectively: Athena, OSF/Motif, and TK.

The basic data structure of ZICLIB is a main memory workspace named *zicimage*, which is used as an abstract device to send graphic requests. This resource will be referenced as a *Virtual Graphic Device* in this paper.

Each *Virtual Graphic Device* has its own characteristics, including the size of a screen in pixels, its depth, and the availability of a software *Z-buffer*. Three predefined depths are available: 1 for monochrome, 8 for pseudo colors and 24 for true colors. A projection type – perspective or parallel – and its definition is also associated to a *Virtual Graphic Device*. Another characteristic is the definition of an illumination model [FVDFH90], including intensity of an ambient light and a light sources list. All the characteristics of a *Virtual Graphic Device* can be given at creation time or modified dynamically.

The three dimensional *graphic elements* which are the bricks of constructions, are curves and surfaces. Curves are defined by sequence of points, and surfaces are defined by structured or unstructured meshes. Graphic requests are available to draw these *graphic elements* "onto" a *Virtual Graphic Device*. Including text is also possible.

2.1 The surfaces

We give here the main characteristics of the *surfaces*:

- Each surface *graphic elements* is supposed to have all its surface elements oriented such that a direct and a back side may be defined for the surface.

- Data for the illumination model [FVDFH90, pages 721–741] is attached to each surface: the reflection coefficient for each component of the ambient light, the diffuse reflection coefficient for each component of the source lights used in the *Lambertian* illumination model, and the specular reflection coefficients used by the *Phong* illumination model.

- To each surface is also associated mode flags defining how to draw it. For example, drawing only the mesh for the direct side of the surface, or drawing the direct and the back side using ambient light and Lambertien model illumination without specular reflection. These flags allow to specify the drawing model illumination using data stored in the *Virtual Graphic Device* and the surface *graphic element*. Another flag allows to specify *flat shading* or *Gouraud shading*.

Once a surface *graphic element* has been created and its drawing mode flags setup, a graphic procedure may be used to draw it on a *Virtual Graphic Device*.

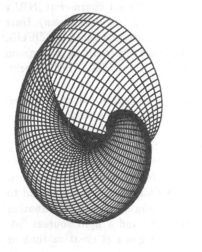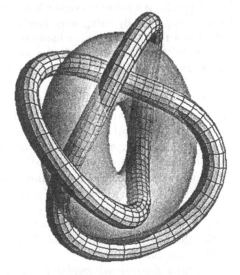

Figure 1: Visualization of structured meshes created by ZICLIB

2.2 The ZIC format

A simple language is also available to ease the definition of the *Virtual Graphic Device*, to use it with all its characteristics, and to construct the *graphic elements*. Keywords are used for each kind of *graphic element*. For example, the keyword *surface* introduces the definition of a surface defined by a structured mesh, and the keyword *surface_g* a structured mesh with an intensity color component known at each node.

The following example uses this language to define colors with the keyword *couleurs* for a 256 entries colormap, the world space coordinate with the keyword *domaine*, the background color with the keyword *bg_col_index*, the name of the next *graphic element* with the keyword *nom*, the transparency coefficient of the next *graphic element* with the keyword *transparence* (default is no transparency) and the *graphic elements* of type *triangles* defined by the number of nodes (600), the world coordinates of theses nodes, the number of triangles (1200) and the index of vertices of each of them.

```
couleurs
# predefined colors
3    # number of colors
0    0. 0. 0.    # first color is black
1    0. 0. 1.    # second color is blue
2    1. 0. 0.    # third color is red

# colors for surfaces
1    # number of ranges
0    # index of the range
3    0. 0. 0.    # the index of the first color (black) of this range is 3
255  1. 1. 1.    # the last one is 255 and corresponds to white.
# Here the color on the surfaces will vary linearly from black [0 0 0]
# to white [1 1 1].

domaine -1.5 1.5 -1.5 1.5 -1.5 1.5
# box corresponding to the view, in world coordinates.

bg_col_index 255   # The color of the background is white

nom tore    # Name of the next graphic element

transparence .4    # coefficient of transparency

# description of the surface by a mesh of triangles
triangles
# list of points
600
0.9 0 0
0.885317 0 0.0927051
0.842705 0 0.176336
...
# list of triangles
1200
0 20 21
0 21 1
1 21 22
...
```

This example describes a surface meshed by triangles, which can be manipulated interactively with a reasonable response-time.

3 Turning ZICLIB into a command interpreter

They are different ways to transform an existing C library, like ZICLIB, into a command interpreter. An obvious solution consists in creating a special-purpose command language, which implies defining the language syntax, implementing a parser, and linking this parser to the application built on top of the library. Another solution consists in defining some or all of the functions implemented in the C library as new commands of an existing language. Obviously this language is supposed to be interpreted, lightweight, simple, and to provide a reasonably large set of predefined commands. Since a few years, different languages have been specifically designed to provide these functionalities, i.e. to allow application programmers to extend the language and embed it into the application as its command language. Examples of such *embeddable languages* include ELK, WOOL, and TCL.

3.1 TCL and TK

To build IZIC, our interactive version of ZICLIB, we use TCL [Ous90], probably the most popular embeddable language at this time. Developed at the University of California at Berkeley, TCL is clearly designed to simplify composition of software modules running as TCL interpreters. Each TCL interpreter is an independent program written in C which make available a well defined set of functionalities through specific TCL commands. These commands extend the generic set of commands provided by TCL, which include assignment, loops, string and array manipulations, input-output, etc.

TCL also comes with TK [Ous91], a graphic toolkit allowing fast development of user interfaces for TCL-based applications. As any X11 toolkit, TK can be used at two different levels, dealing respectively with composing existing graphic interactors – called *widgets* – and defining new types of widgets. TK also includes a simple widget set offering a dozen of predefined widgets types such as button, label, frame, menu, scale, scrollbar, etc. However, an essential difference between TK and most other toolkits resides in the way widgets may be composed dynamically, using TCL, rather than statically in C or C++ as in Athena, OSF/Motif, Interviews, etc.

3.2 IZIC

As shown below, the implementation of IZIC is made of three parts: a new TK widget type called *zic*, the *izic* command interpreter, and a menu-button user interface for *izic*. Then IZIC is based on four C libraries: ZICLIB, X11, TCL, and TK.

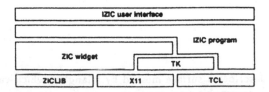

The zic widget

In TK, defining a new widget type is made in C. As any TK widget, zic is created and manipulated dynamically from TCL. The syntax used to create a zic widget is:
zic *pathname* [-zicimage *on|off*] [-doublebuffered *on|off*].

Once created a zic widget is refered by its pathname which defines a new TCL command. This TCL command can then be used to dynamically manipulate the ZIC widget according to the following parameters:

parameter	argument(s)
data	[*adr*]
light	source [*value*]
light	position [*alpha* [*beta*]]
light	ambient [*value*]
refresh	
rendering	phong [*k n*]
rendering	gouraud [*on*	*off*]
rendering	painter [*on*	*off*]
rendering	inside visible	mesh	light [*on*	*off*]
rendering	outside visible	mesh	light [*on*	*off*]
rotate	[*alpha* [*beta* [*gamma*]]]
show	[*adr*]

The IZIC interpreter

An IZIC interpreter can be started from the Shell by typing: izic &. This creates an IZIC server which can be accessed from any other TCL-based application. The IZIC server is linked with the TCL and TK libraries allowing the use of any standard TCL or TK command. Obviously, IZIC is also linked with the zic widget type definition which allows the use of the zic command described above. A second TCL command, named zicdata, is also defined in IZIC to build zic data by parsing either a file or a string, according to the format described in § 2.2.

command	argument(s)
zicdata	readfile *filename*
zicdata	readstr *string*

As we will see in the following, TCL scripts can then be executed by the IZIC server, either from a linked command editor, a menu-button user-interface, or remotely, from any TCL-based application. This possibility to send pieces of TCL code to a running IZIC process allows many interesting manipulations such as animating displayed surfaces or using of fading for switching between two surfaces.

Also, to simplify the driving of IZIC by non TCL-based applications, we provide a simple Unix command named sendizic. sendizic can be used from the Shell, or from any Unix application, to send TCL+TK scripts to a remote IZIC server, as shown on the figure below:

The user interface

The whole capabilities of IZIC are also available through a menu-button user interface made of a set of panels visible on the figure 2. In these panels, the scales allow rotating

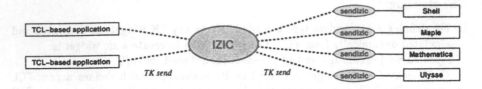

the displayed objects, and positioning the source of light. Other interactors in the panels are build to select between the different modes: painter, gouraud, etc. Also, a line editor allows the user to enter TCL commands that are immediately interpreted by the IZIC server.

Clearly, this menu-button user interface is a default one defined by a TCL file loaded during at initialization time. Any end-user of IZIC may design its own menu-button interface by using the zic widget and standard TCL+TK commands.

Technical considerations

As we already mentioned, ZICLIB is both hardware and software independent by using only ordinary memory manipulations. Up to now, ZICLIB has been successfully compiled and tested on SparcStation, DecStation, and IBM RS6000. Concerning IZIC, its portability is tied to TCL+TK which is now widely available.

The size of the source code of IZIC is composed of: 1800 lines of C for the zic widget, 600 lines of C for the izic program and 350 lines of TCL for the menu-button user interface. This does not include the 14,000 lines of C code of the ZICLIB.

Also, when started on a SparcStation, the IZIC program is about 250 KiloBytes large. When running, this size may increase according to the memory workspaces needed to manipulate the images loaded by the user. For instance, the IZIC process was about 3 MegaBytes when we displayed the surface of the figure 2.

4   Using IZIC from one or more Computer Algebra Systems

In this section, we describe the kind of geometrical objects we are going to manipulate and how our graphic tool can be driven from another process, without modifying the working environment, in order to draw them. So, on one hand, manipulating of equations, derivations, substitutions, numerical evaluations, etc, is done in the Computer Algebra System (here Maple). On the other hand, IZIC draws any curve described by a polygon or any surface described by a mesh, in a very accurate and nice way.

4.1   Formal geometric objects

We choose to manipulate the geometric objects under Maple [CGG+88], in the form of a table which gives the type of the objects (either a curve or a surface) and optional informations such as the points of a mesh, the color of objects (which could be a function of the coordinates), equations or a parameterization, etc. We give here an example of such object, created by the function surface : This dynamical structure allows manipulating patches of surfaces and portions of curves, and treating them as

```
-------------------------------------------------------------------------------
> enneper := surface([u-1/3*u**3+u*v**2, v-1/3*v**3+v*u**2,u**2-v**2],u=-1..1,
v=-1..1, grid=[10,10],color=[1,0,0]);

enneper :=
table([
    _type = surface
    _paramet =
                    3       2          3      2   2    2
    [[u - 1/3 u  + u v , v - 1/3 v  + v u , u  - v ], u = -1 .. 1, v = -1 .. 1]
    _points =
        [[-1.66667, -1.66667, 0], [-1.30667, -1.42933, .360000],
          ...
            [1.66667, 1.66667, 0]]
    _color = [1, 0, 0]
    _box = [-1.66667, 1.66667, -1.66667, 1.66667, -1., 1.]
])
-------------------------------------------------------------------------------
```

formal object with differents levels of evaluation, going from the description by a name
and a type down to the computation of the list of points and polygons for the mesh.
Runtime operations such as modifying structures, moving, scaling, etc, ..., can also be
done with these representations. For example, the intersection of two surfaces could be
done, either using the equations (or the parameterizations) or computing numerically
the intersection of the meshes. Objects can also be colored according to a function of
intensity, for instance to visualize the trace of a surface over another surface. Saving
these objects in files and reusing them in other contexts is also possible.

This approach, which could be implemented in most possible Computer Algebra
Systems, helps us to deal directly with geometric objects and can be freely extended
for specific needs. Moreover, freedom is left to the user to define its own structure.
For instance, one can assemble these objects in a new structure called *solid*, or setup
functions that build special surfaces such as spheres, cylinders, etc.

4.2   The connection with IZIC

Let us see how IZIC is linked with Maple. The objects are loaded in IZIC with the help
of files, where the numerical informations needed for the drawing (see ZICLIB format)
are written. This is not the best way (but an easy way) to transfer great amount of
numerical data and we plan to establish connections based on *Unix sockets*. Commands
corresponding to each graphical interaction on the interface can be send from Maple
to IZIC (in the syntax TCL) but groups of operation such as as opening a window,
refreshing it, rotating the scene, changing the light position, ... are also available.
They correspond to "scripts" which are directly interpreted under IZIC. It uses the
interesting possibilities of the TCL-TK interpreters to send commands to one another,
via the X11 protocol. Thus, collaborating between differents systems, computing on
different machines and displaying their results in the same window is allowed. For
instance, Maple and Mathematica can collaborate on a same plotting. Moreover, this
communication with izic also allows simple animation of surfaces as in the example
below:

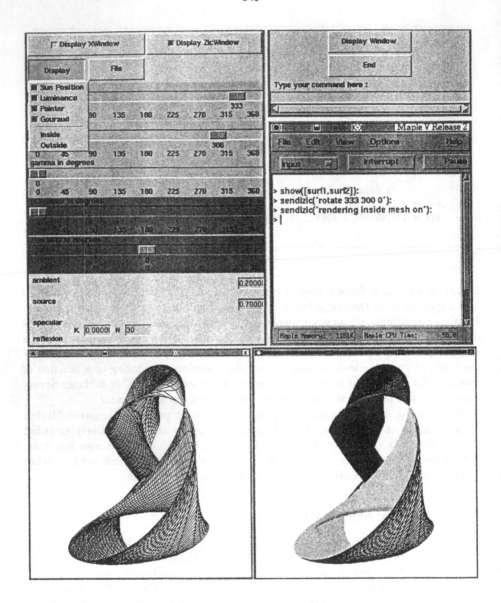

Figure 2: Using IZIC from a Maple session

```
sendizic('for {set i 1} {$i < 36} {incr i}
              {izic rotate [expr "10*$i"];sleep 1}'
         );
```

Macros can be defined on Maple and/or IZIC sides to ease the call of some commonly
used facilities. Here is an example of a surface generated by a point $M(x, y)$ of the
mechanism of 4 bars (described below) when the parameter a is varying and the others
are fixed.

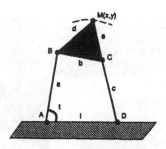

Manipulating distances or angles, solving formally equations of degree 2, etc, is needed for computing the two parts of this surface (the red one surf1 and the blue one surf2) which is of degree six and whose sections correspond to this curves described by the point $M(x,y)$ (when the length of the bars are fixed). The figure 2 shows a Maple session in which we use IZIC to visualize the sextic corresponding to this problem. Obviously, the commands show and sendizic are defined in a file previously loaded in Maple.

In order to have an idea of how much time is spent for computing or drawing, in the previous example the computation of the 5356 points (defined by a huge formula) lasted about 1798 seconds (CPU) on a SparcStation 10, while the time needed for the drawing was about 6.5 seconds. In another example where 15 objects are drawn, requiring 6100 points, the time needed for the computation was about 68 seconds while the visualisation required 7.5 seconds. The simpler surface of figure 3 required about 2.2 seconds for the computation and about 1.1 seconds for the drawing. The time used for sending a command to izic is not significant in these cases.

5 Using IZIC from CAS/PI

CAS/PI is a graphic user-interface allowing simultaneous use of different scientific tools. Right now, three Computer Algebra Systems (Maple, Sisyphe, Ulysse) and two plotters (Gnuplot and IZIC) are connected to it. However the extensible nature of CAS/PI allows the connection of new classes of tools, as well as various others adaptations as described in [Kaj92b].

Basically, the connection of IZIC under CAS/PI follows the same principles as exposed in the §4.2, i.e. we first use a Computer Algebra System to perform data computations, then save data on a file and eventually send a TCL script to IZIC to read the file, display the surface, animate it, etc. However the setting of these successive tasks can be combined in a much more flexible way using CAS/PI. First, communications between tools connected through CAS/PI are made through the concept of *services*. Examples of existing services in CAS/PI are: Evaluation, Factorization, Help, Edition, etc. Then, each tool such as Maple, IZIC, or a Matheditor, is a *server* which provides to the others tools a set of well defined services, and/or may request a set of services. For instance, Maple provides a dozen of services including Evaluation and may request Edition. Conversely, a Matheditor may request Evaluation and provide Edition. At last, links between tools are established dynamically on a service basis, according to the state of the software bus on which the different tools may be plugged and unplugged. Obviously, this architecture enforces tool independence as inter-tool references are replaced by abstract services relations.

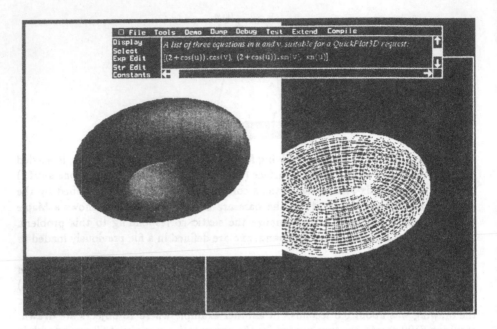

Figure 3: using IZIC from CAS/PI

According to this scheme, we added two services to both **Maple** and **Ulysse** tools:
Plot3D and QuickPlot3D. A Plot3D request includes a list of three functions in two
variables and the bounds and number of points for each of these variables. A Quick-
Plot3D includes only a list of three functions in two variables, the other parameters
being set to default values. The effect of a Plot3D or QuickPlot3D message arriving
for tools like **Maple** or **Ulysse** is implemented as computing corresponding data (as in
§4.2) and then emitting a Display3D request. Obviously, Display3D is provided by
IZIC which displays the surfaces. The advantage of the software bus paradigm is flexibility. Many different combinations can be arranged to link one or more tools providing
Plot3D and/or QuickPlot3D service with one or more instance of IZIC which provide
Display3D. For instance, one may imagine to compare the data computed by the different tools by dispatching Plot3D requests to every tools able to deal with them. One
can also imagine to automatically chain through the bus the result of an Eval service
which is supposed to return a simplified list of three equations with a QuikPlot3D
request.

Figure 3 shows an example of a CAS/PI session in which two IZIC windows display a
same torus respectively as a plain surface on the left and as a mesh surface on the right.
The window on top is a Matheditor, a CAS/PI tool used to edit texts and mathematical
expressions in a graphical way. Obviously, the Matheditor displays three equations in
u and v used for the torus. To display the surfaces the user just selected the list with
the mouse and requested the QuickPlot3D service from the "Tools" menu.

6 Current limitations and future work

The library ZICLIB could be extended for instance, by adding computation of shadows and high-quality output for printing, etc.

Many improvements could be done to the menu-button user interface to make it more user-friendly. However, this could also be done very easily by any end-user who may want to adapt it to its precise needs.

One limitation (up to now) is the use of files in the connection between Maple and IZIC. The transfer of numerical data between the two processes IZIC and Maple could be improved by using sockets and special encoding of the points and elements of ths surface of the meshes.

Computing points could be ameliorated in some cases by using external procedures (C-compiled functions) instead of Computer Algebra Systems once the formula is known. Careful studies of surfaces defined by implicit equations could also be added as well as a larger set of functions for "solids" manipulations.

7 Conclusion

One can observe in different fields of applied mathematics, physics, mechanics, etc, that users of Computer Algebra Systems are not only interested in such systems because of their facilities for doing symbolic computations but also, in a non-negligible way, for extra capabilities such as plotting, generating C, Fortran, or LaTeX codes, etc. Clearly, this shows that people like the possibility of using different kinds of tools from a common software environment. It could be manipulating equations for modeling a problem, generating Fortran Code to approximate the solutions, or the writing of a report presenting the results of experimentations on this problem.

Instead of including all these completely different tools in a same and closed system, we propose to develop specialized, independent, and reusable software components which will be used in what they excel, in the spirit of [Kaj92a]. An advantage of this method is that these tools could be developed, maintained, and composed independently. Drawbacks include the difficulty to link these different tools together and to do so in such a way that the resulting software environment is as homogeneous as a monolithic system.

In IZIC, we applied these principles to the development of a 3D graphic tool, keeping in mind efficiency and flexibility. Efficiency comes from the C-library ZICLIB on which this tool is based. Flexibility is achieved here by embedding a command language into the application. This way, scripts can be written by end-users or generated by other applications to drive IZIC remotely or through a customizable menu-button user interface. Communications with the IZIC server is possible from most of the Computer Algebra Systems, which in this case, are only used for geometrical or algebraic manipulations on the objects we want to visualize.

By this example of graphic server, we claim that independent and specialized tools linked together in a same environment would be more efficient than a general system being able to do all that you want but whose performances, in the field you are interested in, may be much less that what you expect.

Acknowledgements

The authors would like to thank Guillaume Fau who implemented the main part of the TCL code of this software.

References

[CGG+88] B. W. Char, K. O. Geddes, G. H. Gonnet, M. B. Monagan, and S. M. Watt. *Maple Reference Manual, 5th Edition*, 1988. Watcom Publications Limited, Waterloo, Ontario, Canada.

[Fou92] R. Fournier. ZICVIS et la ZIClib – Visualisation de maillages de surfaces. Version 1.6. INRIA, Sophia Antipolis. French INRIA internal report.

[FVDFH90] J. Foley, A. Van Dam, S. Feiner, and J. Hughes. *Computer Graphics – Principles and Practices, 2nd edition*. Addison-Wesley, 1990.

[Kaj92a] N. Kajler. Building a Computer Algebra Environment by Composition of Collaborative Tools. In *Proc. of DISCO'92*, Bath, GB, April 1992.

[Kaj92b] N. Kajler. CAS/PI: a Portable and Extensible Interface for Computer Algebra Systems. In *Proc. of ISSAC'92*, pages 376–386, Berkeley, USA, July 1992. ACM Press.

[Ous90] J. K. Ousterhout. Tcl: An embeddable command language. In *1990 Winter USENIX Conference Proceedings*. Univ. of California at Berkeley, 1990.

[Ous91] J. K. Ousterhout. An X11 Toolkit Based on the Tcl Language. In *1991 Winter USENIX Conference Proceedings*. Univ. of California at Berkeley, 1991.

[Wan90] P. S. Wang. A System Independent Graphing Package for Mathematical Functions. In *DISCO'90*, pages 145–154, Capri, Italy, April 1990. LNCS 429, Springer-Verlag.

The Algebraic Constructor CAC: computing in construction-defined domains

Serge D. Meshveliani

Program Systems Institute,
Pereslavl-Zalessky, 152140, Russia
e-mail: mesh@math.botik.yaroslavl.su

Abstract. We present a symbolic computation program which is able to build new algebraic domains. Many domains in commutative algebra can be built from the ring of integers by applying the constructors of Direct Sum, Polynomial Algebra, Fraction Ring, Quotient Ring. Certain standard arithmetic operations for the elements and ideals of such rings are defined by the system automatically.

1 Introduction

The Commutative Algebra Constructor (CAC) [7] is a program for symbolic computations in algebra. It supplies algebraic *constructors* to define automatically algebraic operations on the domains and builds dynamically new domains with the constructors of

[ ] Polynomial Algebra,	⊕ Direct Sum of rings (modules)
/ Quotient Ring (module),	Fr Fraction Ring,
FreeModPol Free Module over Polynomials,	

For each constructed domain CAC tries to define the functions that we call *K-operations* (*K-functions*):

1) *Arithmetic*: neg, add, sub, mul, power, inv, div, CanAssoc (canonical associated element), CanInv (canonical invertible factor), EucN (Euclidean norm), EucDiv (Euclidean division), GCD (greatest common divisor).

2) *Functions for ideal bases*: CanQ (canonical reduction modulo an ideal), CanB (canonical basis of an ideal), RlB, PSol (general solution of a linear system).

3) *Interface functions*: read, write.

Every domain created by CAC is a description that contains definitions for as many *K*-functions as possible.

Once a domain is constructed, the following *interpreters* can be run over it: *Linear Algebra* over a Euclidean ring, *Polynomial Arithmetic*, *Gröbner Bases and Syzygies*, *Symmetric Polynomial Package* and others.

2 Mathematical background

The basic problem is to develop a method for the *canonical simplification* of the elements and ideals of each K-constructed domain.

2.1 K-rings and K-operations

Here are two examples showing the expressive power of the above constructors $\oplus$, [ ], Fr, /.

Example 1. The field of rational functions on an irreducible algebraic surface $\{\mathcal{P} = 0\}$, where $\mathcal{P}$ is a prime ideal in $k[x_1, \ldots, x_n]$, is expressed as

$$(k(t_1, \ldots, t_d)[y_1, \ldots, y_m])/\mathcal{P},$$

where the t-coordinates are chosen in a proper way.

Example 2. The finite field arithmetic: see section 5.

We call a *K-ring* any ring that is built from the ring of integers $\mathbf{Z}$ with the above constructors.

The set of constructors used in CAC agrees with the set of K-functions in the sense that very often a K-function can be automatically generated for a composite domain provided that the necessary K-functions are defined for all the lower level domains. For example, the extended GCD operation in a Euclidean ring A gives rise to the inversion function for the elements of the quotient ring $R = A/(d)$ (returning "failed" if an element is non-invertible).

2.2 Objects and their attributes

CAC represents an *object* as a *term* (unit of FLAC data). An object contains its *type* ("ring", "ideal" ...), *construction, operation definitions, properties*. We fix a set of the main object properties, such as "field", "Euclidean", "NoZeroDivisors", "maximal" (ideal) etc. They determine what K-functions can be defined correctly for the object being constructed over the given objects. If a certain property value is not declared explicitly for the object being constructed, then the constructor tries to *deduce* it from the ring *axioms* and the property values of the other objects. The current CAC version is capable to process a very restricted set of axioms. For example, ("field" *implies* "Euclidean"), (if I is an ideal in A and $maximal(I)$ then $field(A/I)$). Thus the PolAlgebra constructor will find out that the Euclidean division is to be defined in $(A/I)[x]$, if I contains "maximal" in its property value list.

2.3 Gröbner bases techniques

If R is a *Euclidean ring*, we use the Gröbner basis method [1, 5] to express the functions CanQ and CanB for the ring $A = R[x_1, \ldots, x_n]$ and thus generalize the extended GCD-approach. In this way, arithmetical operations can be defined in A/I for any ideal I in A. For example, the inversion inv(a) in A/I is expressed by applying $CanB(a, a_1, \ldots, a_m; q)$, where $(a_1, \ldots, a_m)$ is the canonical basis of I. The result is $((b_1 \ldots b_k), M)$, where M is the transformation matrix to obtain the canonical basis $(b_1 \ldots b_k)$ from the initial one. If $(b_1, \ldots, b_k) = (1)$, then $M_{1,1}$ is the required inverse element, otherwise a is non-invertible in A/I.

The canonical functions CanB, CanQ are generalizations of the Gröbner basis and Gröbner normal form functions for the K-rings; they provide the comparison of ideals and the canonical form modulo an ideal.

2.4 Transforming ring descriptions

Suppose $R = \mathbf{Q}[x] \oplus \mathbf{Z}$. We cannot use Gröbner bases to construct the quotient ring $R[y]/I$ because R is not Euclidean (we consider the method using syzygy solving in R as too expensive; besides, the syzygy function is also to be *built* for R). But there is an evident algorithmic isomorphism to some "simpler" domain:

$$(\mathbf{Q}[x] \oplus \mathbf{Z})[y] \cong \mathbf{Q}[x, y] \oplus \mathbf{Z}[y] \tag{1}$$

The canonical functions CanB, CanQ for $\mathbf{Q}[x, y]$, $\mathbf{Z}[y]$ are expressed in terms of Gröbner bases. Then they are lifted up to the direct sum. Hence these functions are defined for $R[y]$ too. There are some more isomorphisms like (1). Using them one can sometimes reduce the ring descriptions to one form and thus establish the isomorphism between the given rings.

A.V.Bocharov had noticed that the transformations of type (1) can be formulated as rewriting rules.

3 Background language and implementations

CAC is written in FLAC (author V.L.Kistlerov), which is a *Functional Language for Algebraic Computations* [6, 2, 8]. FLAC is a dialect of the REFAL language (that was developed by V.F.Turchin in the sixties [9]) and it is based on the mechanism of *term substitution and pattern matching*.

CAC makes use of the following features of the FLAC system:

— "lazy" copying of the trees representing the subterms during the evaluation,

— input-output *preprocessors — frames*,

— *dialogue* system,

— built-in infinite-length rational arithmetic.

CAC was implemented in FLAC system for MSDOS, IBM PC AT-386 and System V UNIX, Motorola-68020 in the Program Systems Institute, Pereslavl-Zalessky [2, 8].

Memory: CAC occupies about 200K bytes of interpreter code.

Computers and software: The FLAC-CAC system can be compiled under any 32 bit C compiler.

4 The CAC programmer concept

CAC consists of *constructors, interpreters* and *general interpreter* $. Constructors set operation *definitions* and other attributes in the *domain* term. Each definition contains the *interpreter* name &Int and the *parameter* list #p. When the general interpreter is given a K-operation, an argument #arg and a domain &Dom, it applies &Int from the corresponding definition to the list #arg #par. Each interpreter presents the general solution to a certain algebraic task.

Functionality: CAC is purely functional — with the exception of using the global variables (stacks) to organize the *dialogue*.

5 Computing with CAC. Example

To build the field F_{25} of 25 elements we enter the FLAC Dialogue and construct successively (see comments inside /* */) :

```
._ Z5: (QuotientRing  Z~ (5) );        /* build the residue ring */
                                 /* Z/(5) and store it in stack Z5 */
._ Z5X: (PolAlgebra Z5~ (x) (named Z5X));      /* Z5X = Z5[x] */
            /* the coefficient ring term being extracted from Z5 */
._ p: ($ Z5X~ read  x~2+2 );
              /* read expression to internal form as an element */
              /* of the ring Z5X and store it in stack p        */
._ F25: (QuotientRing  Z5X~ ((p~)) ) ;        /* F25 = Z5X/(p) */
              /* F25  is a field of 25 elements for  p */
              /* is irreducible over Z5 and  deg(p) = 2 */
._ (properties F25~)  (construction F25~ short);
@: field  (QuotientRing Z5X (...))
._ interface: F25~;   /* prepare easy reading-writing for F25 */
._ (load ReadWrite AlgExpr);  /* Now do computations in F25: */
._ x. ;          /* read  x  as an element of current domain */
@: ((1 1))         /* this is the internal form, */
._ (W @);          /* write it as an element of current domain */
@: x
._ (W  1. /. (x+1). );      /* inversion in the current domain */
@: 3*x-3        /* Now compute the GCD of polynomials over F25 */
._ B: (PolAlgebra  F25~ (y z) ) ;   (RETOP  interface B~);
._ f: (R  y~5 + z~5 );    g:  (R  (y+z)~2 * (2*y*z+1) ) ;
._ (W  ($ B~ GCD  f~ g~) );
@: y~2 + 2*y*z + z~2
              /* y~5+z~5 = (y+z)~5  in B,  hence GCD = (y+z)~2 */
```

6 Conclusions

FLAC and CAC systems provide a certain *shell* in which an algorithmic solution for an algebraic task can be easily written as a program.

6.1 Future of the project

In the nearest future, we are planning to implement:

— domain reconstructions and standard isomorphisms mentioned in section 2.4,

— a more general version of the *fraction ring* constructor and some other constructors,

— a more sophisticated deduction method for the object properties,

— factorization, computing in arbitrary algebraic extensions, primary decomposition of ideals, etc.

6.2 Comparing with the AXIOM system

A categorical view of algebraic calculation is being developed in the AXIOM project of IBM since 1980 [4, 3]. CAC is many times smaller than AXIOM, it contains few methods and it is only a demonstration of a certain easy and general approach to the programming of algebra.

At present CAC deals mainly with the Construction-Defined Commutative Noetherian rings. It cannot manipulate such objects as *categories* or *abstract data types*.

Acknowledgements

The author is grateful to the Multiprocessor Research Center of the Program Systems Institute for the opportunity to continue the studies during the last three years. He is obliged to Dr. S.V.Duzhin and DISCO referees for their numerous comments and advices.

References

1. Buchberger, B.: Gröbner Bases: An Algorithmic Method in Polynomial Ideal Theory. CAMP. Publ. No. 83–29.0 November 1983
2. Chmutov, S.V., Gaydar, E.A., Ignatovich, I.M., Kozadoy, V.F., Nemytykh, A.P., Pinchuk, V.A.: Implementation of the Symbol Analytic Transformations Language FLAC. Lect. Notes in Comp. Sci., **429** (1990) 276–277
3. Davenport, J.H., Trager, B.M.: SCRATCHPAD's View of Algebra I: Commutative Algebra. Lect. Notes in Comp. Sci. **429** (1990) 40–54
4. Jenks, R.D., Sutor, R.S., et al.: Axiom, the Scientific Computation System. Springer-Verlag, New York-Heidelberg-Berlin (1992).
5. Kandri-Rody, A., Kapur, D.: Computing Gröbner basis of a polynomial ideal over a Euclidean domain. J. of Symbolic Computation **6** (1988) 37–57
6. Kistlerov, V.L.: Design Principles of the Computer Algebra Language FLAC. Preprint of the Institute for Control Science (*in Russian*) (1987) Moscow
7. Meshveliani, S.D.: CAC, the Commutative Algebra Constructor. User's Manual. PSI, Pereslavl-Zalessky (1992)
8. Meshveliani, S.D.: FLAC. Functional Language For Algebraic Computations. Short Manual. Manuscript, PSI, Pereslavl-Zalessky (1992)
9. Turchin, V.F.: Refal-5, Programming Guide and Reference Manual. New England Publishing Co., Holyoke, 1989

Extending *AlgBench* with a Type System

Stephan A. Missura*

Institute for Theoretical Computer Science
ETH Zürich, CH-8092 Zürich, Switzerland
E-mail: missura@inf.ethz.ch

Abstract. This paper presents some design decisions taken in the on-going project of extending the symbolic computation system *AlgBench* with a type system. The only mechanism for structuring values in *Alg-Bench* are classes which are first-class values and anonymous, hence type constructors are simple functions. We show how within this approach abstract datatypes and categories can be modelled. Subtyping by means of embeddings (inclusions) is distinguished from subtyping by coercions to avoid some deficiencies which arise in a coercion-based mechanism. Furthermore, we allow to compute with "declared-only" identifiers to retain the symbolic capabilities of the original *AlgBench*.

1 Introduction

To benefit from the advantages of a type system, the originally untyped interpreter of the core system *AlgBench* [6] was extended by a type system suitable for symbolic computation [8]. This paper describes some design decisions taken in this still ongoing project.

In the next section we give a short overview how to define and specify new datatypes in *AlgBench* and show how categories can be modelled within this approach. Afterwards, we will discuss why we prefer having both *embeddings* (inclusions) and *coercions* parallely available. Finally, we give some arguments for being able to compute in a computer algebra environments with *"declared-only" identifiers*, i.e. identifiers without an explicit value.

2 Abstract Datatypes and Categories

New datatypes have in *AlgBench* *first-class* status and are *anonymous*, in contrast to most object-oriented or abstract datatype specification languages. This allows to write type constructors simply as functions accepting types as arguments or returning types. Furthermore, the creation of them is combined with the *data abstraction* principle [1]. This means that they are automatically abstract datatypes which can be accessed only through a given interface. The simple example of rational numbers shows how this abstraction from the representation is done in *AlgBench* [2]:

---

* Research supported by the Swiss National Science Foundation.
[2] The double colon syntax ("::") is used for declaration.

```
Rational :=
  Class( Integer*Integer,
    [ numerator   (arg::Self) :: Integer      := arg(1),
      denominator (arg::Self) :: Integer      := arg(2),
      (num::Integer) / (den::Integer) :: Self :=
           (num div gcd(num, den), den div gcd(num,den))
    ] )
```

The type constructor `Class`[3] which is used to produce new types, takes as the first argument the *implementation* of the datatype (in our case $\mathbb{Z} \times \mathbb{Z}$) and as a second argument a list of function definitions (the *access functions*). The result of the application can be thought to be a black-box which surrounds the implementation, where the access functions have exclusively the right to access the implementation. The *type variable* `Self` has to be used to be able to name this anonymous black-box.

In the body of these access functions those parameters which are declared as `Self` become the type of the implementation (namely `Integer*Integer`), and those values which we are returning as results get the type `Self`. We call this process "unboxing" and "boxing", respectively, of values. The rational arithmetic can now be written in terms of these access functions.

We treat categories[4] to be classes of classes, where an algebraic structure is a class with operations defined on it. Therefore we need only classes for structuring data. This is reflected in the definition of the class Monoid. Its instances are classes which obey at least the specification inside the definition[5]:

```
Monoid := Class(Class( [ 1 :: Self,
                         * :: Self*Self-->Self,
                         (x::Self) * 1 := x,
                         1 * (x::Self) := x
                       ] ))
```

Strings are an example of a (free) monoid, hence we can declare String to be a monoid, in the same way as we declare any identifier: `String :: Monoid`, compute with its instances in a term-algebraic way and add (optionally) an implementation to it in the same way as we did define the rationals above[6]:

```
String :=
  Monoid( List(Character),
    [ 1 :: Self                          := [],
      (l::Self) * (r::Self) :: Self      := l * r,
      embed (char::Character) :: Self    := [char]
    ] )
```

---

[3] The word `Class` should remind us of the denotation of a type: a class.
[4] Categories in the sense of Axiom [5], i.e. without morphisms.
[5] In addition, the binary operator * has to be associative.
[6] The operator `embed` in the example is explained in the next section.

3   Why Embeddings And Coercions are Both Needed

Subtyping is a relation on types which is used by the compiler or interpreter for converting implicitely elements. It usually defines one of the following two relations on types called the embedding hierarchy and coercion hierarchy respectively:

1. One domain can be embedded into another by means of an *embedding* or *inclusion* (an injective function which usually commutes with some operations, i.e. a monomorphism), used for example in order-sorted algebras [3].
2. Elements of one domain can be coerced into elements of an another with the help of a coercion function (implicit conversion) which needn't to be injective [10].

One can quickly verify that both relations are at least pre-orders[7]. They don't have to be anti-symmetric because we can have bijective coercions between isomorphic domains as the example of the two function spaces $A \rightarrow (B \rightarrow C)$ and $A \times B \rightarrow C$ shows.

The only difference between the two notions is the property of *injectiveness* in the case of embeddings. This allows to move up values in the type hierarchy without any loss of information, in contrast to coercions where we can loose information.

Every embedding is clearly also a coercion. Why shouldn't we therefore unify the two concepts into one as it is done for example in Axiom where only coercions are available? The following arguments show why we abandoned this idea and provide embeddings and coercions in *AlgBench*:

Axiom and *AlgBench* both allow heterogeneous lists[8]. The former uses the coercion hierarchy for converting elements into a "maximal" type, whereas the latter uses the embedding hierarchy for the same purpose. $\mathbb{Z}$ is in Axiom a subtype of the finite field $\mathbb{Z}_5$ because there exists a coercion $m \mapsto m \bmod 5$ of type $\mathbb{Z} \rightarrow \mathbb{Z}_5$. Suppose we have a list of integers in addition to at least one element of $\mathbb{Z}_5$. Then the "maximal" type of the list is $\mathbb{Z}_5$ and all elements of the list are converted into the domain $\mathbb{Z}_5$. This behaviour is usually not intended and not very intuitive. In contrast to Axiom, *AlgBench* uses the embedding hierarchy to "lift elements" and finds the top element in the hierarchy, namely Object.

An embedding f allows a notion of a (total[9]) *projection function* which is the left-inverse of f. This left-inverse, which in general does not exist for coercions, can be used to *simplify* expressions: E.g. if the denominator of a rational number is 1 we can project it back into the integers.

These two arguments lead us to the conclusion that the notion of an embedding is indeed needed. We could now ask the reverse question: Do we need coercions at all? We think yes, because there are a lot of examples where coercion are really useful [2].

---

[7] As it is shown in [4] they have to be restricted more such that they stay decidable.
[8] In contrast to most functional languages.
[9] In contrast to Axiom or OBJ [3] where the so-called *retracts* are partial.

The authors in [7], where one can find all the subtyping rules of embeddings and coercions, too, came essentially to the same conclusion that both notions are needed. But in contrast to [7] we don't think the embedding hierarchy has to be anti-symmetric.

To establish an embedding or coercion between two types we adopted Axiom's notion, i.e., using two special operators embed and coerce respectively. E.g. we can embed the integers into the rationals or we can establish a coercion between $\mathbb{Z}$ and $\mathbb{F}_5$:

```
embed  :: Integer-->Rational  := (n |-> n/1)
coerce :: Integer-->Modulo(5) := (n |-> n mod 5)
```

4   Symbolic computation in computer algebra systems

One of the big advantages of symbolic computation languages over conventional languages is the capability to compute with symbols. So we can build terms like 1+x, 1/sin(x), ... without needing an explicit value for the identifier x. In a computer algebra environment we usually have the algebraic view in mind (a semantics), i.e., what elements in abstract mathematical domains do these terms *denote*? Does the first term denote the real number $1 + x$ or an element in the polynomial ring $\mathbb{Z}[x]$? Therefore we need further information besides the terms: e.g. $x \in \mathbb{R}$.

A type system is very helpful to support this more semantic than syntactic view (especially if overloading of identifiers is available), because types allow an easy distinction between several interpretations of a term. Concretely, we can *declare* identifiers with a type which corresponds to the phrase "let $x \in domain$" in mathematics. Untyped computer algebra languages tend to mix identifiers (syntactical entities) and indeterminates (semantical entities) which results in mixing syntactical and semantic stages.

But this declaration of identifiers is not enough: we need to be able to *compute* with them in a term-algebraic way without giving an explicit value (they are "declared-only") because that is the key feature of symbolic computation! This is not possible in Axiom, so it can't be counted as symbolic computation system. In contrast, logical frameworks like Isabelle [9] or the OBJ system [3], both not designed exclusively for computer algebra, and *AlgBench* can be viewed as (typed) symbolic computation systems.

Examples of declared-only identifiers are π and e, the base of the natural logarithm, which can't be defined with an exact value but with relations they fulfill; say $\sin(\pi) = 0, \pi > 0, e^0 = 1, \ldots$

In *AlgBench* we would simply declare them to be of type Real:

```
pi :: Real ; e :: Real ; sin :: Real-->Real
```

define some facts about them:

```
sin (pi'Real) := 0  ;  pi > 0 := true ; e^0 := 1
```

and compute then in a term-algebraic fashion. For example the expression "sin(sin(pi))-e*pi" which is type correct would evaluate to "-e*pi" with type Real. In fact, this type is also an example of an identifier without a explicit value and would be in *AlgBench* simply declared to be a field: Real :: Field.

The conclusion is that a system which allows to compute with "declared-only" identifiers can model some mathematical facts in a more natural way.

5 Conclusions and Acknowledgements

We have described some design decisions for a type system in a computer algebra environment which can improve the capability of describing mathematical facts, namely: classes as the only way for structuring data, embeddings instead of coercions for subtyping and computing with "declared-only" identifiers.

An experimental version of the type system was built into the the *AlgBench* interpreter using the language C++ [8]. Future work consists of making the concrete syntax of *AlgBench* more flexible (like OBJ's mixfix syntax).

We thank Roman Maeder and Georgios Grivas for reviewing a draft.

References

1. H. Abelson and G. J. Sussman. *Structure and Interpretation of Computer Programs*. The MIT Press, Cambridge, Mass., 1985.
2. L. Cardelli and P. Wegner. On Understanding, Data Abstraction and Polymorphism. *ACM Computing Surveys*, 17:471–522, 1985.
3. Joseph A. Goguen et al. Introducing OBJ. Technical report, SRI International, March 1992.
4. A. Fortenbacher. Efficient type inference and coercion in computer algebra. In *DISCO90 (Lecture Notes in Computer Science)*, pages 56–60. Springer Verlag, 1990.
5. Richard D. Jenks and Robert S. Sutor. *Axiom: the scientific computation system*. Springer, 1992.
6. Roman E. Maeder. Algbench: An object-oriented symbolic core system. In J. H. Davenport, editor, *Design and Implementation of Symbolic Computation Systems (Proceedings of DISCO '92)*. (to appear), 1992.
7. N. Marti-Oliet and J. Meseguer. Inclusions and Subtypes. Technical report, SRI International, December 1990.
8. Stephan A. Missura. Klassenbasierte Umgebung für algebraische Modellierungen in AlgBench. Diploma Thesis, ETH Zurich, 1992.
9. Lawrence C. Paulson. Introduction to Isabelle. Technical report, Computer Laboratory, University of Cambridge, 1993.
10. John C. Reynolds. Using category theory to design implicit conversions and generic operators. In Neil D. Jones, editor, *Semantics-Directed Compiler Generation, Workshop*, volume 94 of *Lecture Notes in Computer Science*, pages 211–258, Aarhus, Denmark, January 1980. Springer-Verlag.

Modeling Finite Fields with Mathematica.
Applications to the Computation of Exponential Sums and to the Solution of Equations over Finite Fields.

Antonio Vantaggiato

Computer Science Program, University of the Sacred Heart
PO Box 12383 San Juan, Puerto Rico 00914-0383

Abstract: This paper proposes an implementation model for finite fields $GF[m^q]$, m prime, based on a hybrid architecture that integrates symbolic programming developed in Mathematica with an imperative C language module. Its aim is to enable the user to write algorithms to perform calculations in GF's by using Mathematica's programming language and built-in math functions. First, the system's architecture is presented and it is shown that the proposed model has linear time complexity ($O(q)$) for all algebraic operations. Finally, we show the developed modules for the computation of exponential sums and the solution of equations over finite fields.

1. Introduction

We considered symbolic programming environments as possible tools for the solution of very complex problems in finite fields such as the computation of character sums in coding theory. We set our goal to design an integrated tool, within one of such environments, that was as efficient, and, at the same time, as flexible and easy to use as possible. Our architecture, embedded in Mathematica, supports finite field structures and operations: the user is able to write algorithms to perform calculations in GF's by using the environment's programming language and built-in math functions.

We designed a *hybrid two-level architecture*: a high-level module within Mathematica that communicates with another module (containing the basic data structures in C) that runs parallel to the first, through the MathLink [5] protocol.

Two (one-dimension) arrays efficiently implement the accessor procedures to the field elements and arithmetic operations. In particular, our model provides for both the accessor function and its inverse, and so allows for an effective implementation of multiplication and reciprocals. We also use a coding mechanism that allows for simple and space-conscious representation and storage of field elements.

2. A Proposed Model for Finite Fields

The architecture of the proposed system is based on two separate levels that

This research was supported by the Puerto Rico Research Center for Science and Engineering and by the Center for Academic Research of the University of the Sacred Heart.

correspond to two processes that run simultaneously and that communicate through the MathLink protocol. The high-level process (where the end-user interacts and works), is implemented in Mathematica. The user will be accessing finite field arithmetics via Mathematica definitions and rules. These rules call their counterparts in the C low-level process, in which most computations occur.

The C process sets up the field once the latter is activated from Mathematica: its elements are computed and stored (in encoded form) into two arrays so to implement the basic selector operator and its inverse. Also, the C-level contains encoding and decoding functions that (uniquely) map field elements to Integers.

2.1 The Finite Field Abstract Data Type

Let t be the primitive element of a Galois Field $GF(m^q)$, (m prime).

1) Constructors.

Constructors for our ADT are entirely closed within the C level program (they are not accessible by the user): the field is constructed and set up by means of two arrays. Once the field is built, the user will need to use only selector operators. Such two arrays implement the following functions:

table1: $N \rightarrow N$ | table1(i) = Code(t^i)

Array table2 implements the inverse function of table1 (a sort of finite field logarithm):

table2: $N \rightarrow N$ | table2(Code(t^k)) = k

2) Selectors.

The selector operator is defined in Mathematica by means of up-values associated with t by overloading the built-in Power function:

```
t/:     t^i_Integer := table[i]   /; i >= q;
t/:     t^i_Integer := Reciprocal[t^(-i)] /; Negative[i];
```

Functions table and Reciprocal are direct calls to the C language functions that are activated through MathLink, as described above. Function table2 (pertaining only to the C level) is not accessible to the user. It is immediately observed that through this data type design, constructor and selector operators have time complexity of the order of $O(q)$[1], where q, even for large fields, is usually small.

3) Arithmetics.

The arithmetic operations between elements of a finite field are also defined in the C level and are activated through mirror functions in Mathematica. They are:

i) Addition: fPlus
    Exclusive Or between the two operands' codes   when m is 2;
    Modular addition   otherwise
ii) Multiplication: fTimes

---

[1]From a macroscopic point of view, access to a field element would require a *constant-time operation* (a direct access to an array component), but given our coding mechanism, every access to a field element implies a decoding operation that requires a loop of q steps. Therefore it is correct to state that its time complexity is *linear*, or $O(q)$.

iii) Quotient: fDivide   pol1 / pol2 = pol1 * reciprocal(pol2)

First we find the value of k such that $t^k = pol2$. Then $t^{(-k)} = t^{(m^q - k - 1)}$.

iv) Power: fPower

Let m = 2 and q = 9 (GF[2^9]); the computations performed when evaluating a product or a power, say $(t^2 + t + 1)^5$ are:

```
Encode[(t² + t + 1)] -> 7
table2[7] -> 223,      i.e. t²²³ = t² + t + 1.
result = (t²²³)⁵ = t¹¹¹⁵ = t⁹³ = Decode[table1[93]] = t² + t⁴ + t⁸.
```

All these steps, of course, are transparent to the user.

Here follow some examples in GF[2^9].

```
In[5]:= k = t^167
                2   3   6    7    8
Out[5]= 1 + t  + t + t  + t  + t

In[6]:= Timing[Reciprocal[k]]
                                  4   6    8
Out[6]= {0.0333333 Second, 1 + t + t  + t  }

In[7]:= fTimes[k, Reciprocal[k]]

Out[7]= 1

In[8]:= (1+t+t^5)^-10
            4   5    8
Out[8]= t + t + t  + t
```

The Galois Field as implemented in our system shows linear time complexity on all algebraic operations. Allocated memory comes to amount to 2 m^q words (a word being the machine-dependent number of bytes used to represent an integer). We observe that the proposed implementation dramatically reduces the memory requirements of such a system [2], in that, by using a coding mechanism, it needs only one $\lceil \log_2 (m^q - 1) \rceil$-bit word (for a field GF[m^q]) to represent and store a field element.

3. Solution of problems in Galois Fields

We concentrated our efforts in two basic problems, both characterized by extensive computational complexity: character sums and solution of polynomial equations in one variable over a finite field.

We designed the second level of abstraction in Mathematica so to add higher-level calculation capabilities and functions to be used by the algorithms for character sum computation and the solution of equations:

•The Greatest Common Divisor[2] between any two field polynomials (implemented

---

[2]It is worth noting that the defined implementation in levels of abstraction has allowed for

with the Euclidean algorithm) [3];

•The Trace function [3, 6]: $$Tr(x) = \sum_{i=1}^{m-1} \left(x^2\right)^i \qquad GF(2^m) \to GF(2).$$

•The Power function.
The ADT fPower function cannot be used, because it is only valid for "constant" polynomials, i.e. elements of the field. The built-in function Power is therefore overloaded using properties of finite fields of characteristic 2.

3.1 Character Sums

Character Sums are defined as follows:
$$\sum_{x \in GF(2^m)} (-1)^{Tr(f[x])} = |GF(2^m)| - 2\sum_{x \in GF(2^m)} Tr(f[x])$$
where $|GF(2^m)|$ indicates the cardinality of GF. Tipically, it has been choosen:
$f(x) = x + x^3 + x^5 + ... + h\,x^{2m-1} = f(x, h)$ in $GF(2^9)$, i.e. m=9.
The definitions of both functions are as follows:

```
f[n_, h_] := fPlus[Fold[fPlus,t^n,
                  {t^3 n,t^5 n,t^7 n,t^9 n,t^11 n,t^13 n,t^15 n}]],
              fTimes[h, t^(17 n)]]];

charSum[f_Symbol, h_]:= 2^q - 2 Apply[Plus, Table[Tr[f[j, h]],
                                            {j,0,2^q-1}]
```

Here is one example with timing (run on a 19 MIPS Sun SPARCstation under Unix).

```
In[11]:= Timing[charSum[f, t^2]]

Out[11]= {105.867 Second, -8}
```

3.2 Solution of a polynomial equation in one variable over a finite field $GF(2^m)$

We implemented a simple algorithm [3], a modification of Berlekamp's, that is very well suited for large fields with small characteristic. It is based on the property that any polynomial $f(z) = \Pi_{i=1\ ->\ n}\ (z - g_i)$, with $g_i \in GF[p^m]$, can be written as:
$f(z) = \Pi_{c \in GF[p]}\ fGCD(f(z),\ Strace(t^j z) - c)$
for $0 \le j \le m-1$. (Strace is the symbolic trace function.) This formula calls for the calculation of only p GCD's.

fFactor takes a polynomial p(z) and a variable symbol, z, and (if possible) factors

---

the use of Mathematica's built-in function PolynomialQuotient (among others) in implementing the finite field GCD operation, with no modifications. Also, operations among field elements carry over other Mathematica's primitives. For instance, one can define a matrix over the field and invert it using Inverse.

the polynomial with respect to z, according to the following steps:

Initialization:

```
listOfFactors = Nil;
```

Computation of congruences z^n Modulo p(z) by repeated squaring, $n = 2,4...2^{m-1}$;
Recursive part:

i)   Normalization of p(z) so that its first coefficient is 1;

ii)   `trace = Strace(t`^{j} `z)`, the symbolic absolute trace function for a value of j (j = 0,1...m-1) such that `trace` is nontrivial;

iii)   Append to `listOfFactors` the lists produced by the two recursive calls to

- `•fFactor[fGCD[p, trace], z]` and
- `•fFactor[fGCD[p, trace + 1], z]`,

through points i), ii), iii).

```
In[6]:= fFactor[t+t^3+ (t+t^4+t^5) x + (t+t^2+t^3+t^5) x^2 + x^3, x]
                     2      3               5
Out[6]= {1 + t + x, t  + t  + x, 1 + t  + x}
```

Example in $GF[2^6]$

Conclusions. We have developed a linear model for finite fields based on a hybrid architecture. This paper shows how the model allows the user to write algorithms to perform calculations in GF's. We feel that the proposed implementation should be easily portable across platforms as soon as some minor adjustments are made to the C portion of the code. Our most immediate plans for the future are to work on the algorithms' efficiency, to expand the developed model to deal with finite fields of whatever characteristic, and to use it to study some functions (including character sums) over very large finite fields.

Acknowledgements. We wish to thank Prof. Oscar Moreno, of the University of Puerto Rico, and Profs. Mayra Alonso and Luciano Rodríguez, of the University of the Sacred Heart. We would also like to express gratitude to a student of ours, José Pérez, who developed most of the C language code of our model.

References

1. A. Cáceres, O. Moreno: On the Estimation of Minimum Distance of Duals of BCH Codes. Congressus Numerantium 81, (1991).
2. J.H. Davenport: Current Problems in Computer Algebra Systems Design. In: A. Miola (ed.): Design and Implementation of Symbolic Computation Systems (DISCO '90). Berlin: Springer-Verlag 1990.
3. R. Lidl, H. Niederreiter. Finite Fields. In: G.C. Rota (ed): Encyclopedia of Mathematics and its Applications. Cambridge: Cambridge University Press 1984.
4. R. Maeder. Abstract Data Types. Tutorial Notes, Mathematica Conference, Boston. Wolfram Research 1992.
5. Wolfram Research, Inc. MathLink Reference Guide. Champain, Ill. 1992.
6. C. Moreno, O. Moreno. Exponential Sums and Goppa Codes I, Proc. of the American Mathematical Society, Vol. 3, No. 2, (Feb. 1991).
7. S. Wolfram. Mathematica. Reading, Mass.: Addison-Wesley 1992.

An Enhanced Sequent Calculus
for Reasoning in a given Domain *

S. Bonamico and G. Cioni and A. Colagrossi

Istituto di Analisi dei Sistemi ed Informatica del C.N.R.
Viale Manzoni 30, 00185 Roma, Italy
e-mail: CIONI@IASI.RM.CNR.IT

1 Introduction

In this paper we present a system able to perform reasoning processes on mathematical objects in a given domain, i.e. a system that furnishes solutions for the three kinds of inferential problems, namely verificative, generative and abductive.

The system implements an enhanced sequent calculus and is an extension of the desk top machine presented in [3].

The problem of reasoning in a mathematical environment has been recently approached in two different ways. Both approaches provide an architecture based on two fundamental components, namely the automated deduction and the computational components. The approaches differ from one another in the roles played by such components. One direction, followed by [1], integrates the deduction tool in a computational system: the reasoning process is, then, activated by the computational system itself. A different approach is followed, for example, by [6]: a deduction tool activates the computational system services only when a specific computation is required.

Our work follows the latter approach. The proposed method is a specialization of the reasoning tool based on a sequent calculus presented in [2, 4] and [3]. This sequent calculus allows to verify the validity of properties of mathematical objects expressed by formulas in first order logic and, in the case the verification fails, to abduce or generate new formulas (expressions of new properties) which, once added to the original formulas, make them valid. This paper presents a sequent calculus to be used for reasoning in given domains of interpretation.

Here, we refer to the sequent calculus based on that proposed in [5] and extended in [4] to the treatment of different kinds of problems, namely verificative, generative and abductive. This sequent calculus is actually implemented in a desk top reasoning machine [3].

In this section we only recall the basic definitions of the sequent calculus given in [4], needed to the development of automated reasoning procedures in a given domain.

---

* This work has been partially supported by Progetto Finalizzato "Sistemi Informatici e Calcolo Parallelo" of CNR

A *sequent* is a pair: $\Gamma \Rightarrow \Delta$, where Γ and Δ are respectively a conjunction and a disjunction of first order wffs, without function symbols. The usual representation of a sequent is as follows:

$$\alpha_1, \ldots \alpha_{k-1}, \alpha_k, \alpha_{k+1}, \ldots, \alpha_n \Rightarrow \beta_1, \ldots, \beta_{h-1}, \beta_h, \beta_{h+1}, \ldots, \beta_m$$

Γ, i.e. $\alpha_1, \ldots, \alpha_n$, is called antecedent and Δ, i.e. $\beta_1, \ldots, \beta_m$, is called succedent. Given an interpretation I, the sequent s as in (1) is satisfied in I if the formula $\alpha_1 \wedge \ldots \wedge \alpha_n \supset \beta_1 \vee \ldots \vee \beta_m$ is satisfied in I. A sequent s as in (1) is *basic satisfied* if either:

- in the antecedent one or more formulas have False as truth value; or
- in the succedent one or more formulas have True as truth value; or
- there exists in the antecedent a formula α which is equivalent to a formula β in the succedent.

On a given sequent a set of decomposition rules can be applied, one for each connective of the sequent. We do not list the rules, which can be found in [3] or [4].

In the next section a proving procedure for satisfiability problems will be described after the introduction of the necessary definitions regarding the interpretation in a specific domain.

2   Proving in a given domain

An interpretation I is defined as a triple: $I = \langle I_D, I_S, I_F \rangle$, where: I_D is the interpretation domain; I_S is a triple: $I_S = \langle S_P, S_V, S_C \rangle$, where S_P, S_V and S_C are sets of predicates, variables and constants, respectively; I_F is a triple $I_F = \langle F_P, F_V, F_C \rangle$ of mappings such that: $F_P : S_P \times I_D^r \to \{True, False\}$, where r is the arity of the considered predicate; $F_V : S_V \to I_D$ and $F_C : S_C \to I_D$ The

mathematical operations to be performed over the elements of I_D, as well as the properties those elements satisfy, are expressed by giving a semantics to the predicates. Then, we associate to each predicate P a *predicate shape* PS, defined as a triple: $PS = \langle \mathcal{I}^P, \mathcal{O}^P, \mathcal{C}^P \rangle$

Given a tuple X of k elements of S_V: $X = \langle x_1, \ldots, x_k \rangle$, where x_i are identifiers for variables of the language, a tuple $T = \langle t_1, \ldots, t_k \rangle$ is called a *total assignment* for X if $t_i \in I_D$ for $i = 1, \ldots, k$ and the term t_i is assigned to the variable x_i, for $i = 1, \ldots, k$ (in short $x_i \leftarrow t_i$). T is called a *partial assignment* for X if only

$r < k$ elements $t_{j_1}, \ldots, t_{j_r}$ of T belong to I_D and $x_{j_m} \leftarrow t_{j_m}$ for $m = 1, \ldots, r$.

On the basis of these definitions we can now consider the predicate shape PS^P. Given n, the arity of the predicate P, we have [2] :

---

[2] For brevity sake, here and in the following we will use the notation $T_i = \langle t_1, \ldots, t_{i-1}, t_{i+1}, \ldots, t_n \rangle$, for $i = 1, \ldots, n$ with the meaning: $T_1 = \langle t_2, \ldots, t_n \rangle$, $T_n = \langle t_1, \ldots, t_{n-1} \rangle$, $T_i = \langle t_1, \ldots, t_{i-1}, t_{i+1}, \ldots, t_n \rangle$, for $i = 2, \ldots, n-1$.

- $\mathcal{I}^P$ is a set of $n+1$ tuples $\mathcal{I}^P = \{\mathcal{I}_0^p, \mathcal{I}_1^p, \ldots, \mathcal{I}_n^p\}$; $\mathcal{I}_i^p$, for $i = 0, \ldots, n$, are called *input variables tuples* and are defined in the following way:
  $\mathcal{I}_0^p$ is a n-tuple of elements of S_V: $\mathcal{I}_0^p = \langle x_1, \ldots, x_n \rangle$;
  $\mathcal{I}_i^p$, for $i = 1, \ldots, n$, are n (n-1)-tuples of elements of S_V: $\mathcal{I}_i^p = \langle x_1, \ldots, x_{i-1}, x_{i+1}, \ldots, x_n \rangle$ for $i = 1, \ldots, n$.
- $\mathcal{O}^P$ is a set of n+1 tuples $\mathcal{O}^P = \{\mathcal{O}_0^p, \mathcal{O}_1^p, \ldots, \mathcal{O}_n^p\}$; $\mathcal{O}_i^p$ are called *output variables tuples* and are defined in the following way:
  $\mathcal{O}_0^p$ is the empty tuple; $\mathcal{O}_i^p$, for $i = 1, \ldots, n$, are tuples of only an element of S_v: $\mathcal{O}_i^p = \langle x_i \rangle$, for $i = 1, \ldots, n$.
- $\mathcal{C}^P$ is a set of n+1 algorithms $\mathcal{C}^P = \{\mathcal{C}_0^p, \mathcal{C}_1^p, \ldots, \mathcal{C}_n^p\}$ where:
  each $\mathcal{C}_i^p$, for $i = 1, \ldots, n$, is an algorithm that, given as input a total assignment $\langle y_1, \ldots, y_{i-1}, y_{i+1}, \ldots y_n \rangle$ for $\mathcal{I}_i^p$, computes $\langle y_i \rangle$, a total assignment for $\mathcal{O}_i^p$;
  $\mathcal{C}_0^p$ is an algorithm that, given as input a total assignment $\langle y_1, \ldots, y_n \rangle$ for $\mathcal{I}_0^p$, returns a boolean value: True, if for $i = 1, \ldots, n$, all the tuples $\langle y_1, \ldots, y_{i-1}, y_{i+1}, \ldots, y_n \rangle$ are total assignments for $\mathcal{I}_i^p$ and all the tuples $\langle y_i \rangle$ are total assignments for $\mathcal{O}_i^p$; False, otherwise.

Due to the introduction of the interpretation, the decomposition rules concerning quantified formulas show some little differences in comparison with those given in [4]. Let s be a sequent: $Y = \{y_0, y_1, \ldots\}$ is the set of all symbols not occurring in s; $\Theta = \{\langle t_0, \Phi_0 \rangle, \ldots, \langle t_k, \Phi_k \rangle\}$ is a set of pairs where $t_i \in I_D \cup Y$ and Φ_i are sets of quantified wffs occurring in s.
The rules concerning quantified formulas are defined in the following way:
$\underline{\exists A_D}$
given the sequent

$$\Gamma_1, \exists x \alpha, \Gamma_2 \Rightarrow \Delta$$

let $P(x_1, \ldots, x_{i-1}, x, x_{i+1}, \ldots, x_n)$ be a proposition occurring in α and x occurring as the i-th variable in P;
if $\langle x_1, \ldots, x_{i-1}, x_{i+1}, \ldots, x_n \rangle$ is a total assignment for $\mathcal{I}_i^p$,
    then   **begin**
          the following sequent is obtained: $\Gamma_1, \alpha[d/x], \Gamma_2 \Rightarrow \Delta$
          where $\langle d \rangle$ is the total assignment for $\mathcal{O}_i^p$ returned by $\mathcal{C}_i^p$;
          if d is not present as first element of any pair in Θ,
             then add the pair $\langle d, \emptyset \rangle$ to Θ;
    end
    else   **begin**
          the sequent $\Gamma_1, \alpha[y/x], \Gamma_2 \Rightarrow \Delta$ is returned where $y \in Y$; and the
          pair $\langle y, \emptyset \rangle$ is added to Θ and y is removed from Y.
    end

The rules $\underline{\forall S_D}$, $\underline{\forall A_D}$ and $\underline{\exists S_D}$ are defined analogously. The following theorem holds:
Theorem: a sequent is satisfied under a given interpretation if it can be decomposed, by applying the decomposition rules in all basic satisfied sequents.

The proposed procedure $\mathcal{P}_D$ for proving the satisfiability of a sequent under a given interpretation will use the sets Θ and Y and the following *association list* $\Lambda : (\langle y_0, d_0 \rangle \langle y_1, d_1 \rangle \ldots \langle y_l, d_l \rangle)$ whose elements $\langle y_i, d_i \rangle$ are pairs such that y_i are symbols (removed from) of Y and $d_i \in I_D$.

Such a procedure is a modification of that given in [4] and uses two algorithms, namely the *Truth Evaluation and Elimination in a domain* and the *Computation in a domain and Substitution*, which are given in the following.

Algorithm $\mathcal{E}_D$ (Truth Evaluation and Elimination in a domain)

Input: a sequent r, a set L of sequents and an interpretation I.

Output: the sequent r and the set L of sequents purged by atomic component formulas.

Step 1 (Evaluation): for each proposition $P(d_1, \ldots, d_n)$, belonging to r or L, with $\langle d_1, \ldots, d_n \rangle$ a total assignment for $\mathcal{I}_0^p$, compute the total assignment for $\mathcal{O}_0^p$ by the Algorithm $\mathcal{C}_0^p$ giving it as input the total assignment $\langle d_1, \ldots, d_n \rangle$ for $\mathcal{I}_0^p$.

Step 2 (Elimination): delete all the atomic formulas of the antecedents and succedents of r and L which have respectively True and False truth values.

Algorithm $\mathcal{CS}_D$ (Computation in a domain and Substitution)

Input: a sequent r, a set of sequents L, an association list Λ, an interpretation I.

Output: the sequent r, a set of sequents L, an association list Λ, updated with the computed values.

Step 1: for each proposition $P(y_1, \ldots, y_n)$, occurring in the sequent r or in some other sequent of L, such that there are in Λ the pairs $\langle y_{i_1}, d_{i_1} \rangle \ldots \langle y_{i_r}, d_{i_r} \rangle$, with $r > 0$, substitute the d_{i_j} for the y_{i_j} for $j = 1, \ldots, r$.

Step 2: **while** there exists, in the sequent r or in some other sequents of L, a proposition $P(d_1, \ldots, d_{i-1}, y, d_{i+1}, \ldots, d_n)$ such that:
- y has been removed from Y, and
- y doesn't belong to any pair of Λ, and
- $\langle d_1, \ldots, d_{i-1}, d_{i+1}, \ldots d_n \rangle$ is a total assignment for $\mathcal{I}_i^p$;

do

compute the total assignment $\langle d_1 \rangle$ for $\mathcal{O}_i^p$ using $\mathcal{C}_i^p$ giving it as input the total assignment $\langle d_1, \ldots, d_{i-1}, d_{i+1}, \ldots d_n \rangle$ for $\mathcal{I}_i^p$;

substitute d_i for y in all the predicates where y occurs;

insert the pair $\langle y, d_i \rangle$ in Λ.

3   Reasoning in a given domain

When a sequent is proved to be not satisfied under a given interpretation the reasoning process can be started. The reasoning Procedure $\mathcal{R}_d$ furnishes wffs ϕ_i which are either generated wffs, i.e. the sequents $\Gamma \Rightarrow \Delta, \phi_i$ are satisfied under the interpretation I, or abduced wffs, i.e. the sequents $\Gamma, \phi_i \Rightarrow \Delta$ are satisfied under the interpretation I.

In order to generate/abduce the wffs ϕ_i a new set of rules and algorithms, called $\mathcal{A}_1$ and $\mathcal{A}_2$, are needed. The new rules, called recomposition rules are:

$$\mathbf{\underline{S\neg}}\ \frac{\Gamma, \neg\alpha_1, \Delta \Rightarrow \Lambda}{\Gamma, \Delta \Rightarrow \alpha_1, \Lambda} \qquad \mathbf{\underline{A\wedge}}\ \frac{\Gamma, \alpha_1 \wedge \alpha_2, \Delta \Rightarrow \Lambda}{\Gamma, \alpha_1, \alpha_2, \Delta \Rightarrow \Lambda}$$

$$\mathbf{\underline{A\neg}}\ \frac{\Gamma, \Delta \Rightarrow \neg\alpha_1, \Lambda}{\Gamma, \alpha_1, \Delta \Rightarrow \Lambda} \qquad \mathbf{\underline{S\vee}}\ \frac{\Gamma, \Delta \Rightarrow \alpha_1 \vee \alpha_2, \Lambda}{\Gamma, \Delta \Rightarrow \alpha_1, \alpha_2, \Lambda}$$

The Algorithms $\mathcal{A}_1$ and $\mathcal{A}_2$, are described in [4].

As conclusion, the reasoning Procedure $\mathcal{R}_d$ provides a first step which is the application of the Procedure $\mathcal{P}_D$, a second step, if $\mathcal{P}_D$ terminates with a not empty set S of sequents (i.e. the input sequent is not satisfied), which is the application of the Algorithm $\mathcal{A}_1$ and a third step which is the application of the Algorithm $\mathcal{A}_2$. The Procedure $\mathcal{R}_D$ allows to generate or abduce different formulas which, once "added" into the succedent or the antecedent of the sequent given as input of the Procedure $\mathcal{P}_D$, make it valid. In that way generative and abductive logic problems can be solved.

References

1. M.J. Beeson. Logic and computation in mathpert: An expert system for learning mathematics. In E. Kaltofen and S. M. Watt, editors, *Computers and Mathematics*, pages 202–214. Springer Verlag, 1989.
2. S. Bonamico, G. Cioni, and A. Colagrossi. A gentzen based deduction method for mathematical problem solving. In *Proc. of Indian Computing Congress*, 1990.
3. G. Cioni, A. Colagrossi, and A. Miola. A desk top sequent calculus machine. In J. Calmet, editor, *Proc. 1992 Conf. on Artif. Intell. and Symb. Math. Comp.*, LNCS, 1992. Karlshrue.
4. G. Cioni, A. Colagrossi, and A. Miola. A sequent calculus for automatic reasoning. 1993.
5. J.H. Gallier. *Logic for Computer Science*. Harper & Row, 1986.
6. P. Suppes and S. Takahashi. An interactive calculus theorem-prover for continuity properties. *J. Symb. Comp.*, 7:573–590, 1989.

Problem-Oriented Means of Program Specification and Verification in Project SPECTRUM

V.A.Nepomniaschy and A.A.Sulimov

Institute of Informatics Systems
Russian Academy of Sciences, Siberian Division
6, Lavrentiev ave., Novosibirsk 630090, Russia
e-mail: {vnep,sulimov}@isi.itfs.nsk.su

Abstract. In this paper we present an outline of the project SPEC-TRUM based on the problem-oriented method which includes specification languages and knowledge bases. The bases consist of axioms, application strategies and procedures for proving of axiom premises. The axioms are conditional rewriting rules whose premises contain conditions and built-in case analysis. We describe the architecture and special features of the verification system oriented towards sorting, linear algebra and compilation.

1 Introduction

Universal program verification systems based on Hoare's method have advantages as well as shortcomings [1-3]. The essential shortcoming is interactive proving. Indeed, it is difficult to prove interactively large size formulas (verification conditions).

To overcome the difficulty and to extend a space of verification systems application, a problem-oriented method is proposed including:
— problem-oriented specification languages;
— special and modified proof rules for verification condition generation which enable us to simplify specifications;
— problem-oriented knowledge bases consisting of axioms, application strategies and procedures for proving of axiom premises.

Within the framework of the project SPECTRUM the verification system SPECTRUM92 has been designed on the basis of the method. The system is oriented towards such problem areas as sorting, linear algebra and compilation. During experiments a kernel of the knowledge base is extended such that verification conditions proving becomes automatic. Knowledge bases contain both standard axioms represented by conditional rewriting rules and axioms of a new kind including built-in case analysis. The aim of the paper is to describe specification and verification means for the areas mentioned above and their application in the system SPECTRUM92.

2 Specification and Annotation Means

For a problem area the basis of the specification means is the notions intended for description of main program actions and results. The most of the notions are functions and predicates [4,5]. Their properties are described by axioms. Formulas of the specification language are built out of the notions by means of first-order logical operations. The existential quantifier and the negation operation are not used in program annotations. In many cases they may be reduced by means of introducing special notions.

Let $var(F)$ be the set of free variables of a term (formula) F. All axioms are conditional rewriting rules of the form $P \Rightarrow L = R$, where P is the premise, $L = R$ is the conclusion and $var(R) \subset var(L)$. If $R = true$, the axiom is written as $P \Rightarrow L$. The premise P can take one of the following three syntactic forms:

1. A standard premise consisting of conjunctions — $P = P_1 \wedge \ldots \wedge P_n$, where P_i $(i = 1, \ldots, n)$ does not include boolean connectives.
2. A premise with conditions — $P = P_1 \wedge \ldots \wedge P_n$ and P_i $(i = 1, \ldots, n)$ either does not include boolean connectives or contains the only implication.
3. A premise with a marked component — $P = P_1 \wedge \ldots \wedge P_n \wedge \{Q_1 \wedge \ldots \wedge Q_m\}$ where P_i and Q_j $(i = 1, \ldots, n; j = 1, \ldots, m)$ do not include boolean connectives.

Formula e which is a premise of a verification condition is called the environment of axioms being applied in the process of its proving. Now we define the application of the axiom $P \Rightarrow L = R$ in the environment e to a formula f, when $var(P) \subset var(L)$. At first a subformula l of the formula f is searched for matching with the term L. After matching l with the term L the values of all free variables of L are determined. By replacing variables in P and R by their values we obtain the formula $p = p_1 \wedge \ldots \wedge p_n$ (or $p = p_1 \wedge \ldots \wedge p_n \wedge \{q_1 \wedge \ldots \wedge q_m\}$) and the expression r respectively. Our aim is to prove p. Let us consider three cases depending on the kind of the premise P:

1. P is the standard premise. All p_i must be derived from e.
2. P is the premise with conditions. To prove $p_i = c_i \rightarrow q_i$ one would derived q_i from $e \wedge c_i$.
3. P is the premise with marked component. All p_i are proved as above. All q_j are also tried to prove. Let q_j be not proven, then the negation of the formula $q = q_1 \wedge \ldots \wedge q_m$ is tried to prove. If this is successful, the axiom is not applicable to the subformula l of the formula f, otherwise the attempt to prove $l = r$ from $e \wedge q$ is made. Thus the case analysis is used, i.e. the axiom independence of q is verified in environment e.

Let the formula p be proven. The result of the application of the axiom to the subformula l of the formula f is the formula obtained from f by substituting r for l.

Let us consider the axiom $P \Rightarrow L = R$ where there are free variables (called extra) from the set $var(P) \setminus var(L)$. The difference between the application of this axiom and of the axiom written above in the environment e consists only in the way of the extra variables matching. At first the values of all free variables of L are determined as above. In the obtained formula the first conjunct with extra variables is matched with a suitable term of e. After successful matching and replacing extra variables by determined values in P the process continues for rest unmatched extra variables. Further the application of the axiom is continued as above, i.e. the remaining conjuncts of the premise are proved. If the premise is not proven, one tries to match the extra variables with another suitable term of e.

Let us consider examples of axioms A1-A3 from sorting, linear algebra and compilation.

Sorting. Let $ord(A[i..j])$ be the predicate $\forall k(i \leq k < j \rightarrow A[k] \leq A[k+1])$.
A1. $i \leq j \wedge ord(A[k..l]) \wedge \{k \leq i \wedge j \leq l\} \Rightarrow A[i] \leq A[j]$.

Linear algebra. Some of the notions are the following:
— Index sets built out of empty (ϵ) and basic sets by employing union ($\cup$), intersection ($\cap$), and difference ($\backslash$) operations. Some of six basic operations are denoted by $set(k, m)$ (the set of the single element (k, m)), $row(l, m, n)$ (the set describing indices of l-th row of a matrix from m-th to n-th column), $col(m, k, l)$ (the set describing indices of m-th column from m-th to k-th rows).
— Universal quantifiers of the fixed variables u, v bounded by index sets.
— Permutation of segments of matrix's rows and replacement of the matrix's elements whose indices belong to the index set by the value of a given expression.
— Algebraic notions including unitary and inverse matrices, determinant and the column-vector that gives the solution of a system of linear equations.
A2. $l \leq n \wedge n \leq m \Rightarrow col(k, l, m) \backslash set(n, k) = col(k, l, n - 1) \cup col(k, n + 1, m)$

Compilation. The correctness proof of syntax-directed compilers with respect to their specifications is considered. More exactly, such properties as parsing, translation into the intermediate representation, type checking and code generation are specified and verified. A3 is an axiom of the function $FTYPE$ which specifies type checking and semantic constraints.
A3. $(FTYPE(condition, s) \neq' error') \wedge (FTYPE(statement, s) \neq' error') \Rightarrow$
$FTYPE(CON(while, condition, do, statement), s) \neq' error'$,
where s is a set of symbol definitions, CON is the concatenation.

We use two axiom application strategies for sorting and linear algebra. The first strategy consists in the attempts to apply one axiom followed by another to the very end and to repeat this process until no axiom can be applied. The second one also tries to apply the axioms successively and returns to the first axiom immediately after an axiom having been applied. In the latter case every two axioms whose left parts of conclusions are equivalent (they can be matched with the same term) have mutually exclusive premises. Under this condition both efficiency of the axiom application and absence of looping are provided.

This condition is also used to axiomatize compiler notions. In this area we have only three major notions and each of them is defined by a lot of axioms. Compiler axiom system is divided on sets and each set defines a function and consists of groups of axioms whose left parts of conclusions are equivalent. According to the function and its arguments the strategy determines a group of axioms of a corresponding set and applies only one axiom.

3 Verification System

The system consists of three main components: a verification conditions generator, a constructor and a prover. Annotated source program is translated into an intermediate representation. Verification conditions are generated by employing proof rules which include standard VCG rules [6], a modified call procedure rule, file procedures rules and special proof rules applied to *for* loops without invariants [3,4].

The constructor transforms axioms written in a natural notation into program functions (called axiom-functions) written in a implementation language and sets the application strategy. The aim of the transformation is the use of axioms in a compilation mode because it is more efficient than a interpretation one.

The prover tries to determine the truth of verification conditions. There are four components in the prover: a simplifier, a normalizer, a managing module and a axiom-functions modules. The first three components are constant part of the prover and common for all problem areas.

Each verification condition is simplified, normalized and transformed into a set of formulas of the form $p_1 \wedge p_2 \wedge \ldots \wedge p_m \rightarrow p_0$ (where each p_i does not include boolean connectives except negation) by the simplifier and the normalizer. One of the following representations of p_0 is possible:

— elementary linear inequality which contains only simple variables or constants;

— quantified formula containing universal quantifiers bounded by index sets;

— quantifier-free formula with notions of a problem area;

— predicate of membership of the set generated by a grammar;

— relation between terms (i.e. relation which differs from mentioned above).

A verification condition proving is reduced to the proving of mentioned formulas. Some formulas are proved by the simplifier, others are passed to the managing module which determines the axiom-function or the strategy to be applied.

To decide linear inequality systems, Shostak's method for determination of validity and satisfiability for quantifier-free Presburger arithmetic formulas in integers is used. The method is extended by special features. Deciding equality and inclusion of index sets is reduced to the recognition of their emptiness. Index sets are transformed into quantifier-free Presburger arithmetic formulas such that the emptiness of index sets corresponds to nonsatisfiability of the formulas. Decision procedure for quantified formulas of the form $\forall (u \cdot) \in S$ $e(u, v)$ is approximate. It combines the strategies of simplifying the index set S and employing quantified formulas from verification condition premises. Simplifying strategies check the equality of S to one of the sets *set, row* or *col* with suitable parameters by employing axioms like A2. To extend verification condition premises, the strategy of replenishment is used. For example, if there is an inequality of the form $A[i] > A[j]$ in a premise, then the inequality $i \neq j$ is added to it. Quantifier-free formulas with notions of a problem area and predicates of membership of sets generated by grammars are proved by employing axiom-functions.

The source language of the system is Pascal extended by annotations. Note that the constructor and the prover do not depend on the source language. The functional language Refal [7] is used for implementation the system. It runs on IBM PC AT/286. The size of executable modules depends on problem areas. The average size is 200K. Time of proving depends on complexity of the verification conditions and takes up from several seconds to thirty minutes. Experiments are carried out with programs of simple sorting (including insertion with additional array) and of linear algebra oriented towards the gauss method (including matrix inverse) and with Basic and Pascal subset compilers.

Experiments over both correct and incorrect programs have demonstrated the possibility of automatic running of the verification system employing a powerful knowledge base. In spite of using low-powered computer PC AT/286 the system is efficient, since most of the verification conditions take up a few minutes to be proved (time of proving is increased by using the loop invariant elimination method for linear algebra programs).

The verification system has a special module which finds paths in the programs corresponding to the verification conditions. Owing to the module the system determined and localized semantic errors during the experiments.

SPECTRUM92 is the experimental verification system which enables us to investigate proving in logical theories extended by a problem areas specific. The system uses all the proposed verification means: axioms with conditions and built-in case analysis, proof procedures and strategies oriented towards problem areas, special and modified proof rules which enable us to simplify program specification, the axiom-function constructor which permits us to introduce new axioms written in a natural notation and to set the strategy of axiom application. It is intended to apply the system to new problem areas including network protocols and file editors.

References

1. D. Craigen: Strengths and weaknesses of program verification systems. Lecture Notes in Computer Science 289, Berlin:Springer 1987, pp.396-404
2. P.A.Lindsay: A survey of mechanical support for formal reasoning. Software Engineering J. 1 (1988) 3-27
3. V.A. Nepomniaschy and O.M. Ryakin: Applied methods of program verification. Radio and Svjaz, Moscow, 1988 (in Russian).
4. V.A.Nepomniaschy, A.A.Sulimov: A problem-oriented verification system and its application to linear algebra programs. Theoretical Computer Science 119 (1993) (to appear).
5. A.A.Sulimov: Functional means of compiler specification and verification and their application in problem-oriented system of program verification. Ph.D. thesis, Novosibirsk, Institute of Informatics Systems, 1991 (in Russian).
6. S.Igarashi.,R.L.London and D.C.Luckham: Automatic program verification I: A logical basis and its implementation. Acta Informatica 4(2) (1975) 145-182.
7. V.F.Turchin: A supercompiler system based on the language REFAL. SIGPLAN Notices 14(2) (1979) 46-54.

General Purpose Proof Plans[*]

Toby Walsh
Department of Artificial Intelligence
University of Edinburgh
T.Walsh@edinburgh.ac.uk

Abstract. One of the key problems in the design and implementation of automated reasoning systems is the specification of heuristics to control search. "Proof planning" has been developed as a powerful technique for declaratively specifying high-level proof strategies. This paper describes an extension to proof planning to allow for the specification of more general purpose plans.

1 Introduction

The heuristics used to control search in automated reasoning systems are often implicitly encoded in the implementation. This has many disadvantages. For example, it is difficult to reason about their properties, or to modify or generalise them. Proof planning has been developed to tackle some of these problems [Bun88]. Proof planning provides a declarative specification of high-level proof strategies. This paper describes an extension to proof planning to improve its flexibility and generality. This extension allows us to represent explicitly more flexible and general purpose proof strategies. Such general purpose proof strategies have been used, for instance, to describe (families of) decision procedures [Bun91].

2 Proof Planning

Proof plans are built in terms of **methods**. Methods are meta-level descriptions of **tactics**, computer programs which build parts of an object-level proof. Methods encode discrete proof "chunks". For example, the **remove** method removes a defined symbol from a formula by replacing it with its definition. Every method has a set of preconditions; that is, a set of conditions necessary for the method to apply. The preconditions of the **remove** method test that a given symbol occurs in the formula and is a defined symbol. Every method also has a set of postconditions that describe the result of the method. The **remove** method's postconditions state that the defined symbol no longer occurs in the formula. A particular method can capture many different proof "chunks" as it includes parameters that need to be instantiated to give an object-level proof (*eg.* the symbol to be removed). For reasons of efficiency and brevity, methods are often only partial specifications of problem solving strategies; the associated tactic therefore needs to be executed before we are certain that we do indeed have a proof. The tactic associated with the **remove** method thus performs the exhaustive object-level rewriting necessary to remove the defined symbol from the formula.

A proof planner, called CLAM, puts such methods together instantiating their parameters to give a **proof plan** [BvHHS90]. Methods are composed together using the "then" methodical. Each proof plan built by CLAM is tailored to the particular theorem being proved, with the preconditions of methods at the end of the plan being satisfied by the postconditions of methods earlier in the plan. I shall therefore call it a **special purpose proof plan**. CLAM

---

[*] Current address: INRIA-Lorraine, 615 rue du Jardin Botanique, B.P. 101, F-54602 Villers-les-Nancy, France. Email: walsh@loria.fr. This research was supported in part by a SERC Postdoctoral Fellowship. I am grateful to Renato Busatto-Neto for his feedback and for implementing several general purpose proof plans for normalisation. I would also like to thank Alan Bundy and all the other members of the Mathematical Reasoning Group.

uses conventional planning techniques to build special purpose proof plans (*eg.* depth first, iterative deepening, breadth first, or best first search). At each state in the search, CLAM finds a method that is applicable (that is, whose preconditions hold). In 1 to 1 correspondence with each special purpose proof plan is a **special purpose tactic** which, if executed successfully, builds the corresponding object level proof. Special purpose tactics are built from atomic tactics composed together using the "then" tacital.

3   General Purpose Proof Plans

Special purpose proof plans are not, in general, applicable to other formulae. They cannot, for instance, describe the rewriting of two *different* formulae into a *common* normal form. By extending the methodical language, however, we can explicitly represent more general purpose proof plans which are applicable in a variety of different situations.

Special purpose proof plans are built from (fully instantiated) atomic methods and other special purpose proof plans using the then methodical. By comparison, **general purpose proof plans** are built from a larger class of methodicals and (possibly uninstantiated) atomic methods. The following are general purpose proof plans:

Method1 or Method2	complete(Method)	progress(Method)
if(Cond,Method)	if(Cond,Method1,Method2)	repeat(Method)
until(Cond,Method)	while(Cond,Method)	exists(X,Method)

where Method, Method1 and Method2 are general or special purpose proof plans, Cond is some boolean condition, and X is a variable. We will also use a simple definition mechanism for proof plans.

Informally, general purpose proof plans are interpreted as follows:

Method1 or Method2	applies Method1 or Method2.
complete(Method)	only succeeds if Method completes the proof.
progress(Method)	only succeeds if Method makes some progress.
if(Cond,Method)	if Cond holds then apply Method.
if(Cond,Method1,Method2)	if Cond holds then apply Method1 else Method2.
repeat(Method)	repeat Method until it fails to apply.
until(Cond,Method)	repeat Method until Cond holds.
while(Cond,Method)	repeat Method while Cond holds.
exists(X,Method)	succeeds if there is some X for which Method succeeds.

The interpretation of general purpose proof plans is provided by a **proof plan interpreter** which, when given a goal, unpacks a general purpose proof plan into a special purpose plan. For example, the general purpose proof plan until(Condition,Method) is unpacked into a sequence of applications of Method,

$$\text{Method1 then Method2 then } \ldots \text{ Methodn}$$

where Methodi is an application of Method and Methodn is the last application of Method for which Condition held.

The methodicals proposed here bear more similarities to the tacticals proposed by Felty in [Fel89] than to the tacticals introduced in LCF [GMW79]. In particular, these methodicals pass information around by the sharing of logic variables. In addition, as with Felty's tacticals, failure of execution of a compound method leads to backtracking search.

4 Some Examples

To illustrate the sorts of general purpose proof plans which can be represented using these extensions, consider the following examples. The first is a fragment of a general purpose proof plan for normalisation. Suppose we have a formula to prove which contains (defined) symbols not in a given decision class. By unfolding the definition of these defined symbols, we can rewrite the formula to give an expression within the decision class. This rewriting is represented by the general purpose proof plan,

```
repeat(exists(Symbol,if(deviant(Symbol,Class),remove(Symbol))))
```

where deviant(Symbol,Class) is true if Symbol occurs in the given formula but not in Class, the given decision class and remove(Symbol) is a method which removes Symbol from a formula by exhaustive rewriting. Note the use of the existential methodical to allow for different symbols to be removed on successive iterations of the repeat loop. For example, when conjunctive normal form is the given decision class and *a iff b* is the given formula, the proof plan interpreter generates the special purpose proof plan, remove(iff) then remove(imply). This general purpose proof plan also represents a proof strategy for removing the predicate symbols $\geq, >$ and $\leq$ in Hodes' algorithm [Bun91].

The second example is the recursive definition of a general purpose proof plan for removing a list of defined symbols from a formula,

```
remove_list(Symbols) ≡def if(Symbols=[], id_mthd,
                             if(Symbols=[First|Rest],
                                remove(First) then remove_list(Rest)))
```

where id_mthd is the identity method which always succeeds but does nothing. Consider, the general purpose proof plan, remove_list([setequal,superset,subset]). This is evaluated by the proof plan interpreter to give the special purpose plan, remove(setequal) then remove(superset) then remove(subset) which replaces the predicates setequal, superset and subset by their definitions in terms of set membership.

5 General Purpose Tactics

As with special purpose proof plans, we can associate a **general purpose tactic** with every general purpose proof plan, and write a tactic interpreter (analogous to a proof plan interpreter) for executing general purpose tactics. Such an interpreter searches through the space of tactic instances to find an instantiation of a tactic's arguments for which the tactic succeeds.

The special purpose tactic resulting from the execution of a general purpose tactic can be seen as a *trace* of the general purpose tactic's execution. In practice, the execution of a general purpose tactic would return both the special purpose tactic (the execution trace) and the object level proof (the result of the execution).

Starting from a general purpose proof plan EITHER we can interpret the general purpose proof plan building a special purpose proof plan from which we construct a special purpose tactic (the *proof planning route*), OR we can construct the general purpose tactic associated with the general purpose proof plan and execute this (the *tactic route*). On the proof planning route, search is at the meta-level, whilst on the tactic route, search is at the object level. The shift to a meta-level will often cause a large reduction in search as the preconditions of the methods can greatly restrict the methods' applicability whilst tactics, by comparison, lack preconditions. The proof planning route is therefore usually much more efficient than the tactic route. In this sense, *the diagram does not commute.*

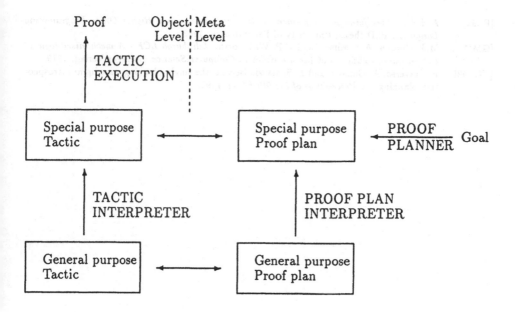

6 Related Work

The initial idea for this work came from the MRG robot planning language [TCS92]. This is a LISP like language for implementing a wide variety of planning systems. One significant difference is that plans in MRG are strictly linear whilst proof plans have a branching tree structure. Another difference, reflecting their different logic and functional programming foundations, is that failure in MRG does not lead to a backtracking search and that logic variables and unification are not used in MRG. As mentioned earlier, these methodicals bear some similarities to the tacticals proposed by Felty in [Fel89]. However, since methods are meta-level descriptions of their corresponding tactics and are restricted in their application by preconditions, a proof planning approach is potentially more powerful than a tactic based one.

7 Conclusions

This paper has described an experiment to extend the proof planning approach with general purpose proof plans. General purpose proof plans improve our ability to represent high-level proof strategies explicitly, to control proof search, and to reason about plan success and failure. Some simple general purpose proof plans have been written, mostly for the normal forming performed in several different decision procedures [Bun91]. Experiments are currently underway with some more sophisticated and powerful general purpose proof plans.

References

[Bun88]   A. Bundy. The use of explicit plans to guide inductive proofs. In R. Lusk and R. Overbeek, editors, *9th Conference on Automated Deduction*, pages 111–120. Springer-Verlag, 1988.

[Bun91]   A. Bundy. The use of proof plans for normalization. In R.S. Boyer, editor, *Essays in Honor of Woody Bledsoe*, pages 149–166. Kluwer, 1991.

[BvHHS90] A. Bundy, F. van Harmelen, C. Horn, and A. Smaill. The Oyster-Clam system. In M.E. Stickel, editor, *10th International Conference on Automated Deduction*, pages 647–648. Springer-Verlag, 1990.

[Fel89]   A. Felty. *Specifying and implementing theorem provers in a Higher Order Programming Language.* PhD thesis, University of Pennsylvania, 1989.

[GMW79]   M.J. Gordon, A.J. Milner, and C.P. Wadsworth. *Edinburgh LCF - A mechanised logic of computation*, volume 78 of *Lecture Notes in Computer Science.* Springer Verlag, 1979.

[TCS92]   P. Traverso, A. Cimatti, and L. Spalazzi. Beyond the single planning paradigm: introspective planning. In *Proceedings of the 9th ECAI*, 1992.

Index of Authors

Springer-Verlag
and the Environment

We at Springer-Verlag firmly believe that an international science publisher has a special obligation to the environment, and our corporate policies consistently reflect this conviction.

We also expect our business partners – paper mills, printers, packaging manufacturers, etc. – to commit themselves to using environmentally friendly materials and production processes.

The paper in this book is made from low- or no-chlorine pulp and is acid free, in conformance with international standards for paper permanency.

Lecture Notes in Computer Science

For information about Vols. 1–650
please contact your bookseller or Springer-Verlag

Vol. 685: C. Rolland, F. Bodart, C. Cauvet (Eds.), Advanced Information Systems Engineering. Proceedings, 1993. XI, 650 pages. 1993.

Vol. 686: J. Mira, J. Cabestany, A. Prieto (Eds.), New Trends in Neural Computation. Proceedings, 1993. XVII, 746 pages. 1993.

Vol. 687: H. H. Barrett, A. F. Gmitro (Eds.), Information Processing in Medical Imaging. Proceedings, 1993. XVI, 567 pages. 1993.

Vol. 688: M. Gauthier (Ed.), Ada-Europe '93. Proceedings, 1993. VIII, 353 pages. 1993.

Vol. 689: J. Komorowski, Z. W. Ras (Eds.), Methodologies for Intelligent Systems. Proceedings, 1993. XI, 653 pages. 1993. (Subseries LNAI).

Vol. 690: C. Kirchner (Ed.), Rewriting Techniques and Applications. Proceedings, 1993. XI, 488 pages. 1993.

Vol. 691: M. Ajmone Marsan (Ed.), Application and Theory of Petri Nets 1993. Proceedings, 1993. IX, 591 pages. 1993.

Vol. 692: D. Abel, B.C. Ooi (Eds.), Advances in Spatial Databases. Proceedings, 1993. XIII, 529 pages. 1993.

Vol. 693: P. E. Lauer (Ed.), Functional Programming, Concurrency, Simulation and Automated Reasoning. Proceedings, 1991/1992. XI, 398 pages. 1993.

Vol. 694: A. Bode, M. Reeve, G. Wolf (Eds.), PARLE '93. Parallel Architectures and Languages Europe. Proceedings, 1993. XVII, 770 pages. 1993.

Vol. 695: E. P. Klement, W. Slany (Eds.), Fuzzy Logic in Artificial Intelligence. Proceedings, 1993. VIII, 192 pages. 1993. (Subseries LNAI).

Vol. 696: M. Worboys, A. F. Grundy (Eds.), Advances in Databases. Proceedings, 1993. X, 276 pages. 1993.

Vol. 697: C. Courcoubetis (Ed.), Computer Aided Verification. Proceedings, 1993. IX, 504 pages. 1993.

Vol. 698: A. Voronkov (Ed.), Logic Programming and Automated Reasoning. Proceedings, 1993. XIII, 386 pages. 1993. (Subseries LNAI).

Vol. 699: G. W. Mineau, B. Moulin, J. F. Sowa (Eds.), Conceptual Graphs for Knowledge Representation. Proceedings, 1993. IX, 451 pages. 1993. (Subseries LNAI).

Vol. 700: A. Lingas, R. Karlsson, S. Carlsson (Eds.), Automata, Languages and Programming. Proceedings, 1993. XII, 697 pages. 1993.

Vol. 701: P. Atzeni (Ed.), LOGIDATA+: Deductive Databases with Complex Objects. VIII, 273 pages. 1993.

Vol. 702: E. Börger, G. Jäger, H. Kleine Büning, S. Martini, M. M. Richter (Eds.), Computer Science Logic. Proceedings, 1992. VIII, 439 pages. 1993.

Vol. 703: M. de Berg, Ray Shooting, Depth Orders and Hidden Surface Removal. X, 201 pages. 1993.

Vol. 704: F. N. Paulisch, The Design of an Extendible Graph Editor. XV, 184 pages. 1993.

Vol. 705: H. Grünbacher, R. W. Hartenstein (Eds.), Field-Programmable Gate Arrays. Proceedings, 1992. VIII, 218 pages. 1993.

Vol. 706: H. D. Rombach, V. R. Basili, R. W. Selby (Eds.), Experimental Software Engineering Issues. Proceedings, 1992. XVIII, 261 pages. 1993.

Vol. 707: O. M. Nierstrasz (Ed.), ECOOP '93 – Object-Oriented Programming. Proceedings, 1993. XI, 531 pages. 1993.

Vol. 708: C. Laugier (Ed.), Geometric Reasoning for Perception and Action. Proceedings, 1991. VIII, 281 pages. 1993.

Vol. 709: F. Dehne, J.-R. Sack, N. Santoro, S. Whitesides (Eds.), Algorithms and Data Structures. Proceedings, 1993. XII, 634 pages. 1993.

Vol. 710: Z. Ésik (Ed.), Fundamentals of Computation Theory. Proceedings, 1993. IX, 471 pages. 1993.

Vol. 711: A. M. Borzyszkowski, S. Sokołowski (Eds.), Mathematical Foundations of Computer Science 1993. Proceedings, 1993. XIII, 782 pages. 1993.

Vol. 712: P. V. Rangan (Ed.), Network and Operating System Support for Digital Audio and Video. Proceedings, 1992. X, 416 pages. 1993.

Vol. 713: G. Gottlob, A. Leitsch, D. Mundici (Eds.), Computational Logic and Proof Theory. Proceedings, 1993. XI, 348 pages. 1993.

Vol. 714: M. Bruynooghe, J. Penjam (Eds.), Programming Language Implementation and Logic Programming. Proceedings, 1993. XI, 421 pages. 1993.

Vol. 715: E. Best (Ed.), CONCUR'93. Proceedings, 1993. IX, 541 pages. 1993.

Vol. 716: A. U. Frank, I. Campari (Eds.), Spatial Information Theory. Proceedings, 1993. XI, 478 pages. 1993.

Vol. 717: I. Sommerville, M. Paul (Eds.), Software Engineering – ESEC '93. Proceedings, 1993. XII, 516 pages. 1993.

Vol. 718: J. Seberry, Y. Zheng (Eds.), Advances in Cryptology – AUSCRYPT '92. Proceedings, 1992. XIII, 543 pages. 1993.

Vol. 719: D. Chetverikov, W.G. Kropatsch (Eds.), Computer Analysis of Images and Patterns. Proceedings, 1993. XVI, 857 pages. 1993.

Vol. 720: V.Mařík, J. Lažanský, R.R. Wagner (Eds.), Database and Expert Systems Applications. Proceedings, 1993. XV, 768 pages. 1993.

Vol. 722: A. Miola (Ed.), Design and Implementation of Symbolic Computation Systems. Proceedings, 1993. XII, 384 pages. 1993.

Vol. 723: N. Aussenac, G. Boy, B. Gaines, M. Linster, J.-G. Ganascia, Y. Kodratoff (Eds.), Knowledge Acquisition for Knowledge-Based Systems. Proceedings, 1993. XIII, 446 pages. 1993. (Subseries LNAI).

Vol. 724: P. Cousot, M. Falaschi, G. Filè, A. Rauzy (Eds.), Static Analysis. Proceedings, 1993. IX, 283 pages. 1993.

Vol. 725: A. Schiper (Ed.), Distributed Algorithms. Proceedings, 1993. VIII, 325 pages. 1993.

Vol. 726: T. Lengauer (Ed.), Algorithms — ESA '93. Proceedings, 1993. IX, 419 pages. 1993.